3

Bioanalytical Reviews

Series Editors

Frank-Michael Matysik, Institute of Analytical Chemistry, University of Regensburg, Regensburg, Germany

Joachim Wegener, Department of Chemistry, University of Regensburg, Regensburg, Germany

Bioanalytical Reviews is the successor of the former review journal with the same name, and it complements Springer's successful and reputed review book series program in the flourishing and exciting area of the Bioanalytical Sciences.

Bioanalytical Reviews (BAR) publishes reviews covering all aspects of bioanalytical sciences. It therefore is a unique source of quick and authoritative information for anybody using bioanalytical methods in areas such as medicine, biology, biochemistry, genetics, pharmacology, biotechnology, and the like.

Reviews of methods include all modern tools applied, including mass spectrometry, HPLC (in its various forms), capillary electrophoresis, biosensors, bioelectroanalysis, fluorescence, IR/Raman, and other optical spectroscopies, NMR radiometry, and methods related to bioimaging. In particular the series volumes provide reviews on perspective new instrumental approaches as they apply to bioanalysis, and on the use of micro-/nano-materials such as micro- and nanoparticles. Articles on μ-total analytical systems (μ-TAS) and on labs-on-a-chip also fall into this category.

In terms of applications, reviews on novel bioanalytical methods based on the use of enzymes, DNAzymes, antibodies, cell slices, to mention the more typical ones, are highly welcome. Articles on subjects related to the areas including genomics, proteomics, metabolomics, high-throughput screening, but also bioinformatics and statistics as they relate to bioanalytical methods are of course also welcome. Reviews cover both fundamental aspects and practical applications.

Reviews published in BAR are (a) of wider scope and authoratively written (rather than a record of the research of single authors), (b) critical, but balanced and unbiased; (c) timely, with the latest references. BAR does not publish (a) reviews describing established methods of bioanalysis; (b) reviews that lack wider scope, (c) reviews of mainly theoretical nature.

Tilman E. Schäffer
Editor

Scanning Ion Conductance Microscopy

With contributions by

L. A. Baker · A. Bhargava · M.-H. Choi · I. D. Dietzel · A. Gesper · J. Gorelik · A. Haak · P. Happel · F. Iwata · Y. Korchev · C. W. Leasor · M. V. Makarova · Y. Mizutani · M. Nakajima · P. Novak · T. E. Schäffer · A. Shevchuk · Y. Takahashi · T. Ushiki · H. von Eysmondt

Springer

Editor
Tilman E. Schäffer
Institute of Applied Physics
University of Tübingen
Tübingen, Germany

ISSN 1867-2086 ISSN 1867-2094 (electronic)
Bioanalytical Reviews
ISBN 978-3-031-14442-4 ISBN 978-3-031-14443-1 (eBook)
https://doi.org/10.1007/978-3-031-14443-1

This Springer imprint is published by the registered company Springer Nature Switzerland AG
The registered company address is: Gewerbestrasse 11, 6330 Cham, Switzerland

Foreword

Tilman E. Schäffer is a distinguished leader in the field of Scanning Ion Conductance Microscopy [1]. He was the first, as far as I know, to use the SICM to answer a real scientific question: how abalone nacre forms. He found it was not by heteroepitaxial nucleation, as was commonly believed, but instead by growth through mineral bridges that formed in the pores he imaged with the SICM [2]. Together with his coworkers, he also developed an elegant, non-contact method for measuring the stiffness of cells [3] and high-speed SICM with imaging speeds faster than one second per image [4]. It is especially important to note that his work has been really focused not just on technique development, but on answering important scientific questions. For example, one of his recent papers reveals differential strategies of how two human-pathogenic viruses manipulate infected cells [5]. One of his own chapters in this book, with von Eysmondt, gives not only a comprehensive overview, but a realistic assessment of both the strengths and the limitations of the technique.

Tilman has also done an excellent job of getting some of the best and brightest in the field to contribute chapters to this book. I was particularly interested in potentiometric scanning ion conductance microscopy (P-SICM) in the chapter by Choi, Leasor, and Baker and in investigating cardiac function with SICM in combination with other techniques in the chapter by Bhargava and Gorelik. Combining SICM with other techniques, pioneered by Tilman himself [2] has been particularly fruitful – especially the combination with super-resolved optical microscopy as discussed in the chapter by Happel, Gesper, and Haak. A recent, spectacular example in their chapter is from Georg Fantner's lab (including one of the inventors of SICM, Barney Drake): a 2021 Nature Communications article with amazing correlative 3D images of single cells using super-resolution optical fluctuation imaging (SOFI) and scanning ion-conductance microscopy [6]. The chapter by Ushiki, Iwata, Nakajima, and Mizutani describes another fruitful combination: SICM with scanning electron microscopy.

One of the real highlights of the book is the chapter by the distinguished pioneers Novak, Shevchuk, and Korchev. They do a wonderful job of summarizing their seminal contributions that helped transform SICM into a practical instrument for widespread applications.

The chapter by Dietzel, Happel, and Schäffer gives a thorough overview of the intellectual context in which the SICM exists. Though I was, of course, well aware of the scanning probe microscopy component of that intellectual context, I must confess that I was unaware of almost all the rest at the time I invented the SICM.

The path to the invention of the Scanning Ion Conductance Microscope started with a sabbatical I spent studying the vacuum/solid interface with the wonderful surface scientist Gabor Somorjai. Near the end, I asked him if he thought I should continue in that type of surface science. He said no! He said that there were many, many tools for studying the vacuum/solid interface, but very few for studying the much more important liquid/solid interface. He thought that I seemed like an inventive person and that I should use my skills to develop tools for studying the liquid/solid interface. Probably the best career advice I ever received!

After numerous marginally successful attempts to do this, my group was able to design a custom scanning tunneling microscope that got the first atomic resolution image in water [7]. One of the problems with this microscope was noise from ion currents that interfered with measuring the tunneling current.

There is an old saying among physicists that one person's noise is another person's signal. I started wondering if it would be possible to build a microscope based on ion currents instead of electron currents. With the long-term goal of imaging a polymer like DNA on an insulating surface, I sketched this idea on Feb. 18, 1986: gluing a ¼" Teflon rod to a plate on the bottom of a glass beaker and then scanning a sharpened, insulated stainless steel rod vertically and laterally with a micropositioner.

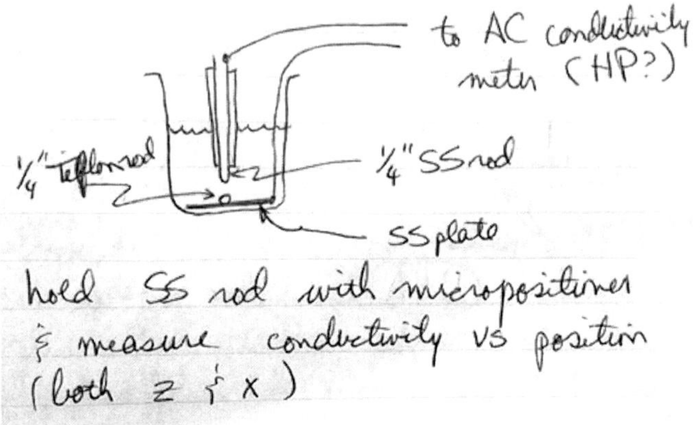

The idea was to lower the rod until the ion current dropped a little, record the micropositioner position, then lift the rod and translate it laterally and lower it again, record the micropositioner position again, and repeat to get a line scan. After trying insulated sharpened rods, flat rods and rounded rods with some, but very limited success, I tried recessing the rod into the insulation. This worked better, but there was too much drift [8] from electrical changes due to electrochemical reactions on the small area of the electrode. Next, I tried a large area electrode inside an eyedropper. This worked much better.

We then went from eyedroppers to micropipettes and from mechanical micropositioners to the scanners we were using for Atomic Force Microscopy. It was slow going at first, especially because the Atomic Force Microscope was occupying most of our attention, but by Feb. 3, 1989, we had our first publication [1] with images of gratings, a polymer, and ion currents through pores. The drawing of the SICM in that publication reminds me of the one above.

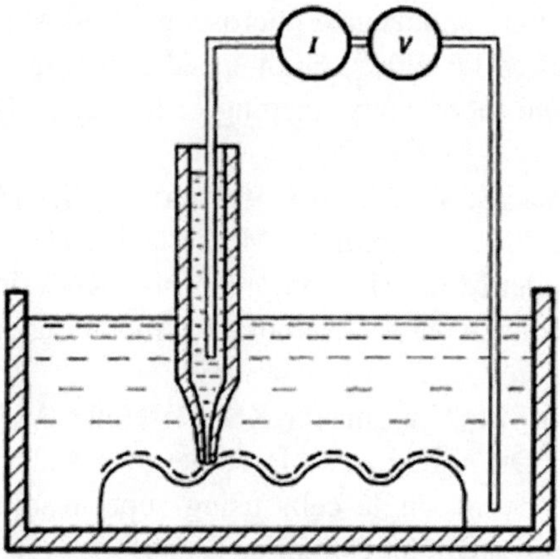

The use of Scanning Ion Conductance Microscopy increased dramatically after the wonderful advances described in the chapter by Novak, Shevchuk, and Korchev and the chapter by von Eysmondt and Schäffer. Now, over 30 years later, it is in a period of rapid growth. A quick search of Google Scholar with only the search term "SICM" revealed over 500 publications on Scanning Ion Conductance Microscopy in the last 4 years [9].

Given this rapid growth, it seems like an ideal time for this book to introduce new researchers to the field and to help existing researchers see the breadth of applications. I am grateful to Tilman, who I am proud to say was my graduate student, and the chapter authors for this wonderful book.

I believe that the future is bright for Scanning Ion Conductance Microscopy because of the dedicated work of the scientists, physicians, and engineers who have advanced the technology and its applications.

Department of Physics
University of California at Santa Barbara
USA

Paul Hansma

References

1. Hansma PK, Drake B, Marti O, Gould SAC, Prater CB (1989) The scanning ion-conductance microscope. Science 243(4891):641–643
2. Schäffer TE, Ionescu-Zanetti C, Proksch R, Fritz M, Walters DA, Almqvist N, Zaremba CM, Belcher AM, Smith BL, Stucky GD, Morse DE (1997) Does abalone nacre form by heteroepitaxial nucleation or by growth through mineral bridges? Chem Mater 9(8):1731–1740
3. Rheinlaender J, Schäffer TE (2013) Mapping the mechanical stiffness of live cells with the scanning ion conductance microscope. Soft Matter 9(12):3230–3236
4. Simeonov S, Schäffer TE (2019) High-speed scanning ion conductance microscopy for sub-second topography imaging of live cells. Nanoscale 11(17):8579–8587
5. Businger R, Kivimäki S, Simeonov S, Vavouras Syrigos G, Pohlmann J, Bolz M, Müller P, Codrea MC, Templin C, Messerle M, Hamprecht K, Schäffer TE, Nahnsen S, Schindler M (2021) Comprehensive analysis of human cytomegalovirus-and HIV-mediated plasma membrane remodeling in macrophages. Mbio 12 (4):e01770–21
6. Navikas V, Leitao SM, Grussmayer KS, Descloux A, Drake B, Yserentant K, Werther P, Herten DP, Wombacher R, Radenovic A, Fantner GE (2021) Correlative 3D microscopy of single cells using super-resolution and scanning ion-conductance microscopy. Nat Commun 12(1):1–9
7. Sonnenfeld R, Hansma PK (1986) Atomic-resolution microscopy in water. Science 232(4747):211–213
8. Perhaps ironically, this “drift” became the signal for the Scanning Electrochemical Microscope, which is well described in the chapter in this book by Makarova and Takahashi
9. As well as about 100 using the acronym for other thing such as Safety Index Computing Module, Sepsis-induced cardiomyopathy, Simplified Interface to Complex Memory and supervised intelligence committee machine

Preface

Scanning ion conductance microscopy (SICM) has undergone a remarkable development since its invention in 1989. Originally devised as a technique for spatially resolving the topography and ion permeability of a sample surface, SICM can now measure many other sample properties such as surface charge, electrochemical activity, or modulus of elasticity. SICM has thereby evolved into a multisensory, versatile measurement platform with a wide range of applications in physics, biology, chemistry, medicine, and materials science.

This book, *Scanning Ion Conductance Microscopy* in Springer's book series *Bioanalytical Reviews*, provides an introduction and overview of SICM technology and applications. It is written by pioneers in the field and is intended for both beginners and experts.

The first chapter, written by Prof. Dietzel, Dr. Happel (Ruhr-University Bochum, Germany), and Prof. Schäffer (University of Tübingen, Germany), gives a historical overview of techniques that paved the way for the development of SICM, starting from the discovery of bioelectricity two centuries ago.

In the second chapter, von Eysmondt and Prof. Schäffer (University of Tübingen, Germany) offer a detailed introduction to the technology of SICM and its strengths and limitations for biological applications in comparison with atomic force microscopy (AFM).

The next chapter, by Dr. Choi, Leasor, and Prof. Lane Baker (Indiana University, Bloomington, USA), provides a comprehensive overview of electrochemical imaging based on the measurement of ions and electrons with SICM.

In the fourth chapter, Dr. Novak, Dr. Shevchuk, and Prof. Korchev (Imperial College London, UK) portray the smart patch-clamp technique, which combines single-channel recording with selective probe positioning on a cell membrane with nanoscale resolution.

Prof. Bhargava (Indian Institute of Technology, Hyderabad, India) and Prof. Gorelik (Imperial College London, UK) illuminate in the following chapter how SICM can be used to associate cell signalling pathways with cellular surface structures in cardiac research.

The sixth chapter, by Prof. Makarova and Prof. Takahashi (Kanazawa University, Japan), presents scanning electrochemical cell microscopy for imaging electrochemical reactions of a variety of materials and for creating patterned microstructures on electroactive surfaces.

In the seventh chapter, Prof. Ushiki, Prof. Nakajima (Niigata University, Japan), Prof. Iwata (Shizuoka University, Japan), and Prof. Mizutani (Hokkaido University, Japan) compare SICM with scanning electron microscopy for imaging cells and tissues.

The final chapter, written by Dr. Happel, Haak, and Dr. Gesper (Ruhr-University Bochum, Germany), introduces different approaches to super-resolved fluorescence microscopy and discusses ways to combine them with SICM.

It is with great sorrow that we have lost our dear friend and colleague Dr. Patrick Happel, who died in 2021 during the making of this book, after a long illness. Patrick was one of the first to work on SICM and he dedicated his life to the development of SICM and its combination with high-resolution optical microscopy. In 2017, he organized, together with Dr. Astrid Gesper, the first dedicated symposium on SICM "Novel tools to investigate cellular physiology at the nanoscale" as a satellite symposium of the Meeting of the German Society for Biochemistry and Molecular Biology at the Ruhr-University Bochum. This symposium was attended by many of the authors of this book. We will greatly miss Patrick and remain grateful for his contributions to the SICM community.

This book could only be written because of the excellent work of all the authors, to whom I am deeply thankful. I would also like to thank Prof. Paul Hansma from the University of California at Santa Barbara, USA, for writing a foreword. Last but not least, I would like to acknowledge the patient editorial support provided by Springer Nature, especially by Dr. Charlotte Hollingworth (Editorial Director, Chemistry and Materials Science) and Ms. Alamelu Damodharan (Project Coordinator, Books).

Institute of Applied Physics
University of Tübingen
Germany

Tilman E. Schäffer

Contents

BIOREV (2022) 3: 1–22
https://doi.org/10.1007/11663_2022_14

Published online: 28 July 2022

The Evolution of Scanning Ion Conductance Microscopy

Irmgard D. Dietzel, Patrick Happel, and Tilman E. Schäffer

Contents

Abstract This chapter recollects the historical evolution of the techniques that set the stage for the development of scanning ion conductance microscopy (SICM). We elaborate how techniques evolved that finally resulted in instruments that now allow researchers to obtain contact-free, three-dimensional images of the surface of living cells with a resolution in the range of a hundredth of a micrometer. The starting point for this as well as for other bioelectric techniques was the discovery of bioelectricity a little more than 200 years ago. After the introduction of the first galvanometers to detect bioelectrical signals in the nineteenth century, in the early decades of the twentieth century, extracellular techniques were developed to record electrocardiographic and electroencephalographic potential changes using chlorinated silver

I. D. Dietzel (✉)
Department of Biochemistry II, Ruhr University of Bochum, Bochum, Germany

P. Happel
Central Unit for Ion Beams and Radioisotopes (RUBION), Ruhr University of Bochum, Bochum, Germany

T. E. Schäffer
Institute of Applied Physics, University of Tübingen, Tübingen, Germany

electrodes. These were miniaturized to allow for the detection of signals from individual cells in the middle of the twentieth century by the development of glass microelectrodes with sub-micrometer tip openings as well as appropriate current and voltage clamp amplifiers. The development of operational amplifiers based on transistors finally led to the detection of the tiny currents flowing through single transmembrane proteins. In the 1980s of the last century, scanning techniques were developed taking advantage of computer-guided piezo actuators to control the fine positioning of microelectrode tips on a scale smaller than the resolution of light microscopy. By combining amplifiers originally developed for electrophysiological recordings with nanoscale scanning techniques, it is now possible to quantitatively analyze the topography of the membranes of living cells with a resolution of down to 10 nm. Combinations with other microanalytical techniques, such as fluorescence microscopy, set the stage for future analysis of protein dynamics in moving membranes at unprecedented accuracy.

Keywords Bioelectricity · Electrophysiology · Instrument development · Scanning probe microscopy · SICM

1 Introduction

Scanning ion conductance microscopy (SICM) [1] is an evolving technique that detects the ionic conductance of the space between the tip of a micropipette and a sample surface. SICM allows researchers to quantitatively image the topography of samples that can be immersed in an electrolyte solution, with a lateral resolution of millimeters down to several nanometers. In SICM, changes of the ionic conductance occurring under the tiny opening of an electrolyte-filled glass pipette that is scanned across a sample surface by computer-controlled piezoelectric actuators are detected to visualize the three-dimensional topography of the sample. Since the tip of the scanning pipette ideally does not touch the sample surface, SICM is particularly well suited for tracing the surface of membranes of living cells, which would deteriorate if touched by sticky probes. The resolution of SICM, which is limited by the size of the pipette tip and by the noise of the amplifiers used to detect the conductance changes, usually exceeds the resolution limit of light microscopy [2].

The emerging technique of SICM can be combined with further microanalytical techniques such as fluorescence microscopy to identify the location of specific protein species or patch-clamp recording to analyze the function of individual ion channels [3]. We can therefore expect combinations of SICM with other techniques to result in so far unsurpassed capabilities in exploring, for instance, the dynamics of membrane-cytoskeleton interactions [4].

Before the invention of SICM and before the researchers who assembled this book entered the stage, many slow and arduous steps of progress had to be accomplished. This chapter reviews some of the many little evolutionary steps that paved the way to make SICM possible.

2 Detection and Registration of Potential Changes in Living Tissues

SICM combines pipettes and amplifiers originally developed for electrophysiological recordings with methods of scanning probe microscopy. SICM can thus be regarded in the tradition that started with Galvani's and Galeazzi's discovery of bioelectricity in the late eighteenth century [5]. Most remarkably, the first observations that indicated that intrinsic electricity governs our abilities to move and think reach back only about 230 years. However, the original observations that the contraction of frog legs could be induced by touching a nerve with the dissected nerve from another leg (Fig. 1) was just the starting spark to understand bioelectricity. It took until 1952, about 160 years later, to convince the scientific community that transient reversals of the membrane potential, called action potentials (first registered by intracellular recordings in 1939 [6]), are in fact the signals underlying information transmission and storage in our brains and during muscle contraction [7]. To put it in other terms: Whatever we do and think is orchestrated bioelectrically. From then on it took roughly another 40 years to satisfactorily understand, at the molecular level, how ion channel proteins, which underly the changes in electrical membrane potential, work [8, 9].

2.1 Discovery of Bioelectricity and of Macroscopic Electrophysiological Recording Techniques

How painfully and slowly the development of bioanalytical tools and our understanding of signals from living matter evolved becomes evident if we trace back the history of the discovery of bioelectricity in more detail:

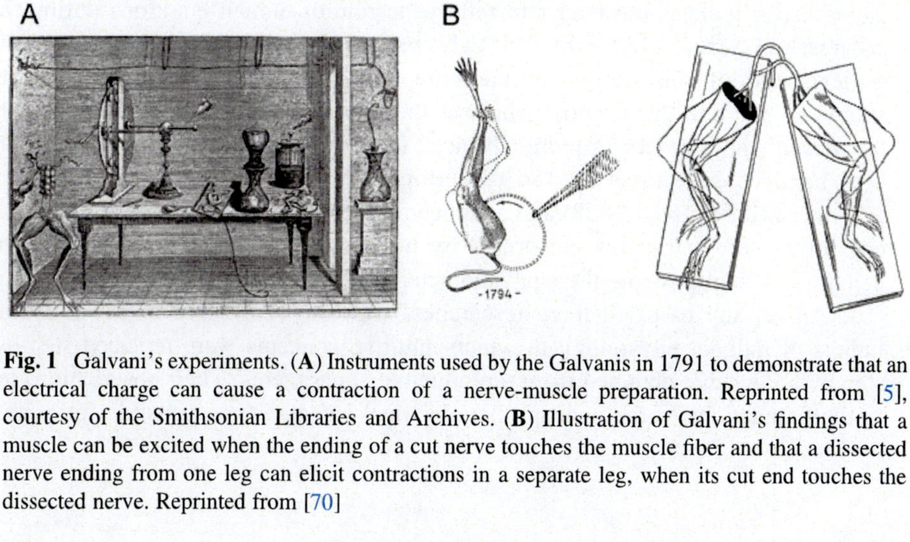

Fig. 1 Galvani's experiments. (**A**) Instruments used by the Galvanis in 1791 to demonstrate that an electrical charge can cause a contraction of a nerve-muscle preparation. Reprinted from [5], courtesy of the Smithsonian Libraries and Archives. (**B**) Illustration of Galvani's findings that a muscle can be excited when the ending of a cut nerve touches the muscle fiber and that a dissected nerve ending from one leg can elicit contractions in a separate leg, when its cut end touches the dissected nerve. Reprinted from [70]

Luigi Galvani and his wife, Lucia Galeazzi, were the first to observe that nerves and muscles can be excited by electrical sparks and by metal arcs connecting nerves and muscles (Fig. 1A). In the ensuing scientific discourse, Volta, who had invented the battery, rejected Galvani's interpretations and interpreted the twitching induced by sparks or by contact with cut endings from other nerves as generated from charges on the metals involved. Galvani then further observed that cut nerves can excite muscles without metals involved (Fig. 1B) and that, with an increase in electrical stimulus strength, the response saturates and the muscles fatigue, which are indications that animals generate intrinsic electricity. Despite Galvani's convincing arguments, the higher reputation of Volta led to a higher rating in the scientific community, and Galvani's interpretation was disregarded for a long time. Galvani died in poor circumstances after an episode of losing his income (because he refused to take an oath of atheism required by an edict by Napoleon). Thus 30 years were to pass in which his nephew Giovanni Aldini and also Alexander von Humboldt maintained the idea of intrinsic animal electricity [10–13].

A milestone in the development of measuring devices was the observation by Hans Christian Oersted (1777–1851) in Kopenhagen in 1820 that a magnetic needle was able to move in the presence of electricity passing through a wire loop [14]. On the basis of this observation, Johann Salomo Christoph Schweigger (1779–1857) in Halle built an instrument in 1820 to measure current strength by the deflection of a magnetic compass needle in a coil (named galvanometer in the memory of Galvani). In a very fruitful scientific period, Carlo Matteucci (1811–1868) discovered the nervous structures responsible for the discharge of the electrical organ in torpedo fish [15, 16] and pioneered the use of galvanometers to detect potential differences between lesioned and intact surfaces of muscles [17], which was the first glimpse of what we now call the resting membrane potential. A detailed historical account of Matteucci's discoveries and the controversy of that time, which arose with Emil du Bois-Reymond (1818–1896), concerning the interpretation of what we now know to be extracellular components of action potentials is given by Moruzzi [16], who is known for the discovery of the role of the reticular formation in arousal. Moruzzi's paper also stimulates the reader to reflect the role of human emotions during the progress of science. Obviously, it has always been a challenge to acknowledge the achievements of competitors. Furthermore, the question arises, whether priority should be given to the scientist who was the first to mention a phenomenon or to the person who arrived at a more advanced interpretation. Although Matteucci had seen negative deflections by 1843, du Bois-Reymond, student and successor of Johannes Müller (1801–1858) and a pioneer of electrophysiology in Berlin, considered only the experiments performed by himself as reliable observations of the negative deflections of the galvanometer needle during muscle contraction [18, 19]. In any case, all three researchers are now considered as the founding fathers of rational physiology in which intuitive concepts were replaced step-by-step by knowledge obtained from reproducible experiments. Their observations led to the foundations of the concept that electrical signals are responsible for animal movements.

A further milestone in technical development was the invention of Lippmann‘s capillary electrometer, which uses the movement of a mercury cylinder in a glass tube in contact with a diluted sulfuric acid solution when a potential in the millivolt-range is applied (Fig. 2B) [20]. Using such an instrument connected to a photographic plate, Augustus Waller was the first to succeed in documenting reproducible potential changes occurring between the front and the back of the chest during the heartbeat of humans and animals (trace *h* in Fig. 2A) [21]. Due to the slow time constant of the capillary electrometer, however, the measured electrocardiogram (ECG) is considerably low-pass filtered. Einthoven (1860–1927) conducted further experiments with this type of electrometer and first developed an algorithm to predict the correct time course of the extracellular electrical potential changes accompanying the heartbeat. In fact, the lettering P to T introduced by him to replace the corresponding lettering A to D of the capillary electrometer records are still in use today to denote the potential deflections corresponding to the contractions of the atria and the de- and repolarizations of the ventricles of the heart (lines A to D in Fig. 2C) [22]. Einthoven then dedicated his further life to improve galvanometry in order to detect undistorted electrical potential changes and bring this technique into clinical practice. He developed an instrument that was based on a string galvanometer, proposed in 1897 by Adler, which uses the deflection induced by current flow in a metal-coated quartz fiber that is exposed to a magnetic field, described more thoroughly in [23] (Fig. 3A,C). The signals were detected on a moving photo plate, which preceded the invention of the Brownian tube that was later used in oscilloscopes, as shown in the first successful recordings reproduced in Fig. 3B [24]. Figure 3D shows a photograph of a typical ECG recording setting. As reference electrode the feet of the subject were immersed in saturated salt solution, and the potential difference was recorded between the hands, which were also immersed in saturated salt solution. An interesting aspect concerning the hardships of life 100 years ago is revealed by the following episode: When Einthoven received the Nobel Prize in 1924, he could not attend the award ceremony, because he was on a lecture trip in the United States and traveling just a century ago was a slow and tiring procedure compared to nowadays, when traveling across continents is a matter of hours. Einthoven died in Leyden 3 years later at the age of 67.

Further refinements to the resolution of galvanometers were developed in the years to come. Another milestone in the evolution of bioelectrical recording techniques was the introduction of more sensitive instruments such as Edelmann’s string galvanometer and the double coil galvanometer developed by the company Siemens and Halske. These improvements of sensitivity, combined with the use of silver-silver chloride electrodes, fabricated by a procedure published by Proebster [25], allowed Hans Berger in Jena to record potential changes in the 10 μV range. With the help of volunteers among his students, colleagues, family, and especially patients who had lost pieces of their skull because of palliative brain tumor treatments, he was able to unambiguously show that these potential changes originated indeed from the electrical activity of the grey matter of the brain. In his pioneering publications from 1929 to 1938 [26, 27], he called his recordings “Elektrenkephalogramm des Menschen.” Figure 4 shows the first electroencephalography (EEG) recordings, taken from his son Klaus, and a photograph of an EEG machine he used in 1926.

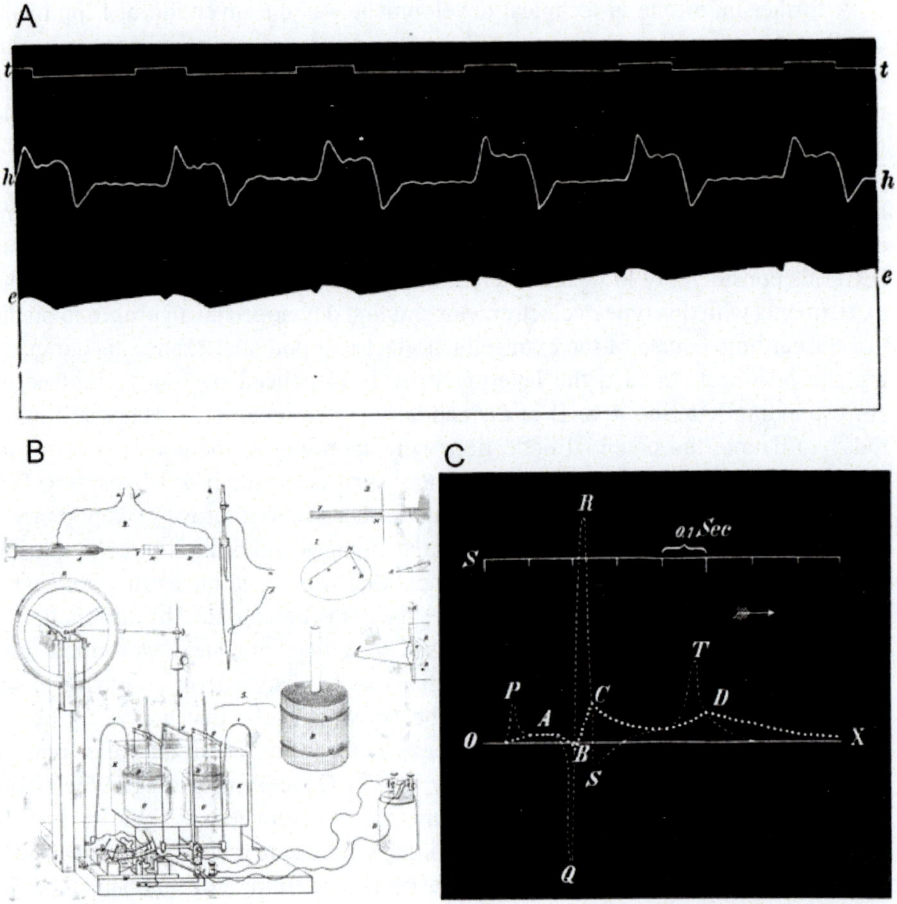

Fig. 2 First recordings of electrical potential changes from the heart. (**A**) Distorted ECG measured between the front and the back of a man's chest. Trace *t* denotes time in seconds, *h* indicates the heart's movements, and *e* the level of the mercury in the capillary. Reprinted with permission from [21]. (**B**) schematic of a mercury capillary electrometer in which fluctuations of a mercury surface in contact with a H_2SO_4 solution induced by voltage changes are visualized by a beam of light projected onto a moving photo plate. Reprinted with permission from [20]. (**C**) Theoretical reconstruction of the expected form of the ECG (with the lettering still used today) from the distorted capillary electrometer readings. Reprinted with permission from [22]

2.2 *Development of Instruments to Record Electrical Activity from Single Cells*

At that stage, it was known that brain and muscle cells operate with electrical signals, but a long way was still to go in our understanding of how neurons and muscle cells generate transient potential changes. From the work of pioneers such as Nernst, Oswald, and Bernstein [28, 29], it was known that cells are negatively charged in their interior and that the generation of the negative transmembrane potential

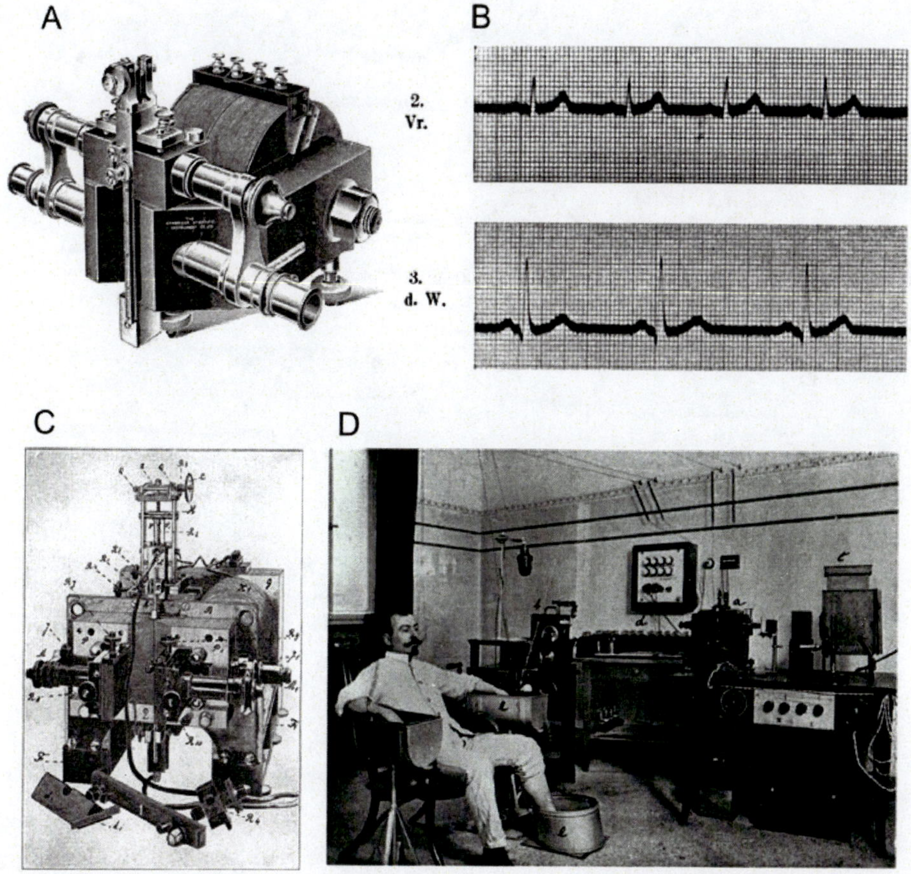

Fig. 3 First recordings of an undistorted electrocardiogram (ECG). (**A**) String galvanometer by "The Cambridge Scientific Instrument Comp.," which is characterized by its compact, solid form. Reprinted with permission from [23]. (**B**) Traces showing the first published electroencephalographic recordings using a string galvanometer. Reprinted with permission from [24]. (**C**) Original construction of the Leyden model of the string galvanometer used by Einthoven. Reprinted with permission from [23]. (**D**) Typical setting of an electrocardiographic recording (at around 1907) used before non-polarizable Ag/AgCl electrodes were introduced (photograph kindly provided by the Siemens Healthineers Historical Institute, Erlangen)

difference requires a selective conductivity of the cell membrane for potassium ions. How the potential difference, however, is transiently reversed during cellular signaling in order to generate the digitally transmitted signals, termed action potentials, still remained in the dark. In addition, the debate, whether electrical signals are constitutive for all nerve and muscle cell actions or whether they are just epiphenomena, was still unresolved and even Berger thought that EEG recordings just represent a reflection of some other phenomenon constituting the underlying excitation process (see pages 195–198 in [27]). These issues were finally settled by the development of several microanalytical tools, allowing researchers to perform

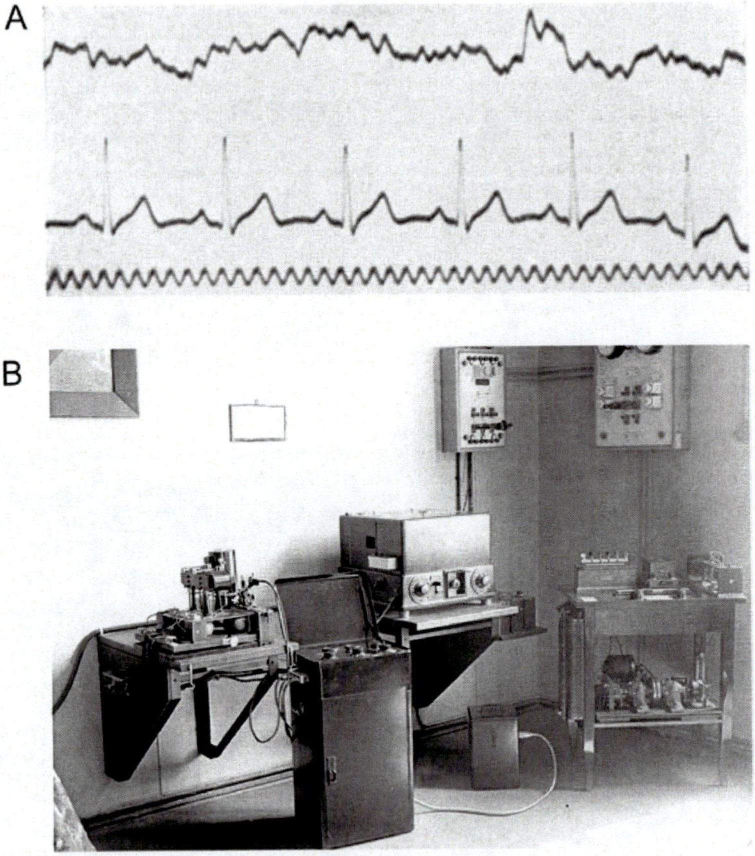

Fig. 4 First electroencephalographic recordings in 1929. (**A**) Recording from the skull of a 16-year-old boy (apparently Hans Berger's son Klaus) using subcutaneous needle electrodes at the front and the back of the skull (upper trace), simultaneously with an ECG recorded with lead foil electrodes from both arms (center trace). Time is calibrated with a sine waveform with a wavelength of 100 ms (bottom trace). Reprinted with permission from [26]. (**B**) Photograph of the typical equipment used by Berger. Courtesy of the archive of the University of Jena

intracellular action potential recordings (Fig. 5A) [6] and analyze the underlying membrane currents [7].

A first important step in this direction was the development of the voltage clamp technique. Action potentials consist of self-sustained reversals of the membrane potential evoked by suprathreshold stimuli that depolarize the membrane to values more positive than about −40 mV (Fig. 5A). Recordings of freely propagating potential differences after the application of suprathreshold stimuli, however, did not allow investigators to find out how the membrane conductance changes in dependence of the membrane potential, because of the self-amplifying propagation process. To answer this question, it was necessary to design an amplifier circuit that could keep the membrane potential constant during the occurrence of the membrane conductance changes after the initiation of the action potential. To do so, the voltage

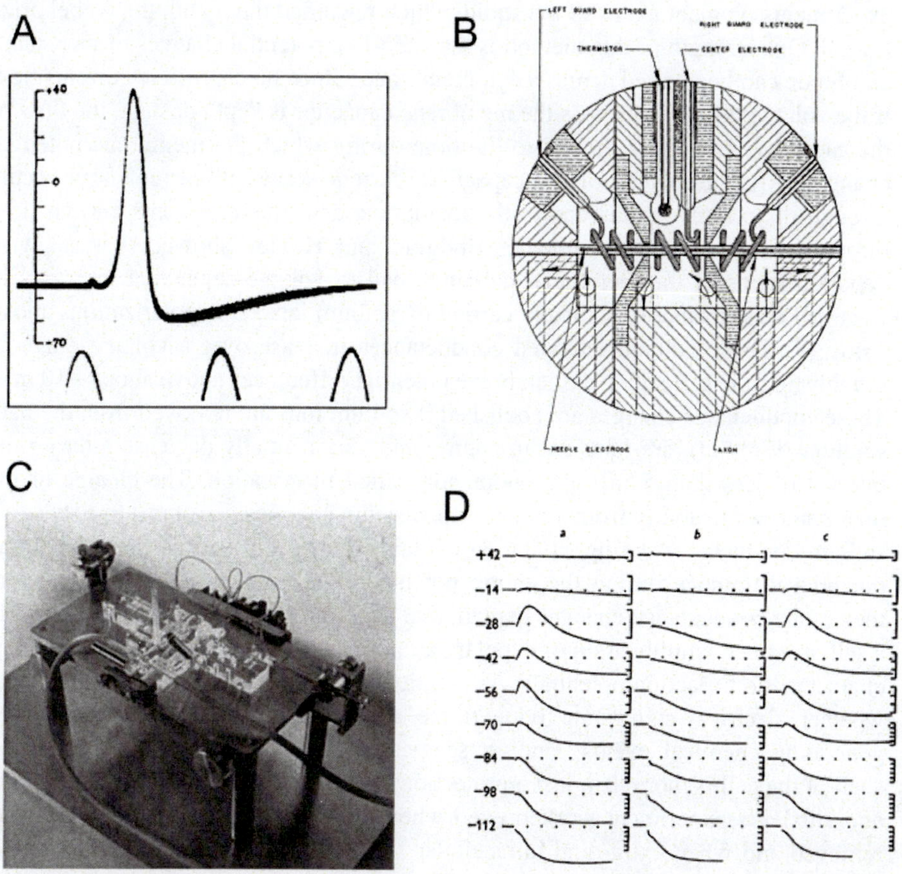

Fig. 5 Early intracellular recordings of action potentials and first description of the voltage clamp. (**A**) Action potentials recorded with a needle electrode from the giant axon of the squid (*Loligo forbesii*). The vertical axis is calibrated in millivolt and the intervals between the sine wave used for time calibration indicate 2 ms. Reprinted with permission from [6]. (**B**) Schematic drawing of the recording chamber containing the giant axon of the squid traversing the middle of the chamber containing a needle electrode, surrounded by coiled stimulation and guard electrodes. Reprinted with permission from [30]. (**C**) Photograph of the recording chamber. Reprinted with permission from [30]. (**D**) Pioneering series of voltage clamped inward and outward current recordings in the presence of extracellular seawater (**a**, **c**) and after replacing sodium by choline (**b**). Note the current is plotted inverted with respect to the current convention. Reprinted with permission from [32]

clamp investigations initiated by Marmont [30] and by Cole and Curtis [31] in 1949 (Fig. 5B, C) used at least two electrodes. One measured the potential difference, while the second one served to inject currents into the cell in order to balance potential changes elicited by current flow out of the cell induced by the opening of additional conductance pathways across the membrane. Their experiments can thus be regarded as the first application of the principle of current measurement induced by defined changes of voltage, and thus the monitoring of conductance changes in investigations of living cells. In 1952, Hodgkin and Huxley used this method in their experiments on giant axons of the squid, which rewarded them with the Nobel prize [7, 32]. One strength of this method is that step-like potential changes of increasing amplitude can be applied across a cell membrane. Since no capacitive currents flow if the voltage difference across the membrane capacitor is kept constant, the flow of the capacitive current is limited to the time during which the membrane potential changes. After the decay of the capacitive current surge, all current flow occurs through the resistive paths across the membrane and thus represents current flow through the permeable membrane. Hodgkin and Huxley showed that stepwise depolarizations of the membrane potential lead to voltage-dependent increases in electrical conductance specifically carried by sodium ions, at depolarizations above −40 mV. These voltage-activated conductances increase over a voltage range of roughly 30 mV, plateauing at membrane potential differences above about −10 mV. These conductance changes are abolished if sodium ions are removed from the bath solution (Fig. 5D), are transient in nature, and exponentially decrease after a time interval of less than 5 ms, a phenomenon called inactivation. The change of the membrane permeability from a selective potassium conductance at rest to a selective sodium conductance during excitation explains the reversal of the transmembrane potential difference during the action potential. The voltage and time-dependent membrane conductance changes explain that suprathreshold depolarizations lead to a self-sustained amplification of depolarization that shuts down on its own, which is characteristic for action potentials. As mentioned above, in 1950 it was still unclear whether electrical signals in living tissue are just epiphenomena to the signal-generating chemical events, since it seemed difficult to conceive that the outer shell of the cell is more than just a protection for the internal events. Since Hodgkin and Huxley's experiment worked even when the cytoplasm had been completely removed and when extra- and intracellular solutions contained only physiological salt solutions, it became more and more clear that, in cells, most biochemical reactions serve the aim to keep the membrane functional, enabling cells to selectively take up nutrients, expel waste molecules, migrate to target cells, respond to signals by electrically activating network oscillations, and change their connections.

The cells investigated in the pioneering work of Hodgkin and Huxley were giant axons from the squid with diameters of 1 mm, large enough to insert Ag/Ag chloride or platinum black coated reversible needle electrodes. Since it was far from clear whether all excitable cells work in a similar manner, it was necessary to miniaturize the recording techniques even further. This was accomplished by the advent of electrolyte-filled glass needle electrodes, into which the larger chlorinated silver wires were inserted (Fig. 6) [33]. In the years to come, several small companies developed electrode pullers of all kinds to manufacture electrode tips of multiple shapes with opening diameters in the range of 20 nm to 10 μm. These electrode pullers allowed researchers to manufacture reliable glass electrode tips and to record from various types of muscle cells and neurons. In the following 20 years, it turned out that all excitable cells function more or less in the same manner, using negative membrane potentials and action potentials for signaling, but that there is a large variability in what we now know to be the relative membrane expression of diverse

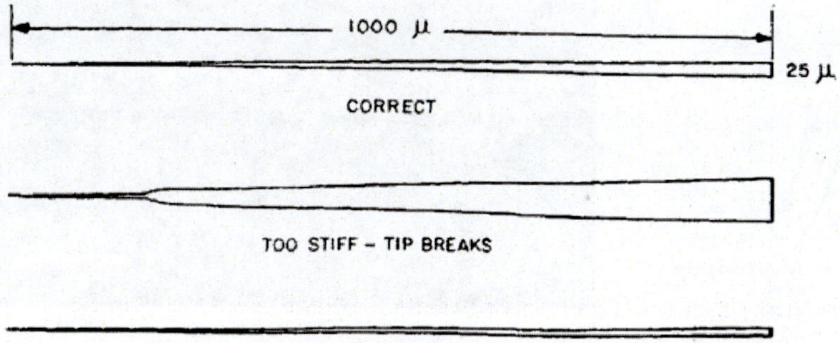

Fig. 6 Shapes of first glass needle electrodes fabricated by Ling and Gerard in 1949. Reprinted with permission from [33]

voltage-selective ion channels leading to cell-specific action potential kinetics and firing frequencies in neurons of different function, even within the same species [34].

After this period, in which it was shown that intrinsic membrane excitability governed by voltage-dependent changes in membrane conductivity is what keeps animals moving, the next question to be asked concerned the mechanisms of how such regulated permeability changes arise. In the 1970s, further progress had been made in the development of operational amplifiers and transistors, and current-voltage-converter circuits had been designed, which enabled investigators to record tiny currents in the range of nano- and picoamperes. Stimulated by the new technical opportunities and the puzzling question whether selective membrane permeability might arise from conformational changes of ion channels in the membrane, Erwin Neher and Bert Sakmann were the first to design an amplifier for single-channel recording and successfully place a fire-polished micropipette on the cleaned surface of a denervated muscle fiber, such that a sufficiently tight seal of more than a gigaohm formed between the rim of the electrode and the membrane. This allowed to record the tiny currents in the picoampere range. In 1976, they published the first direct recordings, showing that membrane permeability changes occur via a sudden opening of channels of high conductance for defined ion species (Fig. 7) [8]. In the years to come, the technique was extensively applied to all kinds of ion channels in a vast number of species and organs and it became clear that all bioelectrical activity works via the transient opening and closing of protein-embedded ion pathways, which can be gated by mechanical strain, chemical transmitters, or changes in membrane voltage. Slower and long-lasting modulations of channel activity can be additionally imposed by more complicated, second-messenger-regulated changes in single-channel opening probabilities, but the final outcome is always a conductive path for either sodium or calcium ions for the generation of action potentials, more unspecific cation channels for the stimulation of cells, and chloride and potassium channels to speed repolarization and to inhibit excitation [35–37]. The remaining fundamental puzzle of the design of a protein channel, which combines a high

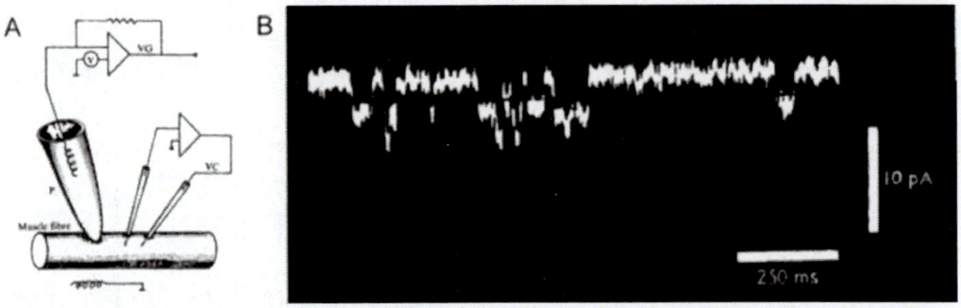

Fig. 7 First patch-clamp recordings. (**A**) Schematic circuit diagram for current recording from a patch of membrane with an extracellular pipette. VC, standard two-microelectrode voltage clamp circuit to locally set the membrane potential of the fiber to a fixed value. P, pipette, fire polished, 3–5 μm diameter opening, containing Ringer's solution and agonist at concentrations between 2×10^{-7} and 6×10^{-5} M. dc resistance of the pipette: 2–5 MΩ. The pipette tip is applied closely on to the muscle fiber within 200 μm of the intracellular clamp electrodes. VG, virtual ground circuit, using a function modules model 380K operational amplifier and a 500 MΩ feedback resistor to measure membrane current. The amplifier is mounted together with a shielded pipette holder on a motor-driven micromanipulator. V, bucking potential and test signal for balancing of pipette leakage and measuring pipette resistance. (**B**) Oscilloscope recording of current through a patch of membrane of approximately 10 μm^2. Downward deflection of the trace represents inward current. The pipette contained 2×10^{-7} M suberyldicholine in Ringer's solution. The experiment was carried out with a denervated hypersensitive frog cutaneous pectoris (*Rana pipiens*) muscle in normal frog Ringer's solution. The record was filtered at a bandwidth of 200 Hz. Membrane potential: −120 mV. Temperature: 8°C. Figure and caption reprinted with permission from [8]

single-channel conductance with a high selectivity for, e.g., the larger potassium ions, excluding the smaller sodium ions, was finally solved using X-ray crystallography. In 1998, Roderick McKinnon's group [9] managed to resolve the structure of a bacterial potassium channel at a resolution of a few Ångstroms: The channel-forming tetramer contains a bulkier water-filled internal cavity with a negatively charged entrance attracting cations into the channel. Only at the uppermost end of the channel pathway the four pore loops, which show negative polarizations toward the aqueous cavity, combine in such a manner, that the carbonyl oxygen atoms from the amino acids of the protein backbone in the signature sequence replace four water molecules of the hydration shell of a permeating potassium ion. Potassium ions thereby slip into the selectivity filter of the channel, pushing a remaining water molecule in front. The electrostatic repulsion then expels the potassium ion out of the channel. Since the hydrating water molecules around a sodium ion sit closer to its (smaller) ionic radius, this more tightly bound hydration shell of sodium ions prevents their entrance into the channel.

3 Development of Scanning Probe Microscopy

3.1 Scanning Tunneling Microscopy and Atomic Force Microscopy

While ion channel researchers were busily characterizing channel properties and while X-ray crystallography was being refined more and more to be able to visualize the structures of increasingly complex proteins, surface scientists and microscopists wondered whether they could visualize details beyond Abbe's limit without the constraints of specimen preparation for electron microscopy.

A major breakthrough was first published in 1982, when Gerd Binnig and Heinrich Rohrer invented scanning tunneling microscopy (STM) and managed to visualize, for the first time, the atoms on the surface of a gold crystal (Fig. 8) [38]. In this technique, a tiny metal tip is scanned across an electrically conducting sample surface at a close distance, whereby the tunneling current between tip and sample is measured and used to control the tip-sample distance. When scanning the tip relative to the sample with x, y piezo scanners, three-dimensional images of the tunneling

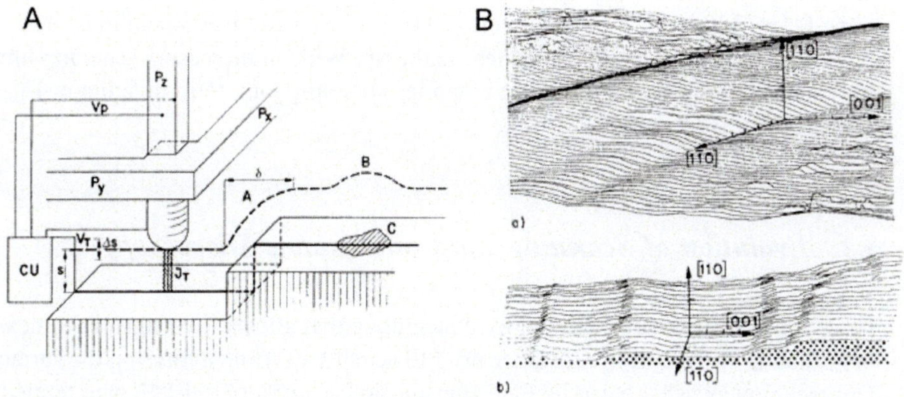

Fig. 8 First images obtained with scanning tunneling microscopy (STM) in 1982. (**A**) Principle of operation (schematic: distances and sizes are not to scale): The piezo actuators P_x and P_y scan the metal tip over the surface. The control unit (CU) applies the appropriate voltage V_p to the piezo actuator P_z for constant tunneling current J_T at constant tunneling voltage V_T. For constant work function, the voltages applied to the piezo actuators P_x, P_y, and P_z yield the topography of the surface directly, whereas modulation of the tunneling distance s by Δs gives a measure of the work function. The broken line indicates the z displacement in a y-scan at (**A**) a surface step and (**B**) a contamination spot, C, with lower work function. (**B**) Two examples of scanning tunneling micrographs of an Au (110) surface, taken at (**a**) room temperature and (**b**) 300°C after annealing for 20 h at the same temperature (and essentially same work function). Sensitivity: 10 Å/div. Because of a small thermal drift, there is some uncertainty in the crystal directions in the surface. In (**a**), the surface is gently corrugated in the [001] direction, except for a step of four atomic layers (≈2 atomic radii) along the [1$\bar{1}$0] direction, as indicated by the discontinuity of the shaded ribbon. The steps in (**b**), which were always found along the [1$\bar{1}$0] direction, are visualized by the possible positions of the Au atoms (dots). Figure and caption reprinted with permission from [38]

current and of sample topography can be obtained. Using this strategy, the three-dimensional topography of the atomic arrangement on a metal surface became directly visible for the first time (Fig. 8B).

STM was limited, however, to visualizing the surface of electrical conductors (mainly metals). To allow for the imaging of electrical insulators, the next step in the methodological development was the invention of atomic force microscopy (AFM) by Binnig, Quate, and Gerber in 1986 [39], in which a sharp tip on a flexible, tiny cantilever is scanned over a sample surface (Fig. 9A). The initial instrumental setup used the tip of an STM to measure the cantilever deflection, which was induced by the force exerted by the sample on a diamond tip (Fig. 9B). This force was kept at a constant level in the range of 10 nN with a feedback mechanism, which allowed the tip to follow the contour of the surface when moved in lateral direction. The first contours obtained in this manner were from an Al_2O_3 sample scanned at forces near 10 nN (Fig. 9C, trace set A) and near 5×10^{-8} N (Fig. 9C, trace set B). These first recordings set the stage to monitor contours of insulating surfaces at molecular and atomic resolution. Soon later, it was discovered that AFM even works in liquids [40]. The use of force as the underlying imaging mechanism, however, also has some limitations, especially for soft and fragile biological samples. For example, the membrane of a cell is indented to some extent by the tip, thus often revealing the structure of the underlying cytoskeleton rather than that of the unperturbed membrane [41]. Furthermore, glycoproteins, which extend from the membrane surface into the extracellular space, can adhere to the tip. With an increasing scanning time, this can lead to increasing tip contamination, impeding long-term imaging of living cells [42, 43].

3.2 Evolution of Scanning Ion Conductance Microscopy

These limitations were overcome by a technique that allows investigators to detect an insulating surface from a distance prior to actually exerting a force on the surface. This technique was invented by Paul Hansma and coworkers in 1989, who termed it scanning ion conductance microscopy (SICM) (Fig. 10A) [1]. Technically, SICM combines electrodes and amplifiers, such as used for electrophysiological recordings, with computer-controlled scanning piezo actuators, such as used in scanning probe microscopy. When an electrolyte-filled glass micropipette immersed in electrolyte solution is approached to an insulating sample surface, the conductance between two chlorinated silver wire electrodes, one inside the micropipette and one in the bath solution, increases when the tip-sample distance gets into the range of the opening diameter of the pipette. When the z-piezo is vertically adjusted in a manner to keep the conductance constant, the topography of the insulator can be imaged with a lateral resolution of the order of the pipette inner opening diameter [1, 2, 44]. The first SICM image of pores in a nylon membrane using this scanning principle is shown in Fig. 10B. In 1996, the Hansma group developed a combined TappingMode AFM and SICM setup, based on a bent glass pipette that was used

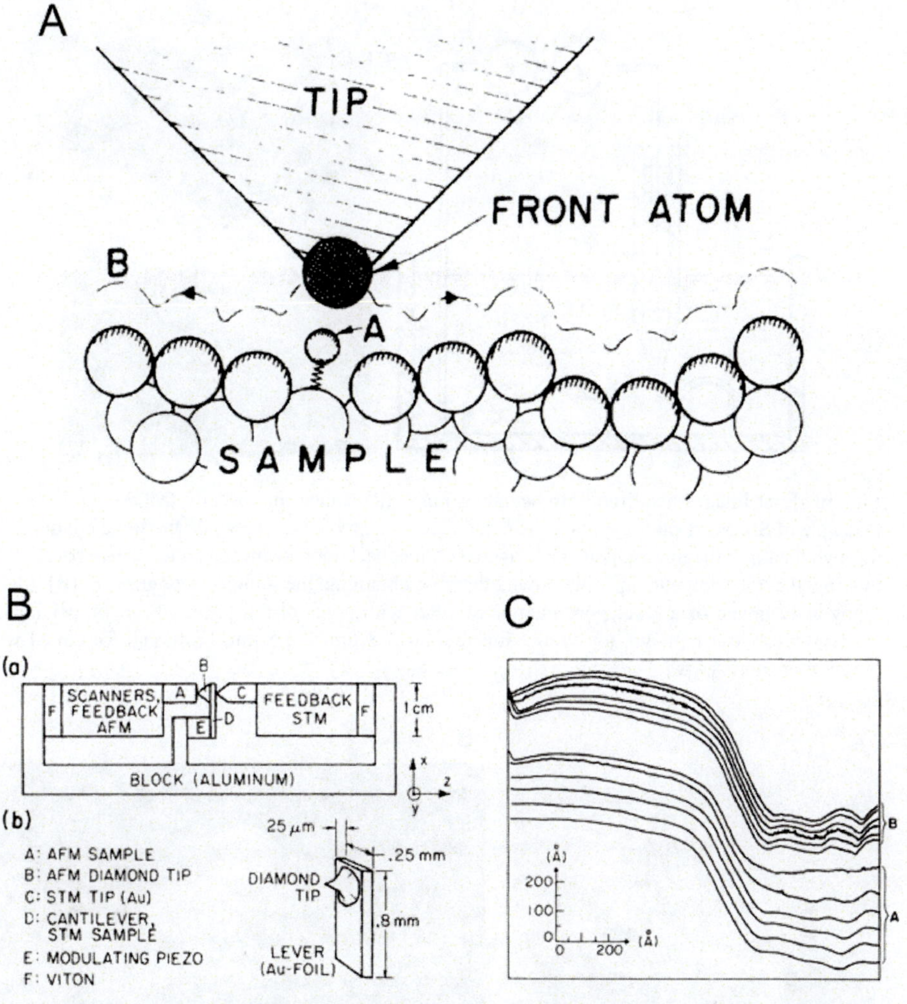

Fig. 9 First images obtained with atomic force microscopy (AFM) in 1986. (**A**) Principle of operation. The tip follows contour **B**, maintaining a constant force between tip and sample by keeping the STM tunneling current constant. (**B**) Experimental setup (**a**) and schematic and dimensions of the cantilever with a diamond tip attached (**b**). The tip is sandwiched between the AFM sample and the STM tip. A small modulating piezoelectric element is fixed to the cantilever, driving it at its resonant frequency. A feedback loop keeps the force acting on the tip constant. (**C**) AFM traces of an area of a ceramic (Al_2O_3) sample. Scanning set in **A**: Imaging force about 10^{-8} N for the highest trace. Scanning set in (**B**): Imaging force 5×10^{-8} N. Reprinted with permission from [39]

both as AFM cantilever and SICM conductance probe [45]. In 1997, they used a combined AFM-SICM setup to reveal conductive channels through interlamellar organic sheets of abalone nacre [46].

In the same year, the group of Yuri Korchev in the laboratory of Max Lab in London developed a SICM setup that enabled investigators to trace the topography

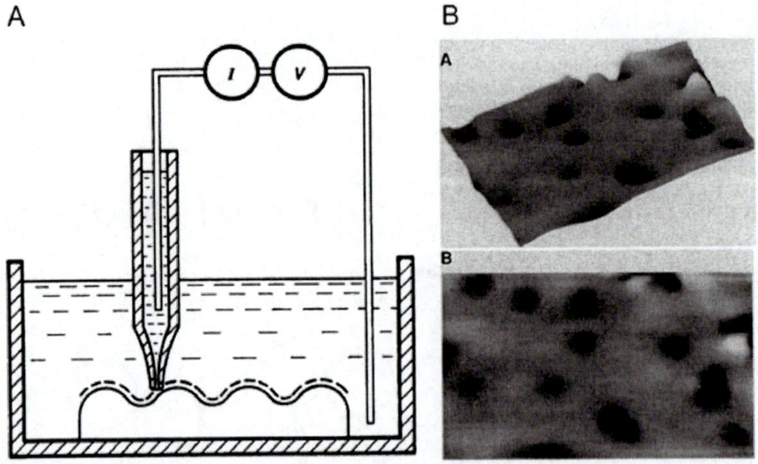

Fig. 10 First images acquired with scanning ion conductance microscopy (SICM) in 1989. (**A**) Principle of SICM. A micropipette is scanned across a sample surface while the ionic conductance through the tip of the micropipette is measured. A feedback loop keeps the conductance constant by moving the micropipette up and down, thereby obtaining the sample topography. (**B**) SICM topography image of a Nuclepore membrane filter with pores of 0.8 μm diameter (upper image: 3D rendering; lower image: top view). Image area: 7.8 μm × 4.5 μm. From [1]. Reprinted with permission from AAAS

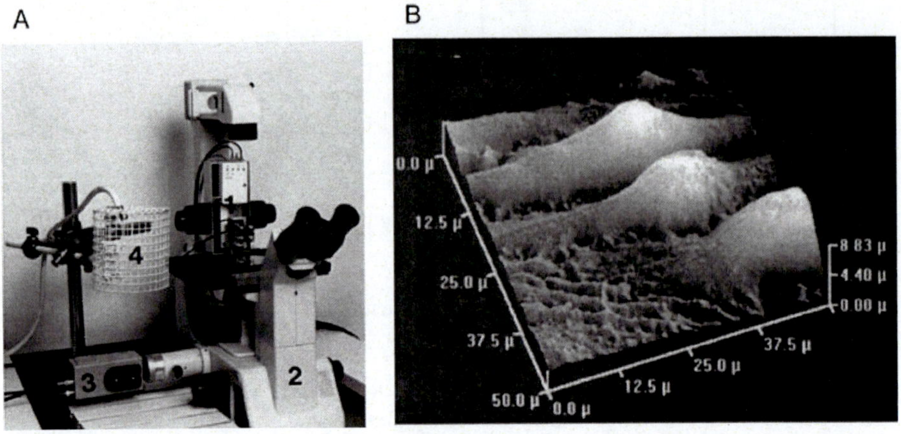

Fig. 11 First living cells imaged with SICM in 1997. (**A**) Photograph of the SICM setup. The SICM head (1) is mounted on top of an inverted light microscope (2). A video camera (3) allows to view the sample and pipette. A Faraday cage (4) is used during SICM operation to shield the SICM head. Reprinted with permission from [47]. (**B**) SICM image of cells from the murine melanocyte cell line melan-b. Reprinted with permission from [48]

of living cells in culture (Fig. 11) [47, 48]. This pioneering work set the stage for many groundbreaking applications of SICM in biology and chemistry in the following years, in a collaboration of the Korchev group with the group of David Klenerman [3, 49–53].

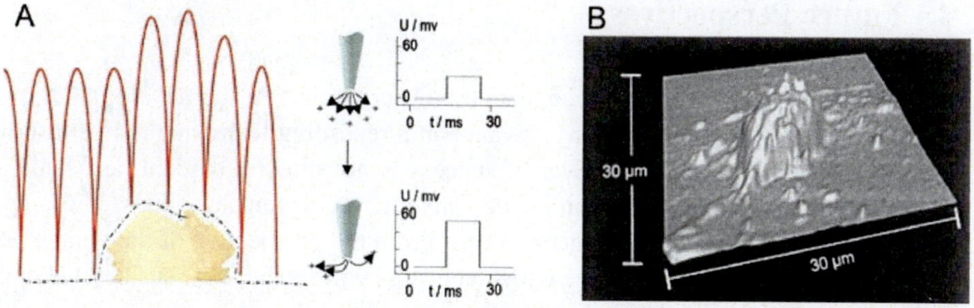

Fig. 12 Backstep/hopping mode SICM. (**A**) Principle of backstep/hopping mode SICM imaging. The pipette is repetitively approached to and retracted from the surface to avoid collisions with sample protrusions (left). In the original backstep mode, pulses of constant current were passed through the pipette, and the surface was sensed by an increase of the resulting voltage drop (right). Partly reproduced with permission from [54]. (**B**) Scan of a hippocampal cell from postnatal rat brain. The pipette was arrested at a resistance change of 3%. The cell body is clearly shown, and neurites start to become visible. Reprinted with permission from [54]

The approach of moving the pipette tip across the sample surface at a constant tip-sample distance by keeping the conductance constant was suitable to obtain topographic scans with high resolution of flat cells. Keeping the tip-sample distance constant or near-constant during imaging, however, caused a major problem for whole cells with steep or even overhanging membranes at the interface between cell and culture dish, because the pipette would laterally contact the cell when moving toward it, thereby possibly damaging the delicate membrane. This problem was solved by approaching and retracting the pipette for every image pixel in addition to applying current pulses to avoid electrode drift (Fig. 12) [54]. While first shown for low-resolution scans, retracting the pipette at every position is the base of the later developed hopping mode [55], the most commonly used SICM imaging mode today.

A major strength of SICM is that it can be used not only for imaging the topography of the sample, but that it can be extended to obtain information, for example, about the stiffness [56, 57], the surface charge [58], or the local conductance [59] of the sample. Furthermore, the SICM pipette can be utilized to apply locally constrained mechanical stimulation to the sample [60] or to locally deposit molecules [51]. By puncturing the cell membrane with the scanning pipette, SICM allows investigators to locally apply molecules to single cells [61] or to extract tiny volumes from selected cells [62]. As elaborated in more detail in other chapters of this volume, SICM has been combined with other techniques to allow, for example, targeted patch-clamping [3] or the correlation of sample topography with fluorescence [50] or electrochemical [63] data.

4 Future Perspectives

This brief historical overview recollects the observation that the development of a small piece of equipment or a tiny advance in a recording technique may consume the work of a whole lifespan. Even if success is not directly evident, as seen, for instance, by the delayed recognition of Galvani's achievements, lines of thought may be picked up by following generations. From the perspective of the contemporary, it is often difficult to foresee which concepts will be the most sustainable ones in the future and in which direction further progress will evolve.

Hence, at the moment, it is difficult to predict how much additional applications SICM will find in its various potential configurations. Even though our understanding of physiology has enormously increased at the cellular and membrane level, our knowledge concerning the signal processes governing membrane reactions to mechanical and chemical stimuli is still incomplete. When it comes to understanding physiology at the membrane and sub-cellular level, a field often termed "nanophysiology" (as introduced by Hans Oberleithner [64]) opens new perspectives. SICM and its various applications have already succeeded in allowing to gather new insights into questions of how proteins arrange in membrane surfaces [65]. Furthermore, SICM investigations, showing that migrating oligodendrocyte precursor cells swell prior to their acceleration [66], have lent support to the hypothesis that ion and water fluxes play an essential role in cellular migration [67]. Dynamical measurements of the morphology and mechanics of cells in motion are a promising route toward understanding the role of the cytoskeleton in cellular physiology, e.g., during haptotactic migration [68].

Another potential next step toward understanding physiology at the nanoscale might be the combination of SICM with techniques from the rapidly evolving field of super-resolution light microscopy [4, 69]. Although several obstacles concerning the poor resolution of super-resolution microscopy concerning the vertical axis still have to be overcome to unambiguously locate protein positions, our concepts of how cells react to environmental challenges will become, step-by-step, more and more complete.

References

1. Hansma PK, Drake B, Marti O, Gould SAC, Prater CB (1989) The scanning ion-conductance microscope. Science 243(4891):641–643
2. Rheinlaender J, Schäffer TE (2015) Lateral resolution and image formation in scanning ion conductance microscopy. Anal Chem 87(14):7117–7124
3. Korchev YE, Negulyaev YA, Edwards CRW, Vodyanoy I, Lab MJ (2000) Functional localization of single active ion channels on the surface of a living cell. Nat Cell Biol 2(9):616–619
4. Hagemann P, Gesper A, Happel P (2018) Correlative stimulated emission depletion and scanning ion conductance microscopy. ACS Nano 12(6):5807–5815
5. Galvani L (1791) De viribus electricitatis in motu musculari commentarius. Press of the Academy of Sciences, Bologna

6. Hodgkin AL, Huxley AF (1939) Action potentials recorded from inside a nerve fibre. Nature 144(3651):710–711
7. Hodgkin AL, Huxley AF (1952) A quantitative description of membrane current and its application to conduction and excitation in nerve. J Physiol (Lond) 117:500–544
8. Neher E, Sakmann B (1976) Single-channel currents recorded from membrane of denervated frog muscle fibres. Nature 260(5554):799–802
9. Doyle DA, Morais Cabral J, Pfuetzner RA, Kuo A, Gulbis JM, Cohen SL, Chait BT, MacKinnon R (1998) The structure of the potassium channel: molecular basis of K+ conduction and selectivity. Science 280(5360):69–77
10. Piccolino M (1997) Luigi Galvani and animal electricity: two centuries after the foundation of electrophysiology. Trends Neurosci 20:443–447
11. Cajavilca C, Varon J, Sternbach GL (2009) Resuscitation great. Luigi Galvani and the foundations of electrophysiology. Resuscitation 80(2):159–162
12. Parent A (2004) Giovanni Aldini: from animal electricity to human brain stimulation. Can J Neurol Sci 31(4):576–584
13. Finger S, Piccolino M, Stahnisch FW (2013) Alexander von Humboldt: galvanism, animal electricity, and self-experimentation part 1: formative years, naturphilosophie, and galvanism. J Hist Neurosci 22(3):225–260
14. Cajavilca C, Varon J, Sternbach GL (2009) Luigi Galvani and the foundations of electrophysiology. Resuscitation 80(2):159–162
15. Matteucci C (1838) Des contrations et des sensations produites dans les animeaux par le courant électrique. 357–364
16. Moruzzi G (1996) The electrophysiological work of Carlo Matteucci. 1964. Brain Res Bull 40(2):69–91
17. Matteuchi C (1843) Mémoires sur l'existence du courant électrique musculaire dans les animeaux vivantes ou réecemment tués 425–462
18. Du Bois-Reymond E (1843) Vorläufiger Abriss einer Untersuchung über den sogenannten Froschrom und über die elektromotorischen. Fischerblatt:1–30
19. Du Bois-Reymond EH (1863) Beschreibung einiger Vorrichtungen und Versuchsweisen zu elektrophysiologischen Zwecken. Abhandlungen der Königlichen Akademie der Wissenschaften zu Berlin, Physikalische Abhandlungen 163:3 Taf
20. Lippmann G (1873) Beziehungen zwischen den capillaren und elektrischen Erscheinungen. Ann Phys 225:546–561
21. Waller AD (1887) A demonstration on man of electromotive changes accompanying the heart's beat. J Physiol 8(5):229–234
22. Einthoven W (1895) Ueber die Form des menschlichen Electrocardiogramms. Pflugers Arch 60(3–4):101–123
23. Einthoven W (1909) Die Konstruktion des Saitengalvanometers. Pflugers Arch 130(5):287–321
24. Einthoven W (1903) Die galvanometrische Registrirung des menschlichen Elektrokardiogramms, zugleich eine Beurtheilung der Anwendung des Capillar-Elektrometers in der Physiologie. Pflugers Arch 99(9–10):472–480
25. Proebster R (1928) Über Muskelaktionsströme am gesunden und kranken Menschen. Zeitschrift für orthopädische Chirurgie. Enke, Stuttgart, p 154
26. Berger H (1929) Über das Elektrenkephalogramm des Menschen. Archiv f Psychiatrie 87:527–570
27. Berger H (1938) Das Elektrenkephalogramm der Menschen. Nova Acta Leopoldina 6(38): 173–309
28. Bernstein J (1902) Untersuchungen zur Thermodynamik der bioelektrischen Ströme. Archiv für die gesamte Physiologie des Menschen und der Tiere 92(10):521–562
29. Seyfarth E-A, Peichl L (2002) Vor 100 Jahren: Julius Bernstein (1839–1917) formuliert seine "Membrantheorie". e-Neuroforum 8(4):274–276
30. Marmont G (1949) Studies on the axon membrane. J Cell Comp Physiol 34:351–382

31. Cole KS (1949) Dynamic electrical characteristics of the squid axon membrane. Arch Sci Physiol 3:253–258
32. Hodgkin A, Huxley A (1952) Currents carried by sodium and potassium ions through the membrane of the giant axon of Loligo. J Physiol 116(4):449–472
33. Ling G, Gerard RW (1949) The normal membrane potential of frog sartorius fibers. J Cell Comp Physiol 34:383–396
34. Nicholls JA, Baylor D (1968) Specific modalities and receptive fields of sensory neurons in CNS of the leech. J Neurophysiol 31(5):740–756
35. Hille B (2001) Ionic channels of excitable membranes. Sinauer, Sunderland
36. Martin AR, Brown DA, Diamond ME, Cattaneo A, De Miguel FF (2020) From neuron to brain. Oxford University Press
37. Kandel ER, Koester JD, Mack SH, Siegelbaum SA (2021) Principles of neural science. McGraw-Hill Education
38. Binnig G, Rohrer H, Gerber C, Weibel E (1982) Surface studies by scanning tunneling microscopy. Phys Rev Lett 49(1):57–61
39. Binnig G, Quate CF, Gerber C (1986) Atomic force microscope. Phys Rev Lett 56(9):930–933
40. Drake B, Prater C, Weisenhorn A, Gould S, Albrecht T, Quate C, Cannell D, Hansma H, Hansma P (1989) Imaging crystals, polymers, and processes in water with the atomic force microscope. Science 243(4898):1586–1589
41. Henderson E, Haydon PG, Sakaguchi DS (1992) Actin filament dynamics in living glial cells imaged by atomic force microscopy. Science 257(5078):1944–1946
42. Braet F, Seynaeve C, Zanger RD, Wisse E (1998) Imaging surface and submembranous structures with the atomic force microscope: a study on living cancer cells, fibroblasts and macrophages. J Microsc 190(Pt 3):328–338
43. Seifert J, Rheinlaender J, Schäffer TE (2019) Scanning ion conductance microscopy and atomic force microscopy for live cell imaging: a comparison. Biophys J 116(3):326a–327a
44. Rheinlaender J, Schäffer TE (2009) Image formation, resolution, and height measurement in scanning ion conductance microscopy. J Appl Phys 105(9):094905
45. Proksch R, Lal R, Hansma PK, Morse D, Stucky G (1996) Imaging the internal and external pore structure of membranes in fluid: TappingMode scanning ion conductance microscopy. Biophys J 71(4):2155–2157
46. Schäffer TE, Ionescu-Zanetti C, Proksch R, Fritz M, Walters DA, Almqvist N, Zaremba CM, Belcher AM, Smith BL, Stucky GD (1997) Does abalone nacre form by heteroepitaxial nucleation or by growth through mineral bridges? Chem Mater 9(8):1731–1740
47. Korchev YE, Milovanovic M, Bashford CL, Bennett DC, Sviderskaya EV, Vodyanoy I, Lab MJ (1997) Specialized scanning ion-conductance microscope for imaging of living cells. J Microsc 188:17–23
48. Korchev YE, Bashford CL, Milovanovic M, Vodyanoy I, Lab MJ (1997) Scanning ion conductance microscopy of living cells. Biophys J 73(2):653–658
49. Shevchuk AI, Gorelik J, Harding SE, Lab MJ, Klenerman D, Korchev YE (2001) Simultaneous measurement of Ca2+ and cellular dynamics: combined scanning ion conductance and optical microscopy to study contracting cardiac myocytes. Biophys J 81(3):1759–1764
50. Gorelik J, Shevchuk A, Ramalho M, Elliott M, Lei C, Higgins CF, Lab MJ, Klenerman D, Krauzewicz N, Korchev Y (2002) Scanning surface confocal microscopy for simultaneous topographical and fluorescence imaging: application to single virus-like particle entry into a cell. Proc Natl Acad Sci U S A 99(25):16018–16023
51. Bruckbauer A, Ying L, Rothery AM, Zhou D, Shevchuk AI, Abell C, Korchev YE, Klenerman D (2002) Writing with DNA and protein using a nanopipet for controlled delivery. J Am Chem Soc 124(30):8810–8811
52. Gorelik J, Shevchuk AI, Frolenkov GI, Diakonov IA, Lab MJ, Kros CJ, Richardson GP, Vodyanoy I, Edwards CR, Klenerman D, Korchev YE (2003) Dynamic assembly of surface structures in living cells. Proc Natl Acad Sci U S A 100(10):5819–5822

53. Rodolfa KT, Bruckbauer A, Zhou D, Korchev YE, Klenerman D (2005) Two-component graded deposition of biomolecules with a double-barreled nanopipette. Angew Chem Int Ed Engl 44(42):6854–6859
54. Mann SA, Hoffmann G, Hengstenberg A, Schuhmann W, Dietzel ID (2002) Pulse-mode scanning ion conductance microscopy-a method to investigate cultured hippocampal cells. J Neurosci Methods 116:113–117
55. Novak P, Li C, Shevchuk AI, Stepanyan R, Caldwell M, Hughes S, Smart TG, Gorelik J, Ostanin VP, Lab MJ, Moss GWJ, Frolenkov GI, Klenerman D, Korchev YE (2009) Nanoscale live-cell imaging using hopping probe ion conductance microscopy. Nat Methods 6(4):279–281
56. Pellegrino M, Pellegrini M, Orsini P, Tognoni E, Ascoli C, Baschieri P, Dinelli F (2012) Measuring the elastic properties of living cells through the analysis of current-displacement curves in scanning ion conductance microscopy. Pflugers Arch 464(3):307–316
57. Rheinlaender J, Schäffer TE (2013) Mapping the mechanical stiffness of live cells with the scanning ion conductance microscope. Soft Matter 9(12):3230
58. McKelvey K, Kinnear SL, Perry D, Momotenko D, Unwin PR (2014) Surface charge mapping with a nanopipette. J Am Chem Soc 136(39):13735–13744
59. Chen C-C, Derylo MA, Baker LA (2009) Measurement of ion currents through porous membranes with scanning ion conductance microscopy. Anal Chem 81(12):4742–4751
60. Sánchez D, Anand U, Gorelik J, Benham CD, Bountra C, Lab M, Klenerman D, Birch R, Anand P, Korchev Y (2007) Localized and non-contact mechanical stimulation of dorsal root ganglion sensory neurons using scanning ion conductance microscopy. J Neurosci Methods 159(1):26–34
61. Adam Seger R, Actis P, Penfold C, Maalouf M, Vilozny B, Pourmand N (2012) Voltage controlled nano-injection system for single-cell surgery. Nanoscale 4(19):5843–5846
62. Actis P, Maalouf MM, Kim HJ, Lohith A, Vilozny B, Seger RA, Pourmand N (2014) Compartmental genomics in living cells revealed by single-cell nanobiopsy. ACS Nano 8(1): 546–553
63. Walsh DA, Fernández JL, Mauzeroll J, Bard AJ (2005) Scanning electrochemical microscopy. 55. Fabrication and characterization of micropipet probes. Anal Chem 77(16):5182–5188
64. Oberleithner H (2008) Nanophysiology: fact or fiction? Pflugers Arch 456(1):1–2
65. Shevchuk AI, Frolenkov GI, Sánchez D, James PS, Freedman N, Lab MJ, Jones R, Klenerman D, Korchev YE (2006) Imaging proteins in membranes of living cells by high-resolution scanning ion conductance microscopy. Angew Chem 118(14):2270–2274
66. Happel P, Möller K, Schwering NK, Dietzel ID (2013) Migrating oligodendrocyte progenitor cells swell prior to soma dislocation. Sci Rep 3:1806
67. Schwab A, Nechyporuk-Zloy V, Fabian A, Stock C (2007) Cells move when ions and water flow. Pflugers Arch 453(4):421–432
68. Seifert J, Rheinlaender J, von Eysmondt H, Schäffer TE (2022) Mechanics of migrating platelets investigated with scanning ion conductance microscopy. Nanoscale 14(22):8192–8199
69. Navikas V, Leitao SM, Grussmayer KS, Descloux A, Drake B, Yserentant K, Werther P, Herten DP, Wombacher R, Radenovic A, Fantner GE (2021) Correlative 3D microscopy of single cells using super-resolution and scanning ion-conductance microscopy. Nat Commun 12(1):1–9
70. Sirol M (1939) Galvani et la galvanisme: Thèse pour le Doctorat en Médecine, Univ. Toulouse. Vigot Frères, Paris

BIOREV (2022) 3: 23–72
https://doi.org/10.1007/11663_2022_15

Published online: 5 August 2022

Scanning Ion Conductance Microscopy and Atomic Force Microscopy: A Comparison of Strengths and Limitations for Biological Investigations

Hendrik von Eysmondt and Tilman E. Schäffer

Contents

Abstract Knowledge of physical properties of cells is vital for many research areas in biology and medicine. Atomic force microscopy (AFM) and scanning ion conductance microscopy (SICM) are two techniques to assess the three-dimensional

H. von Eysmondt and T. E. Schäffer (✉)
Institute of Applied Physics, University of Tübingen, Tübingen, Germany
e-mail: tilman.schaeffer@uni-tuebingen.de

topography and mechanical properties of cells. This chapter introduces the basic working principles and imaging modes of AFM and SICM and then focuses on their similarities and differences. Strengths and limitations in terms of image resolution, imaging speed, and biomechanical applications are discussed. Also, combined applications of SICM and AFM are highlighted. This chapter shows that SICM has emerged as a major addition to the field of biophysics.

Keywords AFM · Biomechanics · Cell morphology · High-speed imaging · Resolution · SICM

1 Introduction

Microscopy has been a key factor for discovery. The emergence and development of new microscopes are strongly intertwined with scientific progress. Imaging cells and molecules have been one of the major drivers in biological research, so it is not surprising that a multitude of different microscopy techniques have been invented [1].

The optical microscope was invented in the sixteenth century, overcoming the limits of the naked eye and opening the path to new discoveries in the world of micrometers. In optical microscopy, visible light is used to illuminate and visualize a sample. However, conventional optical microscopy is limited by the wavelength of the light used for illumination. This so-called optical diffraction limit, first described by Ernst Abbe [2], limits the resolution of conventional optical microscopy to around 200 nm for visible light.

Since then, numerous extensions to optical microscopy have been developed, and a family of microscopy techniques loosely gathered by the term scanning probe microscopy (SPM) has emerged. SPM employs a local probe to measure physical interactions with a sample surface. A prominent member of this family is scanning tunneling microscopy (STM), invented by Gerd Binnig and Heinrich Rohrer in 1981 [3], who were awarded with the Nobel Prize in Physics in 1986. In STM, an electric tunneling current between a sharp tip and a sample surface is used to measure the tip-sample distance, allowing to image the three-dimensional sample surface topography by laterally scanning the tip over the sample surface.

In 1986, Gerd Binnig, Calvin Quate, and Christoph Gerber invented atomic force microscopy (AFM) and demonstrated its capability for measuring the force between single atoms [4]. In AFM, a sharp tip attached to a flexible cantilever serves as a local probe. AFM can be used on both conductive and non-conductive samples in vacuum, air, and liquid environments [5]. This allowed scientists to use AFM for the investigation of biological samples in physiological conditions [5–10]. The resolution of AFM is on the nanometer scale [11], making it ideal for imaging biomolecules such as proteins [12], lipids [13, 14], and DNA [15]. Moreover, AFM is a force sensing device, capable of measuring adhesion forces [16, 17] between molecules or cells [18–22] and mechanical properties of cells [8, 22–24]. A multitude of different biological samples has been investigated with AFM [7, 24–39].

In 1989, scanning ion conductance microscopy (SICM) was invented by Hansma et al. [40]. In SICM, a faradaic current is passed through a nanopipette that is scanned over an electrolyte-submerged sample. The current strongly depends on the tip-sample distance and is used to generate images of sample topography without mechanical tip-sample contact [41–44]. One of the earliest applications of SICM to a biological system was the investigation of the structure of abalone nacre [45]. Korchev et al. [46] first demonstrated that live cells can be imaged with SICM, opening the door to a multitude of biological investigations [43, 47–66].

Today, both AFM and SICM are routinely used in biology. In this chapter, we explain the basic similarities and differences between AFM and SICM. We discuss various aspects of the respective imaging process, such as scanning speed, scan range, and resolution. We introduce instrumental modifications to optimize the performance of these microscopy techniques and present different applications in various research areas. We focus on biological samples, as they are the main application area of SICM. Section 2 introduces the fundamental principles of AFM and SICM. Section 3 explores the capabilities of AFM and SICM as imaging techniques with high spatial resolution. Section 4 compares AFM and SICM in terms of temporal resolution and presents high-speed approaches. Section 5 focuses on biomechanics, especially on the analysis of the elastic and viscoelastic properties of cells. Lastly, Section 6 discusses endeavors in combining AFM and SICM.

2 AFM and SICM Basics

AFM and SICM are complex techniques, but their basic principles of operation are simple to understand. They both use a probe to interact with a sample and visualize its properties, such as surface topography. However, the nature of the probes, their interaction with the sample, and the interpretation of the measured signals are quite distinct.

2.1 Principle of Operation

2.1.1 AFM

A basic AFM setup consists of a cantilever with a sharp tip, a laser, and a photodiode (Fig. 1, left) [4]. Typical cantilever and tip lengths are 100 and 5 μm, respectively. The radius of curvature at the tip apex is usually 5–20 nm, but specialized probes with a smaller or larger radius exist. A laser beam is reflected from the backside of the cantilever and is projected onto a segmented photodiode ("optical lever" principle) [67, 68]. Upon contact of the cantilever tip with the sample surface, the

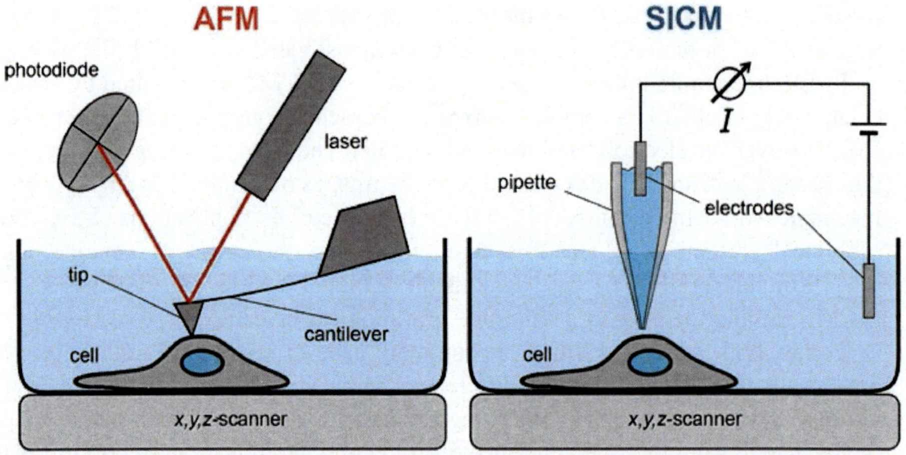

Fig. 1 Schematic of AFM and SICM. In AFM (left), the tip of a flexible cantilever probes the sample (e.g., a cell). The force between the tip and the sample deflects the cantilever. This deflection (and thereby the force) is measured via a laser beam that is reflected from the backside of the cantilever onto a segmented photodiode. An *x*,*y*,*z*-scanner moves the sample relative to the tip. In SICM (right), the tip of a nanopipette serves as the probe. An ion current through the nanopipette is induced by applying a voltage between two electrodes. When the nanopipette approaches the sample surface, the conductive space between the tip and the sample shrinks and thus the conductance and consequently the ionic current decrease. A measurement of the ion current therefore quantifies the tip-sample distance without mechanical tip-sample contact. Adapted with permission from [74]. Copyright 2015 American Chemical Society

cantilever bends, causing the position of the laser beam on the segmented photodiode to change. This change is detected, allowing to measure the deflection of the cantilever and thereby the force on the sample (the force is the deflection multiplied by the cantilever's spring constant). To image the sample topography, the tip is scanned over the sample surface using piezoelectric actuators with sub-nanometer precision. The maximum scan range is typically 100 μm. AFM can be operated in a vacuum [69], a gaseous [4], or a liquid environment [5].

2.1.2 SICM

The probe in a SICM setup (Fig. 1, right) is a nanopipette filled with an electrolytic liquid [40, 70]. A typical inner opening diameter of the nanopipette ranges from 10–1,000 nm. Similar as in the electrophysiological patch clamp technique [71, 72], an electric voltage applied between two electrodes (usually made of Ag/AgCl), one inside the nanopipette and the other one inside the bath solution, induces an ionic current (usually in the range of nanoamperes) through the nanopipette. When the nanopipette approaches a sample, the conductive space between the nanopipette tip and the sample decreases, leading to a reduction of the conductance between the electrodes and thus to a reduction of the measured ion current [40]. This allows for a measurement of the tip-sample distance without mechanical contact between the tip and the sample and can be used to assess sample properties such as topography. The maximum scan range is typically 100 μm, but a large-range scanner with a scan range of 25 mm has been implemented [73].

2.2 Force-Distance and Current-Distance Curves

The dependency of the primary measurement signal (deflection or force in the case of AFM, ion current in the case of SICM) on the tip-sample distance is fundamental to understanding the working principles of these techniques. In this section, we discuss force-distance curves (*FZ*-curves, case of AFM) and current-distance curves (*IZ*-curves, case of SICM) on the example of a mouse embryonic fibroblast (MEF) (Fig. 2a) that is measured both live and fixed with both AFM (Fig. 2b) and SICM (Fig. 2c).

2.2.1 AFM

FZ-curves on a MEF in a liquid environment are shown in Fig. 2b (bottom). To record a *FZ*-curve, the cantilever is lowered toward the sample surface with the help of a piezoelectric z-scanner. When the tip is far away from the surface, the measured

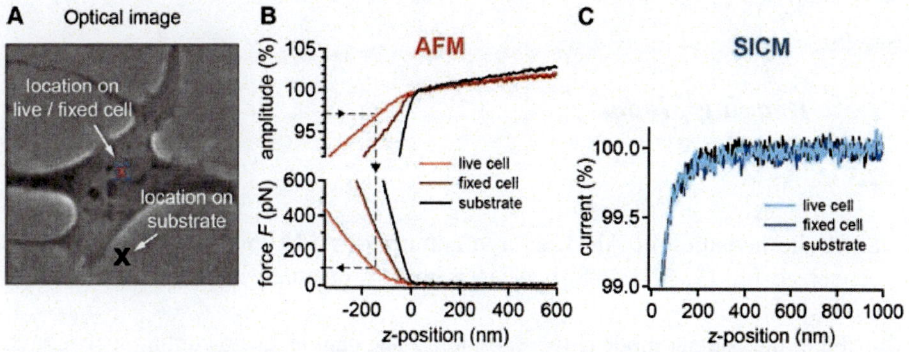

Fig. 2 Vertical approach curves on cells. (**a**) Optical phase contrast image of a live mouse embryonic fibroblast (MEF). (**b**) Amplitude-distance curves (relevant for the AFM tapping mode; top), *FZ*-curves (relevant for the AFM contact mode and the AFM force mapping mode; bottom), and (**c**) *IZ*-curves (relevant for the SICM backstep/hopping mode) on the cell (before/after fixation) and on the underlying substrate, as marked in (**a**). When the cantilever contacts the sample (**b**, $z = 0$), the amplitude decreases and the force increases upon further approach. The dashed line (**b**) indicates a tapping mode setpoint of 97% with an imaging force of roughly 100 pN. (**c**) When the SICM tip comes into close proximity to the sample (here: <200 nm), the ion current decreases. The AFM curves are influenced by the sample stiffness, while the SICM *IZ*-curves are not, indicating that SICM allows for imaging samples at zero mechanical contact force (force curves on the live cell, the fixed cell, and the substrate are shown as examples of increasing stiffness). Adapted with permission from [74]. Copyright 2015 American Chemical Society

force F is zero. When the tip contacts the sample ($z = 0$, for a decreasing z-position), the measured force increases. When the AFM cantilever is additionally vertically vibrated with a small amplitude (as is done in the tapping mode, s. below), an amplitude-distance curve can be recorded (Fig. 2b, top). The vibration amplitude sharply decreases when the cantilever contacts the surface.

The slope of the *FZ*-curve and the slope of the amplitude-distance curve depend on the sample stiffness (Fig. 2b), as can be seen by the different curves for a live MEF (light red), a fixed MEF (dark red), and the substrate (black). We will further discuss this dependency and its use to derive information about the sample's mechanical properties in Sect. 5.

2.2.2 SICM

IZ-curves on the same MEF are shown in Fig. 2c. The ion current I is measured as a function of the z-position of the nanopipette. When the nanopipette is far from the surface, the measured ion current I is constant (at 100% of the saturation current I_0). When the nanopipette approaches the sample surface (decreasing z-position), the ion current decreases. It can be seen that the *IZ*-curves are independent of the sample stiffness, in contrast to the *FZ*-curves in AFM. *IZ*-curves, which can be accurately predicted with a theoretical model [75], depend on the conductivity of the electrolyte, the applied voltage, and the pipette opening size and geometry. *IZ*-curves on a sloped sample surface also depend on sample slope, but this effect can be corrected [76].

2.3 *Imaging Modes*

2.3.1 AFM

Since the invention of AFM, a large number of AFM imaging modes have been conceived [4, 23, 32, 77–82]. AFM is mostly operated in one of three imaging modes.

First, the contact mode (Fig. 3, top left): The cantilever tip continuously contacts the sample surface at a constant force using a feedback loop while scanning laterally across the sample [4]. The contact mode is useful mainly for stiff samples, as the resulting tip-sample forces can influence or even damage soft samples [79], such as live cells [83].

Second, the tapping mode [79] (also called amplitude modulation mode or AC mode) (Fig. 3, top center): The cantilever is vibrated vertically near a resonance, e.g., by driving it with a piezoelectric element, and the resulting vibration amplitude is kept constant by a feedback loop [23, 79, 80, 84]. The vibration significantly reduces the tip-sample contact time and subsequently the lateral tip-sample forces. The

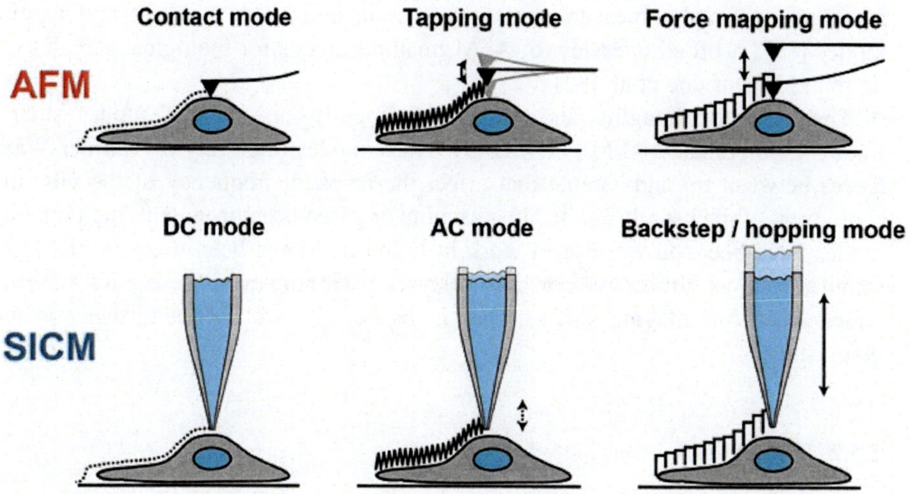

Fig. 3 AFM and SICM imaging modes. AFM contact mode (top left): The cantilever is scanned over the sample while maintaining a constant tip-sample imaging force. A feedback loop controls the vertical (z) position of the cantilever. AFM tapping mode (top center): The cantilever tip vibrates near a resonant frequency and is scanned over the sample while maintaining a constant vibration amplitude using a feedback loop. By intermittently "tapping" the sample, shear forces between tip and sample and thereby possible damage to the sample are reduced. AFM force mapping mode (top right): The cantilever tip is vertically lowered toward the sample until a predefined force threshold is reached (trigger event). The tip is then retracted from the sample and moved laterally to the next location. Recording a force-distance curve (*FZ*-curve) for each location allows the determination of sample properties such as topography, stiffness, or adhesion. SICM DC mode (bottom left): The nanopipette is scanned over the sample by maintaining a constant current through the nanopipette using a feedback loop that controls the z-position of the nanopipette. The nanopipette tip is thereby kept at a constant distance to the sample surface, allowing for the extraction of the sample's topography. SICM AC mode (bottom center): The nanopipette is vertically oscillated and the amplitude of the current through the nanopipette is measured. The amplitude increases with a decreasing tip-sample distance. A feedback loop keeps the amplitude constant during the scanning process. This imaging mode is insensitive to small drifts in the current and is, therefore, more stable than the DC mode. SICM backstep/hopping mode (bottom right): The nanopipette approaches the sample until a predefined current threshold is reached (trigger event), upon which the momentary z-position of the nanopipette is stored and the nanopipette is retracted from the sample. The nanopipette is then moved laterally to a new position over the sample and the process is repeated. The stored z-positions at the trigger events give a map of the sample's topography

tapping mode is less damaging than the contact mode and can be used for soft samples such as live cells [85].

Third, the force mapping mode (Fig. 3, top right): The cantilever is approached toward the sample and is retracted again when the tip-sample force reaches a predetermined level. During this motion, the force acting on the cantilever and the z-piezo position are recorded (*FZ*-curve). The cantilever is then moved to another location on the sample and is approached again. By repeating this process and thereby scanning the cantilever over the sample, its topography or mechanical properties (such as adhesion or elastic modulus) can be mapped [24, 81, 82, 86]. For soft samples, it was shown that the force mapping mode provides more accurate topography measurements when compared to the contact and tapping modes [87]. A broad overview of AFM imaging modes for biological samples can be found in Dufrêne et al. [32].

There are other imaging modes, most prominently "non-contact" modes, such as frequency-modulated AFM [77, 88, 89]. These modes often rely on van-der-Waals forces between tip and sample that affect the resonant frequency of the vibrating cantilever. While usually applied in a vacuum or gaseous environment, such imaging modes have been shown to also work in liquid and have been used on biological samples such as viruses [90] or lipid bilayers [91]. However, these modes are not widely used for imaging soft samples in liquid, and we do not further consider them here.

2.3.2 SICM

Several different SICM imaging modes have been developed [40, 43, 47, 49, 53, 56, 92–96]. An in-depth overview of these developments can be found in Zhu et al. [65]. We will focus on the three most commonly used imaging modes.

First, the DC (direct current) mode (Fig. 3, bottom left): A feedback loop maintains a constant current between the electrodes by controlling the vertical (z) position of the nanopipette. This original operating mode of SICM [40] was later also called constant current mode. The DC mode offers a fast scanning speed, but is best used for flat samples, because the scanning probe may collide with convoluted samples that have abrupt changes in height (like cells usually have). A collision can lead to imaging artifacts or even to sample and probe damage. Imaging at larger tip-sample distances is safer, but decreases the lateral resolution [97]. Additionally, the DC current can drift during prolonged imaging, leading to changes in the tip-sample distance and causing artifacts [42].

Second, the AC (alternating current) mode (Fig. 3, bottom center): This mode is a major improvement over the DC mode. The nanopipette is oscillated vertically (at 100–10,000 Hz) with a fixed amplitude [42, 48, 93]. When the nanopipette is close to the sample surface, this oscillation causes an oscillation of the measured ion current. The amplitude of this current oscillation increases for a decreasing tip-sample distance. A constant oscillation amplitude is maintained by adjusting the z-position via a feedback loop while laterally scanning over the sample. The AC mode is less prone to DC current drift and therefore allows more stable imaging, especially at larger tip-sample distances. Different AC mode variants introduced an AC component into the DC voltage instead of physically oscillating the nanopipette up and down, by modulating bias [98, 99], phase [100], or amplitude [101].

Third, the backstep (or hopping) mode (Fig. 3, bottom right): While the AC mode can be used to visualize samples of moderate surface roughness, steep slopes like the borders of adherent cells are still difficult to visualize in the AC mode. The solution was to alternate vertical and lateral movement of the probe, analogous to the force mapping mode in AFM. Several groups have developed or adapted such an approach, naming it backstep mode [49], hopping mode [43], or standing approach mode [94]. Analogous to the AFM force mapping mode, for each location on the sample, the probe is vertically approached to the sample surface until a predefined current threshold is reached (trigger event). Then the probe height is recorded, and the probe is retracted from the surface again before moving laterally to the next location and repeating this process. Because the lateral motion is performed at a large tip-sample distance, the danger of probe or sample damage is greatly diminished. Compared to the DC or AC modes, however, the backstep/hopping mode is slow.

2.3.3 Comparison of AFM and SICM Imaging Modes

The AFM contact mode is similar to the SICM DC mode. They were the first modes introduced with the respective technique and are the modes that use feedback to keep the primary measurement signal (force or deflection in the case of AFM, ion current in the case of SICM) constant. Both modes allow for fast imaging, but are limited to robust and stiff (AFM) or flat (SICM) samples. While the probe is always in contact with the sample in the AFM contact mode, there are (ideally) no tip-sample forces in the SICM DC mode for flat samples, allowing to image samples as soft as live cells with this mode.

The AFM tapping mode is similar to the SICM AC mode. In these modes, the probe is vertically vibrated while scanning over the sample. Both modes detect the modulation amplitude of the respective technique's primary measurement signal to obtain the topography of the sample. Just as in the AFM contact mode, there are still vertical and lateral forces applied to the sample in the AFM tapping mode that might disrupt soft samples. This is in stark contrast to the AC mode in SICM, where such disruptions were not observed [102].

The AFM force mapping mode is similar to the SICM backstep/hopping mode. In these modes, the probe is lowered toward the sample to a predefined threshold and is then retracted again. These two modes are well suited for investigations of live cells. These two modes can also be used to obtain additional information about the sample, e.g., viscoelastic properties (see Sect. 5).

AFM works in both air and liquid environments and even in a vacuum, while SICM generally only works if the sample is submerged in an electrolyte. However, it was shown that SICM can be operated in the air using a meniscus of electrolyte between the sample and a double-barreled nanopipette [103]. In this configuration, the nanopipette can be used, for example, to deposit fluorescent DNA onto a sample surface, thereby creating paintings at micrometer resolution [51, 103]. A double-barreled nanopipette was also used to produce water droplets with volumes of attoliters inside of organic solutions with high precision [104]. Table 1 summarizes some of the key characteristics of the discussed imaging modes.

Table 1 Comparison of AFM and SICM imaging modes. Modified from [105]. Creative Commons Attribution 3.0 Unported License

Imaging mode	Speed	Sample topography	Characteristics
AFM contact mode SICM DC mode	Fast	Flat	• Exertion of lateral sample forces (AFM) • Possible tip-sample contact for rough samples (SICM) • Prone to DC drift
AFM tapping mode SICM AC mode	Fast	Moderate	• Imaging at larger average tip-sample distances • Reduced lateral forces (AFM) • Less likely tip-sample contact (SICM)
AFM force mapping mode SICM backstep/hopping mode	Slow	Arbitrary	• Best for imaging convoluted/complex topographies such as of entire cells • Low temporal resolution • Additional functional imaging possible (e.g., elastic modulus, surface charge)

3 Resolution

In optical microscopy, resolution can be defined as the minimum distance at which two points can still be distinguished as individual points in the image [2, 106]. In this section, we discuss and compare the resolution and image formation for both AFM and SICM in the context of biological samples.

3.1 AFM Resolution

AFM can achieve sub-Ångström resolution [11, 77, 107–109], both vertically and laterally. Most AFM setups have a vertical detection sensitivity of 0.01 nm or lower [109, 110], but the vertical resolution is mainly limited by mechanical thermal noise of the cantilever. The deflection of a free rectangular cantilever in an AFM setup with an optical lever-based deflection detection [67, 68] typically exhibits a total (root mean square, RMS) vertical thermal noise [110, 111] of

$$\Delta z = \sqrt{\frac{4k_{\mathrm{B}}T}{3k}} \tag{1}$$

where k_{B} is the Boltzmann constant, k is the spring constant of the cantilever, and T is the absolute temperature. For a cantilever with a spring constant of 1 N/m, the total thermal noise at room temperature is 0.074 nm. Stiffer cantilevers display less total thermal noise.

The lateral resolution of AFM on rigid samples mainly depends on the geometry of the cantilever tip and on the vertical resolution [109]. The minimum detectable

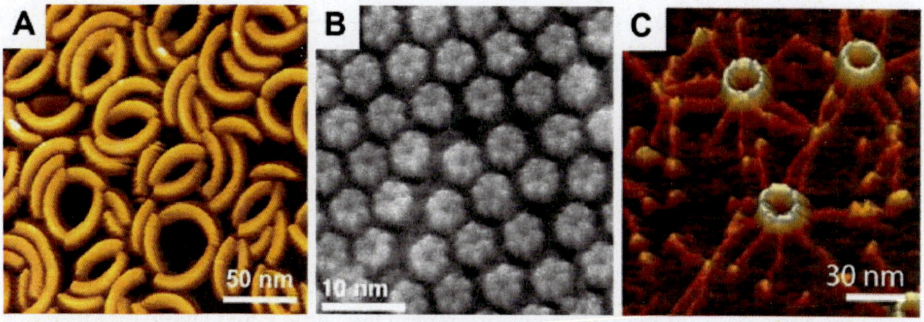

Fig. 4 Examples of high-resolution AFM topography imaging. (**a**) Self-assembled listeriolysin O, a major virulence factor of *Listeria* monocytogenes. Adapted with permission from [127]. Copyright 2015 American Chemical Society. (**b**) Extracellular surface of connexons from rat liver epithelial cells. Adapted from [128], Copyright 2006 with permission from Elsevier. (**c**) SAS-6 oligomers showing their typical cartwheel structures. Adapted with permission from [129]. Copyright 2014 American Chemical Society

distance d between two rigid, sharp objects with a small height difference Δh can be calculated from the radius of curvature R of the tip and from the vertical resolution Δz [112]:

$$d = \sqrt{2R}\left(\sqrt{\Delta z} + \sqrt{\Delta z + \Delta h}\right) \tag{2}$$

In general, the most limiting factor for lateral resolution is tip sharpness. Lateral resolution also depends on tip geometry [113, 114] and on sample stiffness, which sets the tip-sample contact radius [74, 115]. For cells, which are often very soft (0.1—several 100 kPa) [116, 117], the lateral resolution is often majorly reduced. Cell fixation by aldehydes like formaldehyde or glutardialdehyde therefore usually increases the lateral resolution of AFM [118, 119].

AFM has been widely used for single-cell and single-molecule investigations [35, 120] in biology down to nanometer resolution. For example, AFM was utilized to show the polymerization of fibrin, an important protein involved in blood coagulation [5]. The motion of single DNA molecules [121, 122], the activity of RNA polymerase [123] and the dynamic interactions of p53 with DNA [124] were resolved, as well as both strands of the DNA double helix [125]. Bacteriorhodopsin adsorbed to mica was imaged at a resolution down to 0.5 nm [7, 126]. Other examples include the self-assembly of a virulence factor of *Listeria* [127], connexons of rat liver epithelial cells [128], and the structure of SAS-6, a protein vital in centrosome duplication [129] (Fig. 4). Morphological and nanomechanical changes of MCF-7 cells induced by resveratrol, an anti-cancer drug, were detected by AFM [130]. AFM revealed that the outer S-layer of *Corynebacterium glutamicum* consists of hexagonal structures with a diameter of just 16 nm [131]. The architecture of the Gram-positive cell walls of *Staphylococcus aureus* and *Bacillus subtilis* were imaged on the molecular level using AFM [132], and the local stiffness of podosomes was mapped with AFM in combination with super-resolution microscopy [133]. Finally, AFM alongside scanning electron microscopy and confocal microscopy has been used to reveal the viral structure of SARS-CoV-2 in nasopharyngeal tissue from infected humans, revealing that it is similar to Sars-CoV [134].

3.2 SICM Resolution

As for AFM, resolution of SICM is different in the vertical and lateral directions. The vertical resolution of SICM is mainly limited by the electrical noise in the measurement of the ionic current through the nanopipette. The vertical resolution in hopping mode imaging (the mode mostly used for cell measurements) can be estimated by considering the minimal detectable current drop in an *IZ*-curve (Fig. 5) [43]. Expressed as a percentage of the reference current far from the surface (I_{ref}), the minimum detectable current drop was 0.25% for a 100 nm diameter nanopipette and 0.75% for a 30 nm diameter nanopipette. With these values and a chosen current

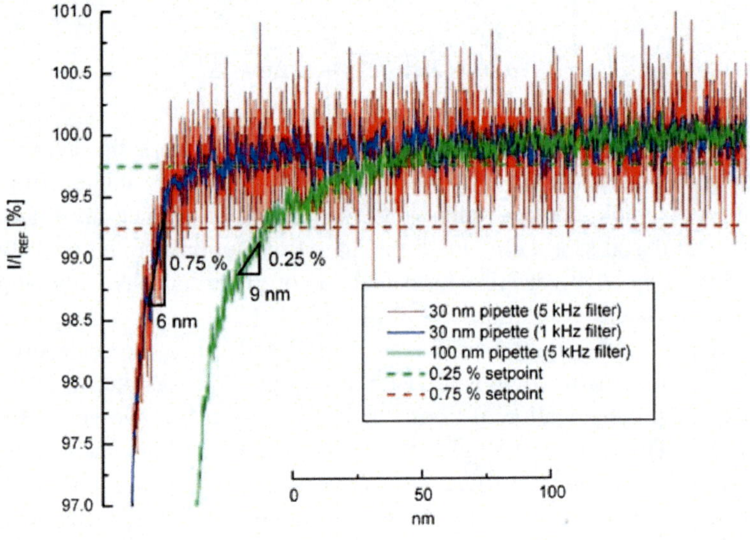

Fig. 5 Vertical resolution in hopping mode SICM. Shown are *IZ*-curves for specified nanopipettes approaching a flat glass surface. Dashed lines denote the minimum detectable current drop for the respective *IZ*-curves in the same color (0.25% for a 100 nm nanopipette, 0.75% for a 30 nm nanopipette, 5 kHz low pass filter), assuming a 3:1 signal-to-noise-ratio for detecting a non-stochastic change of the current. At a 1% trigger point, the vertical resolution can be estimated, using the slope of the *IZ*-curve and the minimum detectable current drop (black triangles), as 6 nm for a 30 nm nanopipette and 9 nm for a 100 nm nanopipette (for this particular case). The signal-to-noise-ratio can be improved by reducing the bandwidth of the low pass filter, enhancing the vertical resolution to $\approx$3 nm for the 30 nm nanopipette (blue *IZ*-curve, 1 kHz low pass filter), but this increases the response time for detecting the trigger event. Reprinted with permission from Springer Nature [43], supplementary information. Copyright Nature Methods 2009

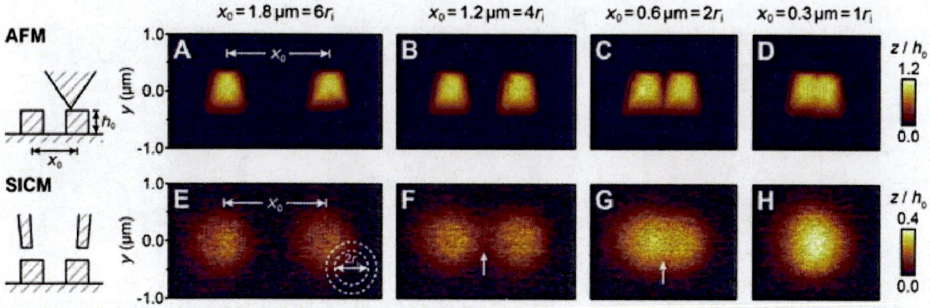

Fig. 6 Lateral resolution in SICM. Pairs of pillars of height h_0 were microfabricated with a varying distance x_0 between the pillars. The height of the pillars was $h_0 = 1.1\ r_i$, where r_i is the inner radius of the nanopipette. **(a–d)** AFM and **(e–h)** SICM images of the pillars. The pillars appear blurred in the SICM images and overlap at a distance of about $3r_i$ and below. Adapted with permission from [135]. Copyright 2015 American Chemical Society

threshold of 1%, a vertical resolution of 9 nm and 6 nm, respectively, was estimated (black triangles in Fig. 5). When reducing the detection bandwidth by a factor of 5, a vertical resolution of 3 nm was estimated (Fig. 5).

The lateral resolution in SICM was investigated by simulation [97] and by experiment [40, 135]. Microfabricated pillars at different distances from each other were imaged with both AFM and SICM (Fig. 6). In AFM, the top faces of the pillars were accurately reproduced, while their side walls appeared sloped ("tip artifact" due to the pyramidal-shaped cantilever tip). In SICM, the resulting images of the pillars were blurred (bell-shaped). Even lower pillars can appear ring-shaped [135]. In these reports, it was shown that the lower limit of lateral resolution d in SICM is roughly three times the nanopipette's inner opening radius r_i:

$$d = 3\ r_i \tag{3}$$

The inner opening radius r_i can be determined, e.g., from an *IZ*-curve [75].

The so far highest lateral resolution (as defined above) of the order of 13 nm has been reported for SICM using slim quartz nanopipettes on a monolayer of S-layer proteins from *Bacillus sphaericus* CCM 2177 [52]. The occasional occurrence of missing protein molecules in the unprocessed image of the protein lattice shows that this resolution is a "true" lateral resolution (i.e., separation of individual points on the sample) and does not only reflect lattice periodicities that can be detected even beyond the fundamental resolution limit [97, 135].

Different processes to easily fill [136, 137] and characterize [138] sub-10-nm nanopipettes have been developed—a difficult task as strong capillary forces normally prevent the filling of the nanopipette tip at these aperture sizes. To achieve sub-10-nm resolution, nanopipettes with an inner opening radius of less than 4 nm are needed. Nanopipettes with an inner opening radius of only ≈2.3 nm have been fabricated [136]. With such nanopipettes, SICM imaging with a lateral resolution below 10 nm might become possible.

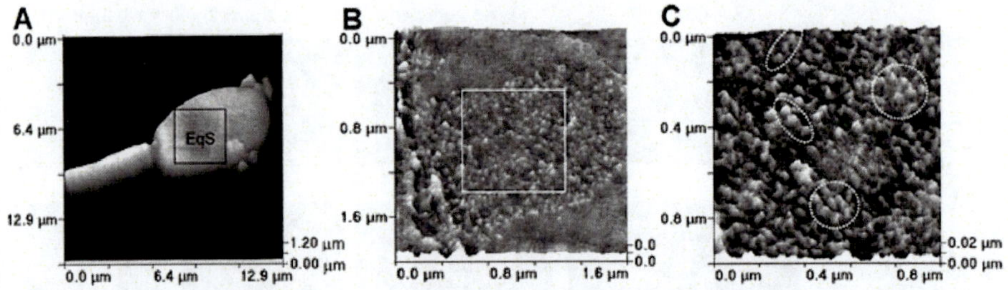

Fig. 7 SICM topography images of protein complexes on the surface of a live boar spermatozoon undergoing a spontaneous acrosome reaction. (**a**) Low-resolution image visualizing the spermatozoon. (**b**) Higher resolution image of the equatorial segment (EqS) region (marked with a box in (**a**)), revealing flat and more convoluted areas (subsegments) of the membrane. (**c**) High-resolution image of the area marked with a box in (**b**). Adapted with permission from [52] Copyright 2006 WILEY-VCH Verlag GmbH & Co. KGaA, Weinheim

SICM has been increasingly used for high-resolution visualization of cellular and sub-cellular structures. Examples include the visualization of microvilli of renal cells [50], murine melanocytes [139], the structure and liquid interaction of leaves [140], amyloid fibrils [55], collagen fibrils, chromosomes and HeLa cells [141], trachea and renal glomerulus of rats observed by both SEM and SICM [142], primary cilia from human retinal pigment epithelium-1 (RPE-1) and Madin-Darby canine kidney (MDCK) cells [143], Z-grooves and t-tubules of cardiomyocytes [144], osteoblasts [145], and a multitude of other studies [65]. Single protein complexes on the surface of a live spermatozoa (Fig. 7) [52] and tip links on hair cells of the inner ear of a mouse (Fig. 8) [43] were imaged by high-resolution SICM.

3.3 *Comparison of Resolution in AFM and SICM*

As illustrated in the previous two sections, resolution of SICM and AFM is not a global constant, but is rather highly dependent on the circumstances. On rigid samples, AFM clearly has higher resolution compared to SICM: AFM can resolve and manipulate even single atoms [11, 77, 107]. Different groups experimentally compared AFM and SICM by imaging the same samples using both techniques. They concluded that AFM generally provides better resolution than SICM, but that AFM can cause damage to delicate features such as cell protrusions [74, 102, 141]. Also, the resolution of AFM on soft samples such as live mammalian cells is reduced, e.g., to 50–100 nm [146].

Figure 9 shows the height profile of a MEF cell scanned both with tapping mode AFM and with hopping mode SICM. For AFM, the shape of the cell strongly depended on the applied imaging force. At low imaging forces (<0.9 nN), AFM overestimated the width of the cell and underestimated its height. At higher imaging forces (>0.9 nN), AFM underestimated both the width and the height of the cell. A deformation of the cell was clearly observed. A deformation of live cells by AFM

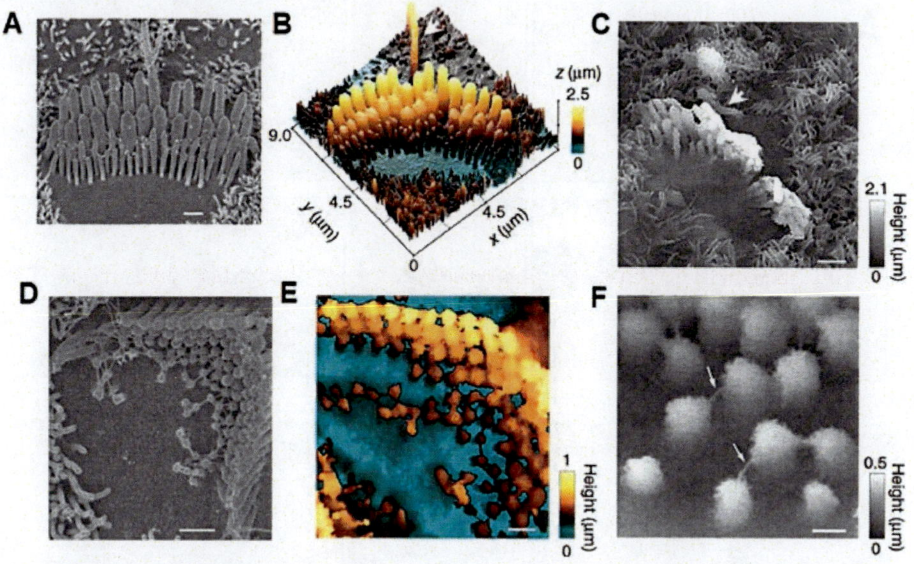

Fig. 8 High-resolution SEM and SICM images of stereocilia of mouse inner and outer hair cells. (**a**) SEM image and (**b**, **c**) SICM images of stereocilia of mouse wild-type inner hair cells including a kinocilium (white arrow). (**d**) SEM and (**e**, **f**) SICM images of stereocilia of outer hair cells in young postnatal Shaker 2 mice. Arrows indicate extracellular filaments interconnecting stereocilia ("tip links"). Only hair cells located approximately at the middle of the cochlea are shown. Scale bars, 500 nm (**a**, **d**, **e**), 1 µm (**c**), and 200 nm (**f**). Adapted with permission from Springer Nature [43]. Copyright Nature Methods 2009

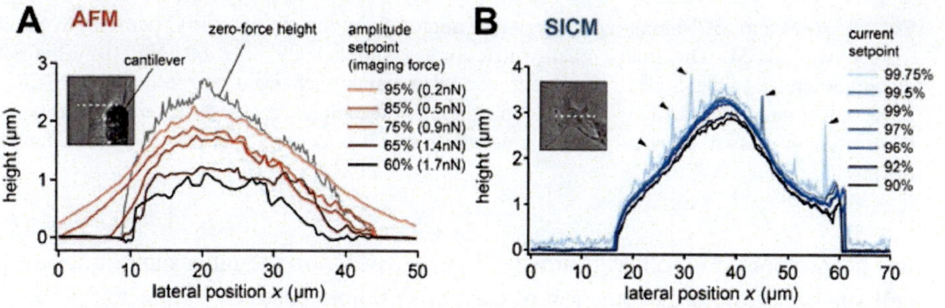

Fig. 9 Influence of the imaging setpoint on sample surface tracking and cell shape in AFM and SICM. (**a**) Height profiles of a live MEF acquired with different amplitude setpoints in AFM tapping mode (red to black lines). The force mapping mode was used to acquire the zero-force height (gray line). For setpoints at 75% and above, the substrate was not imaged accurately, and the cell was slightly indented. For lower setpoints, the substrate was imaged accurately, but the cell showed profound deformations. (**b**) Height profile of a MEF for different setpoints of the ion current. Spikes in the height profile for ion current setpoints of 99.5% and above (black arrowheads) are premature trigger events caused by noise. In contrast to AFM, SICM accurately tracks the substrate and the MEF. Both AFM and SICM setpoints were relative to the free amplitude and the free ion current, respectively. Adapted with permission from [74]. Copyright 2015 American Chemical Society

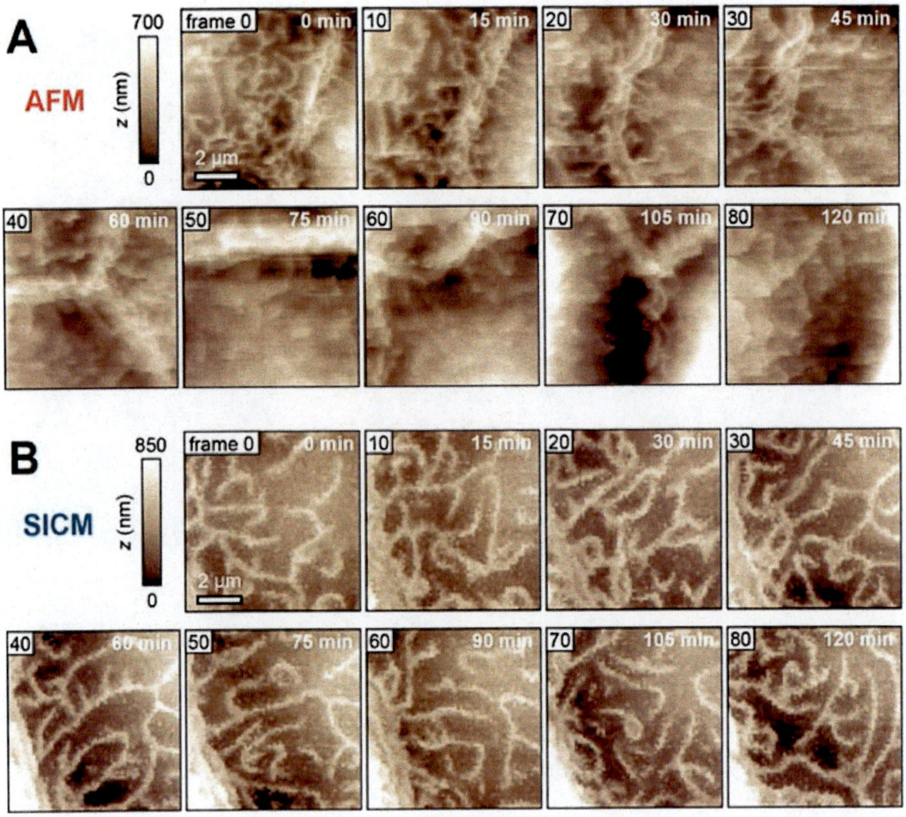

Fig. 10 AFM and SICM long-term imaging of microvilli on live A6 cells. (**a**) Microvilli dynamics imaged with AFM. The image quality deteriorates significantly after 60 min. (**b**) Microvilli dynamics imaged with SICM. The image quality remains unchanged, suggesting that SICM is better suited than AFM for long-term live imaging of soft samples. Image duration 90 s, only every tenth frame is shown. Adapted with permission from [74]. Copyright 2015 American Chemical Society

has been reported by other groups [83, 147]. SICM, on the other hand, tracked the cell surface without deformation at current setpoints between 99.5% and 97%. At setpoints <97%, the nanopipette touched and deformed the cell. At a setpoint of 99.75%, spikes were visible in the height signal due to noise in the current measurement that caused premature trigger events (black arrows in Fig. 9).

As a direct comparison of AFM and SICM imaging, microvilli on live frog epithelial A6 cells were imaged over 120 min with both AFM and SICM (Fig. 10) [74]. Both AFM and SICM revealed the high topographical dynamics of the microvilli (which were first observed with SICM by Gorelik et al. [50, 148]). However, the image quality in AFM gradually degraded with time (significantly after 60 min), while SICM retained a good image quality during the whole experiment. The degradation of the image quality in AFM is likely connected to the non-zero force of the cantilever tip on the cell, causing cell debris to stick and accumulate on the tip [74].

AFM and SICM are comparable in terms of resolution to some other microscopy techniques used in molecular and cell biology (Table 2). However, resolution in AFM strongly depends on sample stiffness [118], while SICM is a "non-contact" technique and its resolution is independent of sample stiffness (Fig. 11) [74]. In summary, SICM can have advantages over AFM in terms of lateral resolution, image quality, long-term imaging stability, and accuracy in topography measurements, especially for soft and delicate samples such as single suspended phospholipid membranes [54] or live cells [74].

4 High-Speed Imaging

Cellular dynamics such as contractions of cardiomyocytes [27, 61], vesicle recycling and structural dynamics of hippocampal neurons [63, 149], or membrane and cytoskeletal dynamics of cells [150, 151] are often on the scale of (milli-)seconds. Several techniques have been devised to resolve these high-speed dynamics at a high temporal resolution. Optical microscopy is still the standard for high-speed live cell imaging. Its main advantage is the acquisition over a wide field at 100–10,000 frames per second using high-speed cameras [152]. Its main disadvantage is the diffraction limit of ≈200 nm. The diffraction limit can be overcome with super-resolution microscopy [106, 153–156], but at the cost of losing imaging speed and having to label samples with fluorophores.

Both AFM and SICM are members of the SPM family, meaning that a probe is scanned over the sample, often taking several minutes per frame in a typical experiment. This slow imaging speed (compared to optical methods) regularly limits the capture of biological dynamics and events that occur on the second and sub-second time scale. There have been developments, however, to increase the scanning speed of AFM and SICM while retaining a high spatial resolution, sprouting the fields of high-speed AFM (HS-AFM) and high-speed SICM (HS-SICM).

4.1 High-Speed AFM

HS-AFM is often done in the contact or tapping mode, as these modes are faster than the force mapping mode (see Sect. 2.3). In the tapping mode, the imaging speed is limited by the cantilever's resonant frequency. Stiffer cantilevers have higher resonant frequencies, but also higher spring constants, making them more likely to damage soft samples. The first reports of HS-AFM for soft samples were published by the Hansma group and included the development of small cantilevers with high resonant frequencies and small spring constants as well as suitable detection

Table 2 High-resolution microscopy techniques used in molecular and cell biology

	Scanning ion conductance microscopy	Atomic force microscopy	Super-resolution microscopy (STED, PALM, STORM)	Transmission electron microscopy	Scanning electron microscopy
Resolution	10–200 nm	$\leq$1–50 nm[a]	20–50 nm	0.2–10 nm	2–10 nm
Working principle	Ion current through nanopipette	Force on cantilever tip	Fluorescence below diffraction limit	Transmission of electrons	Scattering of electrons
Probe	Nanopipette	Cantilever	Fluorophore	Electron beam	Electron beam
Sample preparation and environment	Label-free; physiological conditions	Label-free; physiological conditions, air, vacuum	Fluorescence labeling; physiological conditions, air	Dehydration (negative stain), vitrification (cryo-electron microscopy), vacuum	Fixation, freeze/critical point drying and sputter coating, vacuum
Artifacts	Tip, current, scanning	Tip, force, scanning	Bleaching, toxicity	Dehydration, ice crystal formation, beam damage	Dehydration, beam damage
Advantages	Non-contact imaging; imaging under physiological conditions; no staining, labeling, or fixation necessary; high signal-to-noise ratio; assessment of multiple physical, chemical, and biological parameters	Imaging under physiological conditions; no staining, labeling or fixation necessary; high signal-to-noise ratio; assessment of multiple physical, chemical, and biological parameters	Access to three-dimensional (intra-)cellular structures; high spatiotemporal resolution; monitoring biomolecular processes in live cells	Resolves atomic structures of proteins; conformational snapshots of proteins and complexes; molecular resolution of structures within the cell	Resolves surfaces of tissues, cells, and interfaces at nanometer-scale resolution
Limitations	Restricted to surfaces and ion-conducting solutions	Restricted to surfaces; resolution depends on sample stiffness; sometimes fixation necessary	Needs fluorescent labels	No live processes	No live processes

Resolution values are for nucleic acids, purified proteins, extracted cellular or synthetic membranes, or whole cells

[a] Resolution $\leq$1 nm only on membrane proteins on a solid support; on mammalian cells and microbial cells, resolution of $\approx$50 nm and $\approx$10 nm was achieved, respectively. STED, stimulated emission depletion; PALM, photoactivated localization microscopy; STORM, stochastic optical reconstruction microscopy. Table data partly adapted by permission from Springer Nature [32], Copyright Nature Nanotechnology 2017

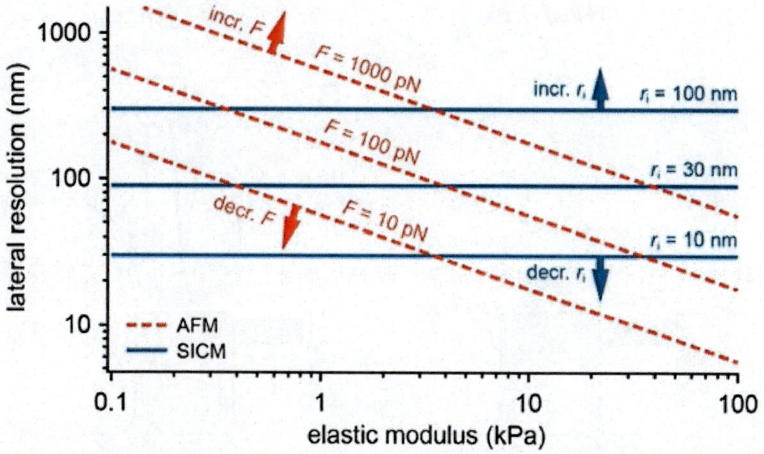

Fig. 11 Models for the lateral resolution in AFM and SICM as a function of elastic (Young's) modulus of the sample. The elastic modulus of a live mammalian cell is usually in the range of 0.1 – several 100 kPa [117]. For AFM, the lateral resolution deteriorates with an increasing imaging force F and with a decreasing elastic modulus of the sample (assuming a conical tip with a half-cone angle of 34°). For SICM, the resolution is independent of the sample's elastic modulus but is dependent on the inner nanopipette radius r_i. In conclusion, the resolution of SICM might be higher compared to AFM for softer samples. Adapted with permission from [74]. Copyright 2015 American Chemical Society

schemes for these cantilevers [157–159]. Further improvements by the Ando group [160–162] led to an HS-AFM design capable of acquiring image sequences at 31 ms/frame (100 scan lines, 240 nm scan range, 323 Hz pixel rate) [163, 164]. HS-AFM has been used, for example, to visualize biomolecule dynamics such as the photo-induced conformational change of bacteriorhodopsin [31], the walking of myosin V along an actin filament [30], interactions of annexin assemblies [165], force-induced conformational changes in the mechanosensitive channel PIEZO1 [166], interactions of human influenza A hemagglutinin with exosomes [167], or the dynamics of intrinsically disordered proteins [168].

The small scan range of HS-AFM is a serious limitation when trying to visualize whole cells. This problem was overcome by using small cantilevers in combination with a large-range flexure stage (Fig. 12a) [169]. One x- and one y-piezo (which can be easily interchanged) elastically displace the flexure stage that is in the form of a 4-legged table, achieving a maximum scan range of 23 μm × 23 μm. The sample stage area (6.5 mm × 6.5 mm) is conveniently large. Collagen fibrils in the air were visualized with up to 46 frames/s (100 × 100 pixels, 460 kHz pixel rate). The dissolution of a calcite crystal in a slightly acidic aqueous solution was imaged with 10 frames/s. Later, a modified version of the same setup was used in the force mapping mode to measure mechanical properties of MEF cells during cytoskeletal reorganization at high speed [170]. HS-AFM for the imaging of biological samples is an actively developing field of research [34, 171, 172].

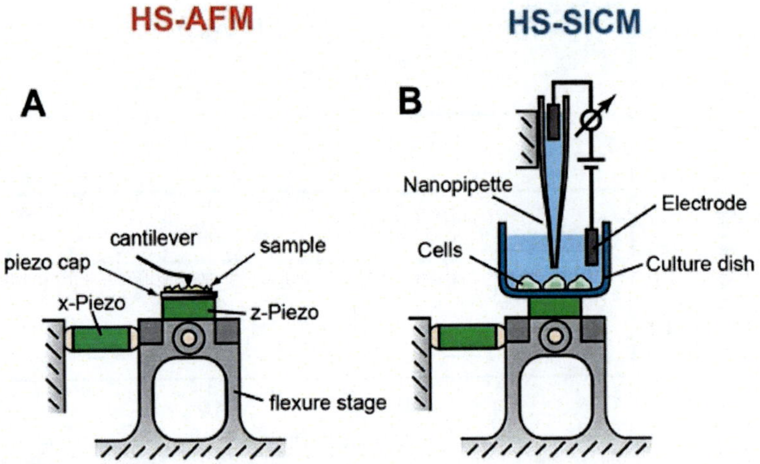

Fig. 12 High-speed AFM and high-speed SICM flexure stage scanners (side view). (**a**) HS-AFM and (**b**) HS-SICM scanner schematics. Both scanners are based on the same principle: An aluminum flexure stage is moved by preloaded *x*- and *y*-piezos (*y*-piezo not shown), while the *z*-piezo is fixed on top of the flexure stage with an adhesive. The sample is likewise fixed to the top of the *z*-piezo. During imaging, the sample rather than the probe moves. The probe is either a cantilever for AFM (**a**) [169] or a nanopipette for SICM (**b**) [96]. (**a**) Adapted with permission from [169], copyright IOP Publishing, reproduced with permission, all rights reserved. (**b**) Reproduced from [96] with permission from the Royal Society of Chemistry

4.2 High-Speed SICM

Early improvements in the scanning speed in SICM included the introduction of a low-resolution pre-scan: A region of interest is first imaged at low resolution to obtain a rough idea of sample topography. For the subsequent high-resolution image, the retract distance is adjusted according to the sample topography from the pre-scan [173]. In another approach, four adjacent pixels are acquired at low resolution and analyzed in real-time to obtain the underlying sample roughness, followed by a measurement of the same area at higher resolution, depending on the determined roughness [43]. This approach effectively reduces the image resolution on the underlying (flat) substrate, thereby speeding up imaging. Another advantage of this approach is the small time delay between the pre-scan and the actual scan, a feature that can become useful when imaging highly dynamic samples.

Further improvements regarding scanning speed were made with the introduction of a hybrid scanning mode for fast SICM [95]. This mode takes advantage of the fact that for many samples, there are only few regions with steep vertical slopes (such as cell borders) and many relatively flat regions (the substrate or the cell apex). For each line of the image, a rough estimation of topography is first gathered on a small number of pixels via the hopping mode. The line is then repeatedly (five–ten times) scanned again without feedback during the scanning motion. Each time, a new corrected path is used that is calculated from the measured ion current during the previous path. The sample topography for each scan line is thereby obtained iteratively.

Shevchuk et al. [174] used a combination of a large-range piezo and a fast, small-range shear piezo with a resonant frequency of 150 kHz to quickly retract the pipette at the trigger event. By this method, mechanical contact of the pipette with the sample was avoided, and the acquisition time was decreased to 7 s per frame (32 × 32 pixels). Seifert et al. imaged the dynamics of activated human platelets at a rate of 4 s per frame (32 × 32 pixels).

A continuous control mode for the nanopipette position enabled approach velocities up to 300 μm/s [175], leading to an HS-SICM imaging mode called transverse-fast scanning mode (TFSM) [176]. TSFM first obtains the highest surface point for each scan line and then chooses an adjusted hopping height for the pixels of that line instead of keeping the hopping height constant throughout the image. In another HS-SICM imaging mode, the retract distance was significantly decreased, and steep slopes were detected by measuring the ion current during lateral movement in hopping mode, thereby avoiding collisions [177]. A closed-loop control strategy for *IZ*-curves including a deceleration phase near the sample surface was devised [178]. In the "automated region of interest" (AR) mode, the region of interest for an image is predicted from the previous one, thereby speeding up imaging [63].

While these developments used software and new measurement algorithms to increase the scanning speed of SICM, other developments focused on improving the SICM hardware. Watanabe et al. presented a fast, tip-scanning *xyz*-nanopositioner [179] and increased the ion concentration inside the nanopipette for an enhanced signal-to-noise ratio [180], achieving a pixel rate of 1,000 Hz on the edge of a HeLa cell and a pixel rate of 2000 Hz on actin filaments. Leitao et al. presented an HS-SICM design with an enhanced *z*-range, achieving a pixel rate of 2,500 Hz on COS-7 cells [181].

An extremely fast HS-SICM design using a flexure stage moved by piezo scanners was reported by Simeonov and Schäffer [96] (Fig. 12b). This design is similar to that of an HS-AFM setup developed previously (Fig. 12a) [169]. The flexure stage is moved laterally with the *x*- and *y*-piezos, while a sample dish (filled with electrolyte) is attached to the stage and moved by the *z*-piezo on top of the stage (Fig. 12b). The electrolyte-filled nanopipette remains stationary during the measurement. This design significantly reduces the inertial mass driven by the *x*- and *y*-piezos, thus increasing the scanning speed. The SICM scanning speed was further increased by employing an immediate sharp retraction of the nanopipette after the trigger event, called "turn step" [96]. In the conventional hopping mode SICM (Fig. 13a), the nanopipette is lowered toward the sample down to a predefined ion current threshold (trigger event), at which the nanopipette is retracted again. Since the setup needs some microseconds to fully stop and retract the nanopipette (due to inertial mass, limited power output of the piezo amplifier, time lag in the control system, and measurement bandwidth), the pipette usually overshoots the trigger position. With an increasing approach velocity, the overshoot distance increases, risking contact between sample and nanopipette. In the hopping mode with turn step (Fig. 13b), the nanopipette is sharply retracted by 10–70 nm immediately after the

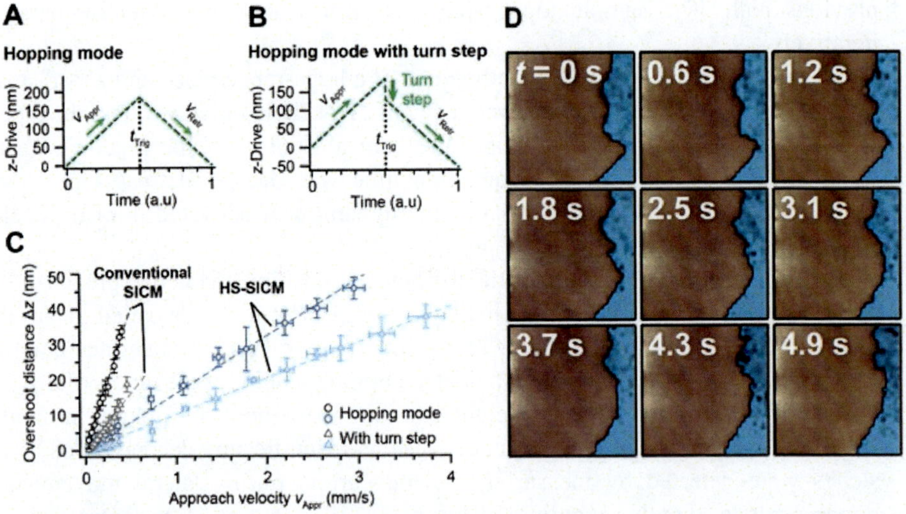

Fig. 13 Reduction of the overshoot distance in hopping mode SICM using a turn step. (**a**) Z-drive voltage during an approach curve before and after the trigger event at t_{Trig} in conventional hopping mode and (**b**) in hopping mode with an additional turn step. In conventional hopping mode, the nanopipette is lowered toward the sample until a trigger event occurs, upon which the nanopipette is retracted again. With the addition of a turn step, the nanopipette is immediately retracted at the trigger event by a certain distance, before gradually retracting as usual. (**c**) The overshoot distance Δz in SICM hopping mode is plotted against the approach velocity v_{Appr} without (circles) and with (triangles) turn step. The turn step significantly reduces the overshoot distance, enabling high-speed imaging with SICM. (**d**) HS-SICM image sequence of the highly dynamic periphery of a human platelet. Protrusions are extending and retracting along the platelet contour in this sub-second image sequence (0.6 s per frame, 32 × 32 pixels, 1.4 mm/s approach velocity). Adapted from [96] with permission from the Royal Society of Chemistry

trigger event, thereby reducing the overshoot distance [96]. Plotting the overshoot distance against the approach velocity (Fig. 13c) reveals a linear dependency (dashed lines: line fits) and the average overshoot time (slope of the fits). It can be seen that the average overshoot time is significantly shortened in the hopping mode with turn step. Using these improvements, this HS-SICM setup achieved an approach velocity of up to 3 mm/s and a pixel rate of up to 11.000 Hz on a calibration grid and 1,000–3,000 Hz during the measurement of live cells. The highly dynamic periphery of a human platelet could be visualized at a rate of 0.6 s per frame (32 × 32 pixels) (Fig. 13d).

Using a combination of SICM with a microelectrode array (MEA), the 3D topography of beating cardiomyocytes was imaged at an ultrafast time resolution of 0.001 s per frame [61]. This was achieved by recording the vertical motion of the cell sequentially at many locations on the cell with SICM and by simultaneously recording the action potential with the MEA. Synchronizing this data by using the action potential as a time reference allowed generating an image sequence of the topography during a full contraction-relaxation cycle with a frame rate of 1,000–5,000 Hz. This is by far the fastest frame rate reported using SICM imaging to date. However, this method works only for samples with highly periodic dynamics.

4.3 *Comparison of HS-AFM and HS-SICM Imaging*

While HS-AFM focuses on the study of fast dynamics of single molecules, HS-SICM focuses on the study of live cells. An overview of pixel frequencies from HS-AFM and HS-SICM studies on biological samples is provided in Table 3.

5 Cell Mechanics

When a mechanical stimulus acts on a material, it undergoes a certain response, depending on its mechanical properties. Biomechanics is the field concerned with mechanical properties in biological systems, from whole organisms to cells. Both AFM and SICM are capable of measuring mechanical sample properties—a capability of special interest in cell biology. In this section, we will explore methods in AFM and SICM to extract (visco-)elastic properties of biological cells and illustrate applications of these methods with examples.

5.1 *Measurement of Elastic Sample Properties*

A cell's compliance or its inverse, the stiffness, are mainly determined by the cell's cytoskeleton, which consists of actin filaments (F-actin), microtubules, and intermediate filaments [183, 184]. F-actin (especially stress fibers) is usually the major factor contributing to cell stiffness [26], while the cell membrane [185] as well as microtubules and intermediate filaments [28, 186] have a smaller impact. The stiffness of a cell changes throughout its life and health status, e.g., during volume change [187], migration [188], cell cycle [189], mitosis [190], and proliferation into a cancer cell [191]. AFM force mapping of the central region of cancer cells revealed that they differ in their mechanical properties in comparison with their healthy counterparts [192], linking biomechanics to cancer research [193], and making cell stiffness an important pathological marker [194].

The elastic (Young's) modulus is a mechanical property that specifies the "stiffness" of a solid material. It is defined as the amount of force per area (stress) per relative deformation (strain) and is measured in units of Pascal. The higher the elastic modulus, the more resistant a material is to deformation. The elastic modulus of mammalian cells typically ranges from 0.1 to several 100 kPa (Fig. 14) [116, 117], with bacteria ranging in the megapascal range [195]. For a detailed review of the different elastic moduli of mammalian cells, see Kuznetsova et al. [117] and Ozkan

Table 3 Examples of HS-AFM and HS-SICM imaging of biological samples in liquid

Technique	Sample	Imaging mode	Scan area (μm^2)	Number of pixels	Time per frame	Pixel rate (Hz)
AFM	IgG antibodies on mica [157]	Tapping mode	0.2 × 0.2	130 × 130	2 s	8,450
	DNA on mica [159]	Tapping mode	1.5 × 1.5/ 1 × 1	256 × 256/ 128 × 128	5.6 s/1.7 s	11,700/9,600
	Myosin V on mica [160]	Tapping mode	0.24 × 0.24	100 × 100	80 ms	125,000
	Myosin V on mica [163]	Tapping mode	0.24 × 0.24	100 × 100	31 ms	323,000
	Bacteriorhodopsin on mica [31]	Tapping mode	0.04 × 0.04	200 × 200	1 s	40,000
	Myosin V walking on actin filament [30]	Tapping mode	0.13 × 0.065	80 × 50	147 ms	27,000
	Mouse embryonic fibroblasts [170]	Force mapping mode	6.4 × 6.4	128 × 128	49 s	333
SICM	A6 cells [95]	Hybrid mode	40 × 40	600 × 1,024	10 min	1,024
	Clathrin-coated pits on COS-7 cells [174]	Hopping mode	1 × 1/ 1.5 × 1.5 2 × 2	32 × 32	7–17 s	60–146
	A431 cells [177]	Hopping mode	10 × 10	64 × 64	18 s	228
	Human platelets [182]	Hopping mode	4 × 4/10 × 10	32 × 32/80 × 80	4 s/17 s	256/376
	Actin filaments [180]	Hopping mode	2 × 2	100 × 100	5 s	2,000
	Platelets [96]	Hopping mode	5 × 5	32 × 32	0.6 s	1,707
	A6 cells [96]	Hopping mode	5 × 5	64 × 64	1.4 s	2,926
	Hippocampal neurons [63]	AR mode	15 × 15	128 × 128	98 s	167
	COS-7 cells [181]	Adaptive hopping mode	3.6 × 3.6	40 × 25	0.5 s	2,500
	Beating cardiomyocytes [61]	SICM-MEA combination	65 × 65	60 × 60	0.001 s	N/A

Please note that the given numerical values do not allow for an abolute comparison of imaging performance, as these values depend, e.g., on the sample and on the imaging mode. SICM data partly reproduced with permission from [65]. Copyright 2020 American Chemical Society. N/A = not available/not applicable

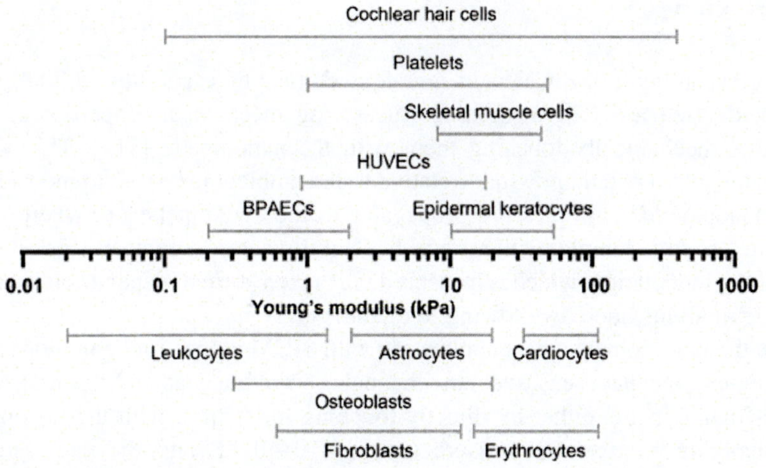

Fig. 14 Typical elastic modulus of mammalian cells measured by AFM. Different types of cells display different ranges of elastic modulus (usually between 0.1 and several 100 kPa). HUVECs = human umbilical vein endothelial cells, BPAECs = bovine pulmonary artery endothelial cells. This figure is based on data from [117]

et al. [195]. It is important to know that different techniques for obtaining mechanical properties can measure different elastic moduli for the same sample [196].

5.1.1 AFM

AFM has been used for biomechanical investigations for 30 years, especially for the measurement of the local elastic modulus [32, 35, 81, 171, 197–204].

In a typical elastic modulus measurement that is done in the AFM force mapping mode (see Sect. 2.1), contact models such as the Hertz model for an indentation with a sphere [205] or the Sneddon model for an indentation with a cone [206] are used to extract the local elastic modulus via a fit to an *FZ*-curve [8, 24, 87, 117, 203, 207].

Mechanosensitive cells such as endothelial cells [208] or platelets [9] react to external forces. The force exerted on a mechanosensitive cell by the AFM cantilever can cause morphological and nanomechanical changes, making it unsuitable to investigate such cells. Measurement of the elastic modulus with AFM is dependent on several parameters such as the indentation, the tip geometry, and the spring constant of the cantilever [209, 210]. This can lead to a variability in the reported elastic modulus and even to the need to standardize AFM investigations of cell mechanics [203].

5.1.2 SICM

The great strength of SICM is its non-contact imaging capability. In 2008, Sanchez et al. devised a SICM method for measuring mechanical properties of samples without mechanically touching them with the nanopipette [211]. This works by applying a hydrostatic pressure (relative to the ambient pressure) to the back end of the nanopipette. This pressure propagates to the nanopipette tip, where it locally indents a soft sample without mechanical tip-sample contact (Fig. 15a). The resulting indentation, which is measured via the ion current, depends on the sample's elastic modulus, thereby allowing to quantify it.

In the first biomechanical application with SICM, dorsal root ganglions (neurons expressing mechanosensitive ion channels) were mechanically stimulated by a SICM nanopipette, either by directly touching the neuron with the nanopipette or by applying an external pressure (around 10–40 kPa) to the back end of the nanopipette. When stimulated, the neuron showed an increase in internal Ca^{2+} concentration and an inward current [41]. The same principle was later used to extract information about the local elastic modulus of a red blood cell, of live and fixed epithelial cells, and of rat cardiomyocytes [211]. Positive pressure applied to the surface of a red blood cell (Fig. 15b) lead to a deformation of the cell that was linear with the applied pressure, allowing to quantify the cell's elastic modulus (Fig. 15c).

In the same study [211], a negative and a positive pressure were sequentially applied to dorsal root ganglions. At a negative pressure or a small positive pressure, the apparent elastic modulus was around 0.5 kPa, whereas for a higher positive pressure, the apparent elastic modulus was 6.9 kPa. It was postulated that the soft elastic modulus at low deformations arises from a sub-cell membrane space without

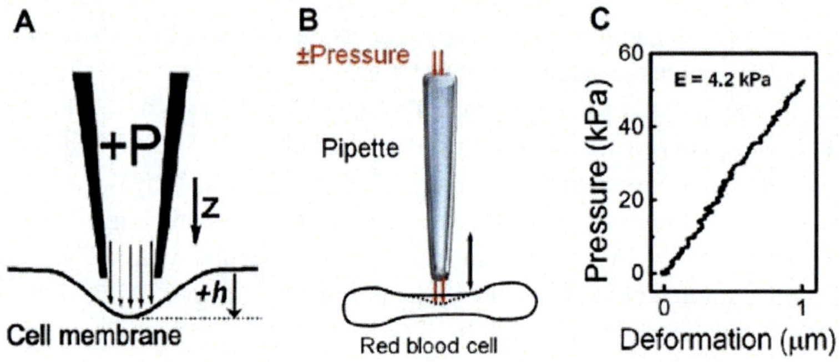

Fig. 15 Measurement of mechanical sample properties with SICM. (**a**) Schematic diagram of the measurement principle. Applying a positive pressure P to the upper end of the nanopipette creates a pressure below the nanopipette tip. This pressure deforms a soft sample such as a cell. The resulting deformation h is measured and quantified via the change of the ionic current through the nanopipette. (**b**) Schematic of the measurement of mechanical properties of a red blood cell. (**c**) Graph showing a linear relationship between the deformation of a red blood cell and the applied pressure, giving an elastic modulus of the cell of 4.2 kPa. Adapted from [211]. Copyright 2008, with permission from Elsevier

cytoskeleton, while the high apparent elastic modulus stems from the rigid cytoskeleton at high deformation, similar to the soft human skin on top of the stiff muscle and skeleton.

In a similar approach, Pellegrino et al. used the weight of the water column in the nanopipette (the filling level) to apply a weak hydrostatic force on a sample [212]. Neuronal growth cones could be guided by this mechanical stimulation [212, 213]. To quantify the force generated by this method, an AFM cantilever was used as a sample: while monitoring the deflection of the AFM cantilever, the applied force was varied by about 8 pN by changing the filling level by 1 mm in a nanopipette with a 1 μm tip diameter [212]. The force sufficient to induce a neuronal growth cone response was about 60 pN [212]. With an expanded setup combining AFM, SICM, and phase contrast imaging to increase visibility and avoid optical artifacts, Pellegrino et al. also demonstrated that it is possible to accurately quantify the force applied on a sample by a pressurized nanopipette [214]. They also demonstrated that *IZ*-curves can be used to indirectly and accurately measure the cantilever deflection and therefore the sample deformation using different values of applied pressure. They used this approach to measure the elastic modulus of live murine fibroblasts [215]. (We will further discuss this combined SICM-AFM setup in greater detail in our discussion on AFM and SICM combinations in Sect. 6).

A disadvantage of these methods introduced by Sanchez et al. and Pellegrino et al. is the time needed to change the pressure, making high-resolution imaging of mechanical properties time-consuming and unfeasible, especially on live cells. Rheinlaender and Schäffer introduced a solution to these issues: The nanopipette approaches a sample surface in the hopping mode, while a constant pressure in the range of several tens to hundreds of kilopascals propels the electrolyte from the nanopipette tip (referred to as pressure-SICM in this chapter, Fig. 16a) [56]. On a stiff sample with an elastic modulus much larger than the applied pressure, such as polystyrene or glass, the pressure from the nanopipette tip does not measurably indent the sample (Fig. 16b, left) and the ion current decreases as usual when the nanopipette approaches the surface. On a soft sample with an elastic modulus on the same order of magnitude as the applied pressure, the sample is indented by the pressure (Fig. 16b, right), causing the ion current to decrease less. The local stiffness is calculated from the slope between 1% and 2% of the current decrease (Fig. 16c) [56], enabling simultaneous topography and elastic modulus measurements from the same *IZ*-curve. This technique has been successfully applied to biomechanics on a range of specimens like mouse embryonic fibroblasts [56], human [58] and mouse [216] blood platelets, adult rat cardiomyocytes [62], and cancer cells [217]. A direct comparison of this SICM technique with AFM force mapping of the same fixed fibroblast revealed a consistent representation of the local elastic modulus at a comparable resolution with both techniques (Fig. 17) [56]. For thin samples, the apparent elastic modulus is overestimated but can be corrected [218].

In another approach, the intrinsic interaction forces between the glass nanopipette and the cell membrane were used to indent a cell at low stress [219]. Two height maps recorded consecutively with different ion current trigger thresholds (e.g., 0.3% and 1.5%) were used to extract maps of cell stiffness with just about 100 nm of

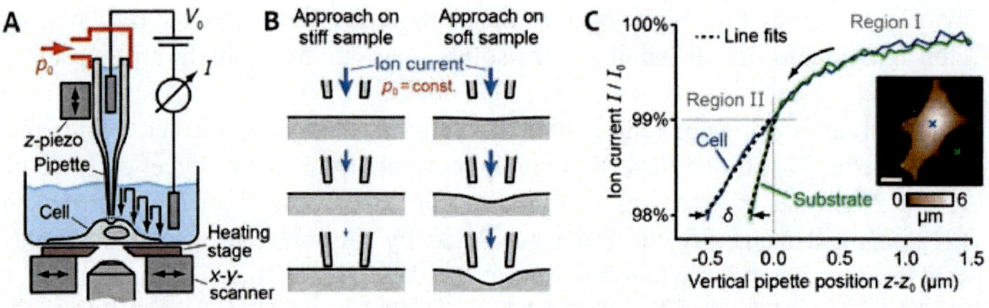

Fig. 16 Measurement of elastic modulus with SICM. (**a**) Schematic of a SICM setup used for measurement of elastic modulus. The tip of an electrolyte-filled nanopipette is approached toward a sample. A static pressure p_0 is applied to the back end of the nanopipette. (**b**) Principle of the elastic modulus measurement. With a decreasing distance of the nanopipette to the sample, the pressure on the sample increases. A softer sample is indented more than a stiffer sample. In both cases, the current decreases, but it decreases less for the softer sample. (**c**) *IZ*-curves on the stiff substrate (green cross in the topography map in the inset, scale bar 15 μm) and on a live MEF (blue cross). In Region I (the initial phase of the approach), the two IZ-curves overlap. At 99% current, the sample is still undeformed, and the piezo position can be used to infer the height of the sample. Below 99% (Region II), the sample is indented by the pressure from the nanopipette and the current drops with a higher rate on the stiff substrate than on the soft cell. The slope between 98% and 99% current can be line-fitted to obtain the elastic modulus of the sample. Adapted from [56]. Republished with permission of The Royal Society of Chemistry

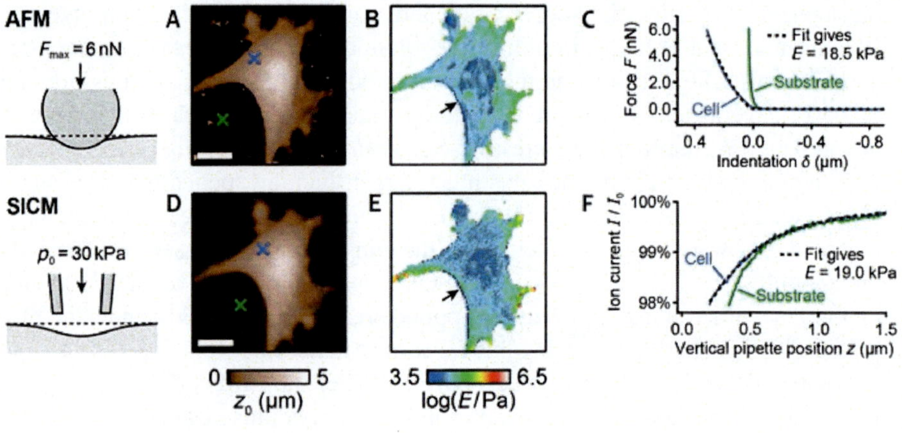

Fig. 17 Direct comparison of the measurement of elastic modulus with AFM and pressure-SICM. (**a**) AFM topography (contact height) and (**b**) elastic modulus images of a fixed fibroblast cell. (**c**) Force vs. indentation curves on the cell and on the substrate (locations indicated by crosses in (**a**)). Fitting the Hertz model for a spherical indenter resulted in a sample stiffness of 18.5 kPa. (**d**) SICM topography and (**e**) elastic modulus images of the same fixed fibroblast as in (**a**). (**f**) *IZ*-curves on the cell and on the substrate (locations indicated by crosses in (**d**)). Fitting the IZ-curve for the cell using the model from Rheinlaender et al. [56] gave an elastic modulus of 19.0 kPa. To approximately match the indentation morphology (schematics on the left in the figure) of both techniques, a cantilever with a spherical tip of radius 0.94 μm, a spring constant of 0.226 N/m, and a maximum applied force of 6 nN was used for AFM, while a nanopipette with an inner radius of 450 nm, a half-cone angle of 4°, and an applied pressure of 30 kPa was used. Reproduced from [56] with permission from the Royal Society of Chemistry

indentation. Advantages of this technique are a reduced stress on the sample and no need for an external pressure [220].

The contact-free manipulation and force measurement capability of SICM [57] is especially interesting for biomechanical investigations of mechanosensitive cells, such as endothelial cells and platelets. Platelets undergo a rapid change in their actin cytoskeleton after activation [221, 222], for example, via adhesion and spreading onto a surface [223–225] or via an endogenous thrombotic agonist such as thrombin [226]. Pressure-SICM revealed an increase in apparent platelet stiffness during surface adhesion and spreading [58]. In contrast, a decrease of the apparent elastic modulus in platelets by a factor of two was observed after activation with thrombin. This decrease in stiffness was distinctly different from treatment with cytochalasin D, a fungal metabolite that depolymerizes actin filaments [227]. A comparison of the change in the sub-cellular elastic modulus (stiffening factor) showed that the softening caused by thrombin is constricted to certain regions, while the softening caused by cytochalasin D is uniform over the whole platelet [58].

5.2 *Measurement of Viscoelastic Sample Properties*

When a mechanical stress is applied to a purely elastic material, it will show an immediate deformation (strain) with a magnitude that depends on the material's elastic modulus (assuming no inertial effects). Once the stress is released, the material immediately returns to its original shape and size: the strain is time-independent and reversible. In purely viscous materials such as some liquids, the strain increases linearly with time when a constant stress is applied. The strain is not reversible, and energy is dissipated. Viscoelastic materials have both elastic and viscous properties: The applied stress causes both an immediate elastic strain and a time-dependent viscous strain. Upon removal of the stress, a viscoelastic material returns to its original shape. Several methods for measuring viscoelastic properties of live cells have been developed [228, 229].

5.2.1 AFM

Most biomechanical investigations with AFM assume biological soft materials to be purely elastic for the sake of simplicity. In truth, cells are viscoelastic, combining both viscous and elastic properties [230]. Rheological studies of viscoelastic materials in the context of AFM exhibit a number of characteristics [183, 229, 231–235]:

- The functional shape of *FZ*-curves depends on the rate of the force application (e.g., on the force curve rate).
- *FZ*-curves on a viscoelastic material show a hysteresis, a "lagging behind" of the retract from the approach curve due to energy loss.

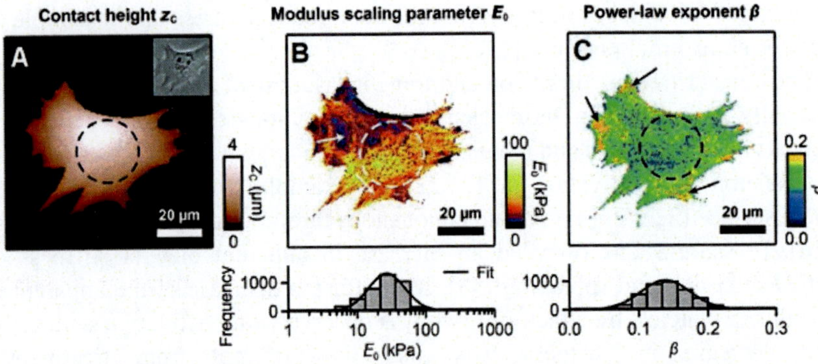

Fig. 18 Force clamp force mapping (FCFM) of a live MEF$^{vin-/-}$ cell. (**a**) Map of contact height z_c with optical phase contrast image as inset. The dashed circle in the contact height map indicates the position of the cell's nucleus. (**b**) Map of the modulus scaling parameter E_0 (measure of stiffness). White arrows indicate stiff fibrous areas, probably cytoskeletal fibers. E_0 is distributed log-normally in the range between 10 kPa and 100 kPa (histogram). (**c**) Map of the power-law exponent β (measure of fluidity). Black arrows indicate areas of high β-values. β follows a Gaussian distribution in the range between 0.1 and 0.2 (histogram). Adapted with permission from [236]. Creative Commons Attribution 3.0 Unported license

- Viscoelastic materials exhibit creep, i.e., the indentation (strain) caused by a constant force (stress) increases with time.
- Viscoelastic materials exhibit stress relaxation, i.e., the force (stress) needed for a constant indentation (strain) decreases with time.
- When a periodic force is applied to a viscoelastic material, a phase lag between force and indentation can be observed that is between 0° (purely elastic material) and 90° (purely viscous material).

Hecht et al. presented force clamp force mapping (FCFM), an AFM-based method for quantifying and mapping the viscoelastic creep behavior of soft samples with sub-micrometer spatial resolution [236, 237]. In FCFM, the cantilever approaches the sample until tip-sample contact occurs. The tip then indents the sample until a predetermined force is reached. This force is then held constant ("clamped") for 50–60 ms by a feedback loop while the indentation (creep) is recorded. Finally, the cantilever is retracted, thereby completing the *FZ*-curve, and the procedure is repeated for the next location on the sample. With a power-law-based model [229, 232, 233, 238, 239], the modulus scaling parameter E_0 (representing the sample stiffness) and the power-law exponent β (representing the sample fluidity) are extracted from each *FZ*-curve. In their work, Hecht et al. [236] used FCFM to image wild-type MEF cells and MEF cells with a genetic knockout of vinculin ($^{vin-/-}$), a cytoplasmic actin-binding protein mainly involved in adhesion [240, 241]. Figure 18 shows three simultaneously acquired maps from an FCFM measurement on a MEF$^{vin-/-}$ cell: (A) the contact height z_c, (B) the modulus scaling parameter E_0, and (C) the power-law exponent β. The map of E_0 revealed stiff fibers presumed to be cytoskeleton fibers (white arrows), while the map of β revealed a high fluidity in the peripheral regions (black arrows).

FCFM has been combined with traction force microscopy and has revealed a correlation between viscoelastic material properties and contractile prestress of fibroblast and epithelial cells [242]. Other examples of rheology measurement techniques with AFM include stress relaxation microscopy [243], multi-frequency force modulation [244], resonance compensating chirp mode [245], and multi-harmonic dynamic AFM with directly excited microcantilevers [246], among others [235].

5.2.2 SICM

Rheinlaender and Schäffer [60] adapted the FCFM approach from AFM [236] to SICM and investigated the correlation between stiffness and fluidity in live MEFs. In this study, the nanopipette (with a constant external pressure applied) was lowered toward the sample until the current decreased by 2%. A feedback loop then kept the current and thus the force acting on the sample "clamped" for $\approx$100 ms, while the sample indentation (nanopipette position) was recorded (creep). Finally, the nanopipette was retracted again, thereby completing the *IZ*-curve, and the procedure was repeated for the next location on the sample. Using a power-law model supported by finite element method (FEM) simulations, E_0 and β were extracted from the *IZ*-curves. This allowed the simultaneous mapping of the height z_0, the modulus scaling parameter E_0 as a measure of "stiffness," and the power-law exponent β as a measure of "fluidity" at sub-cellular resolution (Fig. 19). The imaged MEF cell showed stiff areas with low fluidity (green arrows) and soft areas with high fluidity (orange arrows), supporting the hypothesis that stiffness and fluidity are linearly correlated [229].

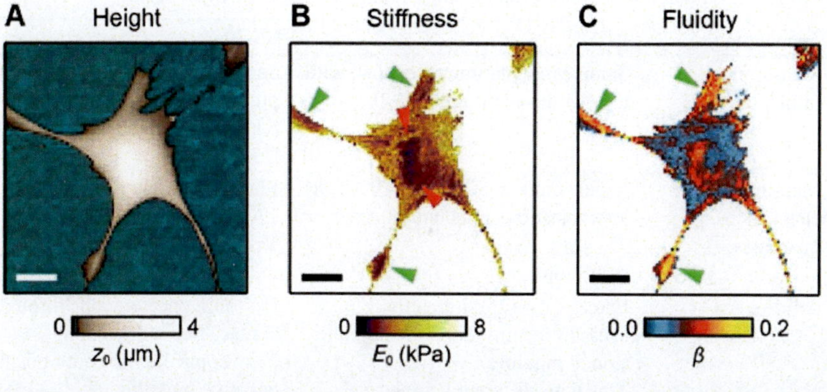

Fig. 19 Viscoelastic SICM measurements demonstrate a correlation of stiffness and fluidity at the sub-cellular level on a live MEF cell. Maps of (**a**) SICM topography z, (**b**) elastic modulus E_0, and (**c**) fluidity β. Areas of high stiffness show a low fluidity (green arrows) and vice versa (orange arrows), suggesting a correlation between stiffness and fluidity. Reproduced from [60] with permission from the Royal Society of Chemistry

5.3 *Comparison of Cell Mechanics Measurements in AFM and SICM*

Both AFM and SICM can be used for both elastic and viscoelastic measurements (Table 4). SICM could be especially useful for long-term measurements of the mechanics of soft or mechanosensitive samples due to its non-contact approach [135, 247, 248].

6 Combinations of AFM and SICM

In the preceding sections, we have drawn a range of comparisons between AFM and SICM and have discussed their advantages and disadvantages. In this section, we will change course and look at the few, yet interesting cases of the combined use of AFM and SICM.

The first combination of AFM and SICM was based on bent nanopipettes and was used to simultaneously measure both deflection and ion current [45, 92]. In the combined setup (Fig. 20a), a bent nanopipette was used as a cantilever from which a laser beam was reflected onto a detector to measure the deflection of the nanopipette due to tip-sample forces. Simultaneously, the bent nanopipette was also a probe for ion conductance. The bent nanopipette was operated in either the contact or the tapping mode (the resolution was higher in the tapping mode) and enabled the simultaneous measurement of force and ion conductance. Using this setup, it was shown that pores in synthetic polycarbonate membranes are of heterogenous conductance [92] and that the build-up of nacre (mother or pearl) of the mollusk *Haliotis*

Table 4 Comparison of cell mechanics measurements with AFM and SICM

Technique	AFM	SICM
Measurement principle	Sample indentation via contact with the cantilever tip [197, 198]	Sample indentation via hydrostatic pressure by the nanopipette tip [211] Sample indentation via intrinsic stress [219]
Measured biomechanical quantities	Elastic modulus [24, 115, 197, 198] Viscoelastic parameters [8, 199, 231–233, 236] Adhesion [16]	Elastic modulus [56, 211, 215, 219, 220] Viscoelastic parameters [60]
Advantages	Force is measured directly Measurement of friction and adhesion is possible Tips with different geometrical shapes are readily available	Less influence on cells through contact-free indentation Easy application of both positive and negative pressure
Disadvantages	Influence on cells and possible damage through mechanical contact	Force is measured indirectly

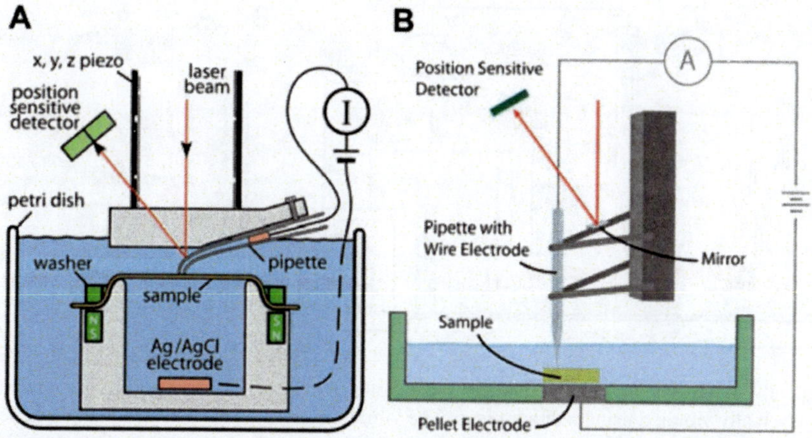

Fig. 20 Simultaneous force and ion conductance measurement. (**a**) Schematic of a combined SICM and AFM setup using a bent nanopipette. An ion current is induced between the electrode inside the nanopipette and the electrode below the sample. This current is recorded while the nanopipette is scanned over the sample (SICM part). The nanopipette is also used as a cantilever for AFM in the tapping mode. The deflection of the nanopipette is measured by reflecting a laser beam from the nanopipette's top surface onto a position-sensitive detector (AFM part). Adapted with permission from [45]. Copyright 1997 American Chemical Society. (**b**) Illustration of ion-sensing deep AFM. A nanopipette is held by two gold-coated triangular flexures. These flexures reflect a laser beam onto a position-sensitive detector (AFM part). The ion conductance is measured between a wire electrode in the nanopipette and a pellet electrode below the sample. Reprinted from [250] with the permission of AIP Publishing

rufescens (red abalone) is facilitated by mineral bridges through interlamellar organic sheets [45].

An AFM setup specifically designed for imaging vertical height differences over 100 μm called "Deep AFM" [249] was adapted to allow for SICM measurements. A nanopipette was mounted to two gold-coated flexures and was immersed in a bath containing the sample (Fig. 20b) [250]. The microscope could be operated in both the AFM contact and tapping mode. A magnetic drive increased the tapping frequency from the probe's resonant frequency of 100 Hz to 4.8 kHz. This allowed the passive measurement of ion conductance while using AFM for imaging of topography. This design also allows imaging in a SICM imaging mode with simultaneous force sensing.

SICM has been combined with an integrated shear force distance control for measuring the ion current independently of sample topography (Fig. 21a) [53]. The thin end of the nanopipette was transversally vibrated at 10–70 kHz using a dither piezo. The vibration amplitude (typ. 50–250 nm peak-peak), which is sensitive to shear forces between the nanopipette and the sample surface, was measured using two air-filled periscopes that were submerged in the bath solution. With this setup, it was observed that the ion current decreased before pipette-sample (shear) forces were detected, showing that SICM can image a sample without mechanical tip-sample forces.

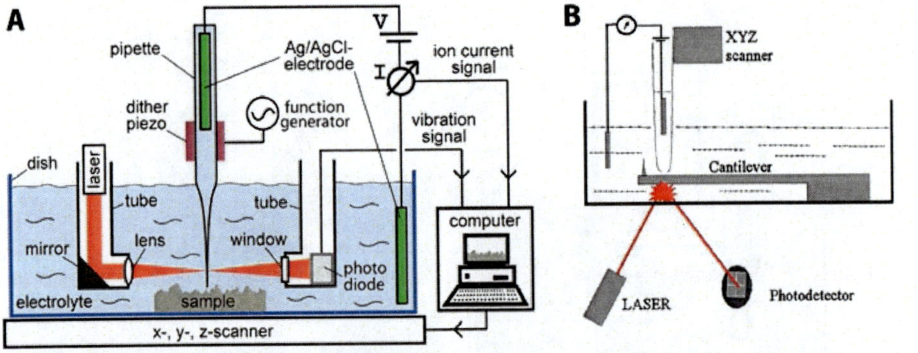

Fig. 21 (**a**) SICM combined with shear force distance control. The thin end of the nanopipette is transversally vibrated using a dither piezo. The vibration amplitude is detected via a laser beam that is focused onto the thin end of the nanopipette and projected onto a photodiode with the help of two air-filled periscopes. When approaching the nanopipette to the sample surface, the vibration amplitude is reduced due to shear forces between the nanopipette and the sample, thereby providing a signal for distance control. Adapted with permission from [53]. Copyright IOP Publishing, reproduced with permission, all rights reserved. (**b**) Combined AFM-SICM setup to measure forces involved in SICM. The cantilever acts as the sample, detecting weak hydrostatic forces from the SICM nanopipette. Adapted from [212]. Copyright 2011 with permission from Elsevier

AFM has been used to measure the forces generated by SICM (Fig. 21b). For this purpose, a SICM nanopipette was positioned directly over a cantilever in a specialized AFM setup [212, 214]. With this setup, it was revealed that the nanopipette exerted a weak hydrostatic force (on the order of 20 pN) on the sample [215]. It was later proven in a similar setup that physical contact of the nanopipette with the sample/cantilever only occurs at large values of current decrease (at roughly 8% as reported in Fig. S3 of [56]).

Yet another approach of combining AFM and SICM evolved from the combination of AFM and nanofluidics, named fluid force microscopy (FluidFM) [251]. Cantilevers containing a nanofluidic channel leading to an opening with an aperture of 100–200 nm diameter in the apex of a pyramidal tip were used to deliver fluorescent dyes into cells or even into sub-cellular regions. The perforation of the cell membrane was monitored via AFM. In a later study, two Ag/AgCl electrodes were added: One in the back of the cantilever's nanofluidic channel, the other one in the sample bath (Fig. 22a–c), allowing simultaneous deflection and ion current measurements with the same probe (Fig. 22d) [44, 252]. Finite element method (FEM) simulations indicated that the different tip geometries (circular in the case of a conventional SICM nanopipette, quadratic in the case of FluidFM) give a similar current behavior upon probe approach. These reports also confirmed that the resolution of SICM is three times the inner nanopipette radius r_i, as it was shown by Rheinlaender and Schäffer [97, 135]. A combined FluidFM/SICM design was also used for force-controlled patch clamping of beating cardiac cells [253]. An extensive review of FluidFM can be found in Li [254].

There are several studies that report the sequential use of AFM and SICM. The elastic modulus of vascular endothelial cells was imaged with AFM and their

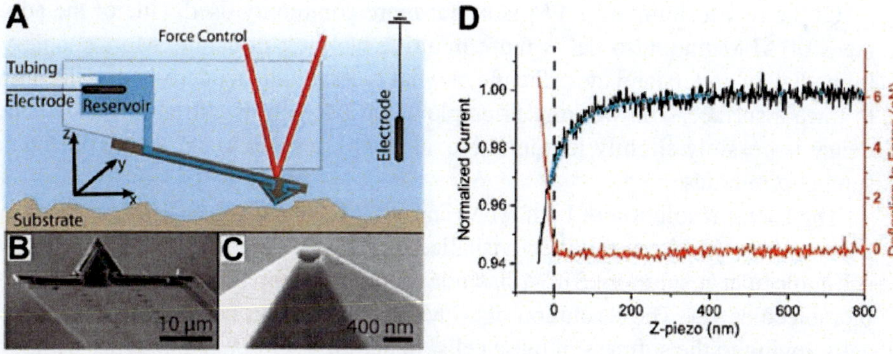

Fig. 22 Fluidic force microscopy (FluidFM) uses an AFM cantilever with an integrated SICM tip. (**a**) Schematic diagram of FluidFM. A microchanneled AFM cantilever with an aperture at its tip is filled with electrolyte. One Ag/AgCl electrode is installed in the cantilever's reservoir, while the other one is submerged in the sample bath. This allows the generation of an ion current through the cantilever. (**b**) Scanning electron micrograph of a cross-section of the cantilever tip, revealing the nanofluidic channel inside. (**c**) Scanning electron micrograph of the tip aperture. (**d**) Simultaneously acquired *FZ*- and *IZ*-curves. The decrease of the current occurs before the deflection increases at the tip-sample contact point ($z = 0$), showing that SICM measurements are indeed "non-contact." Reprinted with permission from [44]. Copyright 2015 by the American Physical Society

topography with SICM [255]. Indentation and elastic modulus maps were constructed by subtracting the SICM topography image from the AFM topography image (see Fig. 4a in [102]). The effects of paraformaldehyde treatment of cells were studied sequentially with AFM (to measure the elastic modulus of the cells) and SICM (to measure the cell surface fluctuation amplitude [256, 257]). Plasma membrane-derived extracellular vesicles were induced and imaged at high resolution with AFM and SICM [258].

7 Conclusion

In this chapter, we discussed the advantages and disadvantages of AFM and SICM, while focusing on biological applications and especially on the investigation of cells. Both techniques allow high-resolution, label-free imaging of soft samples in physiological buffer solutions at ambient and physiological temperatures. While AFM is used also in air and in vacuum, SICM is routinely limited to the aqueous environment. Both techniques allow the simultaneous measurement of topographical and mechanical parameters, providing multi-parametric information on samples.

One limitation both AFM and SICM have in common is the need for an adhered sample. Cells in suspension or non-adhered cells cannot usually be measured using AFM or SICM. Another shared limitation is the slow imaging speed, especially when compared to light microscopy. However, we discussed some approaches of high-speed AFM and high-speed SICM in this chapter, showing that a sub-second frame rate can be achieved.

Of the two techniques, AFM is by far more commonly used. One of the advantages of AFM may also be its more intuitive working principle: Most people will immediately understand the principle of what is essentially a micro-meter-sized lever to trace a surface. The working principle of SICM with its current-distance dependency is possibly slightly less intuitive, also with respect to extracting mechanical sample properties.

The lateral resolution of both AFM and SICM depends on the sharpness of their probe. While AFM can routinely visualize single biomolecules at molecular or even sub-molecular resolution, SICM has not routinely been used for visualizing single biomolecules yet. The resolution of AFM, however, is reduced on live mammalian cells, owing to the softness of these cells. In addition, prolonged imaging with AFM can cause cell damage. SICM offers an elegant solution to solving these probe-sample contact issues. Its non-contact nature significantly decreases the chance of damaging or influencing the sample, its resolution is independent of sample stiffness, and its image quality does not deteriorate over multiple hours of continuous cell imaging.

Overall, we can summarize that AFM is the more known and the more used technique of the two. However, the advantages of SICM discussed in this chapter (and the special capabilities of SICM discussed in the other chapters of this book) make SICM a big competitor to AFM in biological research, especially on live cells.

References

1. Lemon WC, McDole K (2020) Live-cell imaging in the era of too many microscopes. Curr Opin Cell Biol 66:34–42
2. Abbe E (1873) Beiträge zur Theorie des Mikroskops und der mikroskopischen Wahrnehmung. Springer Science + Business Media, pp 413–440
3. Binnig G, Rohrer H, Gerber C, Weibel E (1982) Surface studies by scanning tunneling microscopy. Phys Rev Lett 49(1):57–61
4. Binnig G, Quate CF, Gerber C (1986) Atomic force microscope. Phys Rev Lett 56(9):930–933
5. Drake B, Prater CB, Weisenhorn AL, Gould SA, Albrecht TR, Quate CF, Cannell DS, Hansma HG, Hansma PK (1989) Imaging crystals, polymers, and processes in water with the atomic force microscope. Science 243(4898):1586–1589
6. Egger M, Ohnesorge F, Weisenhorn AL, Heyn SP, Drake B, Prater CB, Gould SAC, Hansma PK, Gaub HE (1990) Wet lipid protein membranes imaged at submolecular resolution by atomic force microscopy. J Struct Biol 103(1):89–94
7. Butt HJ, Downing KH, Hansma PK (1990) Imaging the membrane protein bacteriorhodopsin with the atomic force microscope. Biophys J 58(6):1473–1480
8. Radmacher M, Tillmann RW, Fritz M, Gaub HE (1992) From molecules to cells: imaging soft samples with the atomic force microscope. Science 257(5078):1900–1905
9. Fritz M, Radmacher M, Gaub HE (1993) In vitro activation of human platelets triggered and probed by atomic force microscopy. Exp Cell Res 205(1):187–190
10. Hansma HG, Hoh JH (1994) Biomolecular imaging with the atomic force microscope. Annu Rev Biophys Biomol Struct 23(1):115–139
11. Binnig G, Gerber C, Stoll E, Albrecht TR, Quate CF (1987) Atomic resolution with atomic force microscope. Europhys Lett (EPL) 3(12):1281–1286

12. Weisenhorn AL, Drake B, Prater CB, Gould SAC, Hansma PK, Ohnesorge F, Egger M, Heyn SP, Gaub HE (1990) Immobilized proteins in buffer imaged at molecular resolution by atomic force microscopy. Biophys J 58(5):1251–1258
13. Hansma HG, Weisenhorn AL, Edmundson AB, Gaub HE, Hansma PK (1991) Atomic force microscopy: seeing molecules of lipid and immunoglobulin. Clin Chem 37(9):1497–1501
14. Singh S, Keller DJ (1991) Atomic force microscopy of supported planar membrane bilayers. Biophys J 60(6):1401–1410
15. Hansma HG, Bezanilla M, Zenhausern F, Adrian M, Sinsheimer RL (1993) Atomic force microscopy of DNA in aqueous solutions. Nucleic Acids Res 21(3):505–512
16. Hoh JH, Cleveland JP, Prater CB, Revel JP, Hansma PK (1992) Quantized adhesion detected with the atomic force microscope. J Am Chem Soc 114(12):4917–4918
17. Weisenhorn AL, Maivald P, Butt H, Hansma PK (1992) Measuring adhesion, attraction, and repulsion between surfaces in liquids with an atomic-force microscope. Phys Rev B Condens Matter 45(19):11226–11232
18. Fang HHP, Chan KY, Xu LC (2000) Quantification of bacterial adhesion forces using atomic force microscopy (AFM). J Microbiol Methods 40(1):89–97
19. Razatos A (2001) Application of atomic force microscopy to study initial events of bacterial adhesion. In: Doyle RJ (ed) Microbial growth in biofilms – part B: special environments and physicochemical aspects. Academic Press, pp 276–285
20. Nguyen TD, Gu Y (2016) Investigation of cell-substrate adhesion properties of living chondrocyte by measuring adhesive shear force and detachment using AFM and inverse FEA. Sci Rep 6:38059
21. Ishii S, Yoshimoto S, Hori K (2022) Single-cell adhesion force mapping of a highly sticky bacterium in liquid. J Colloid Interface Sci 606(Pt 1):628–634
22. Dufrêne YF, Pelling AE (2013) Force nanoscopy of cell mechanics and cell adhesion. Nanoscale 5(10):4094–4104
23. Putman CA, van der Werf KO, de Grooth BG, van Hulst NF, Greve J (1994) Viscoelasticity of living cells allows high resolution imaging by tapping mode atomic force microscopy. Biophys J 67(4):1749–1753
24. Radmacher M, Fritz M, Kacher CM, Cleveland JP, Hansma PK (1996) Measuring the viscoelastic properties of human platelets with the atomic force microscope. Biophys J 70(1):556–567
25. Lal R, John SA (1994) Biological applications of atomic force microscopy. Am J Physiol Cell Physiol 266(1):C1–C21
26. Hofmann UG, Rotsch C, Parak WJ, Radmacher M (1997) Investigating the cytoskeleton of chicken cardiocytes with the atomic force microscope. J Struct Biol 119(2):84–91
27. Domke J, Parak WJ, George M, Gaub HE, Radmacher M (1999) Mapping the mechanical pulse of single cardiomyocytes with the atomic force microscope. Eur Biophys J 28(3): 179–186
28. Rotsch C, Radmacher M (2000) Drug-induced changes of cytoskeletal structure and mechanics in fibroblasts: an atomic force microscopy study. Biophys J 78(1):520–535
29. Allison DP, Mortensen NP, Sullivan CJ, Doktycz MJ (2010) Atomic force microscopy of biological samples. Wiley Interdiscip Rev Nanomed Nanobiotechnol 2(6):618–634
30. Kodera N, Yamamoto D, Ishikawa R, Ando T (2010) Video imaging of walking myosin V by high-speed atomic force microscopy. Nature 468(7320):72–76
31. Shibata M, Yamashita H, Uchihashi T, Kandori H, Ando T (2010) High-speed atomic force microscopy shows dynamic molecular processes in photoactivated bacteriorhodopsin. Nat Nanotechnol 5(3):208–212
32. Dufrêne YF, Ando T, Garcia R, Alsteens D, Martinez-Martin D, Engel A, Gerber C, Muller DJ (2017) Imaging modes of atomic force microscopy for application in molecular and cell biology. Nat Nanotechnol 12(4):295–307
33. Stylianou A, Kontomaris SV, Grant C, Alexandratou E (2019) Atomic force microscopy on biological materials related to pathological conditions. Scanning 2019:8452851

34. Uchihashi T, Ganser C (2020) Recent advances in bioimaging with high-speed atomic force microscopy. Biophys Rev 12(2):363–369
35. Dufrêne YF, Viljoen A, Mignolet J, Mathelie-Guinlet M (2021) AFM in cellular and molecular microbiology. Cell Microbiol 23(7):e13324
36. Eid J, Greige-Gerges H, Monticelli L, Jraij A (2021) Elastic moduli of lipid membranes: reproducibility of AFM measures. Chem Phys Lipids 234:105011
37. Main KHS, Provan JI, Haynes PJ, Wells G, Hartley JA, Pyne ALB (2021) Atomic force microscopy-A tool for structural and translational DNA research. APL Bioeng 5(3):031504
38. Pena B, Adbel-Hafiz M, Cavasin M, Mestroni L, Sbaizero O (2022) Atomic force microscopy (AFM) applications in arrhythmogenic cardiomyopathy. Int J Mol Sci 23(7):3700
39. Joshi J, Homburg SV, Ehrmann A (2022) Atomic force microscopy (AFM) on biopolymers and hydrogels for biotechnological applications-possibilities and limits. Polymers 14(6):1267
40. Hansma PK, Drake B, Marti O, Gould SA, Prater CB (1989) The scanning ion-conductance microscope. Science 243(4891):641–643
41. Sanchez D, Anand U, Gorelik J, Benham CD, Bountra C, Lab M, Klenerman D, Birch R, Anand P, Korchev Y (2007) Localized and non-contact mechanical stimulation of dorsal root ganglion sensory neurons using scanning ion conductance microscopy. J Neurosci Methods 159(1):26–34
42. Li C, Johnson N, Ostanin V, Shevchuk A, Ying LM, Korchev Y, Klenerman D (2008) High resolution imaging using scanning ion conductance microscopy with improved distance feedback control. Prog Nat Sci 18(6):671–677
43. Novak P, Li C, Shevchuk AI, Stepanyan R, Caldwell M, Hughes S, Smart TG, Gorelik J, Ostanin VP, Lab MJ, Moss GW, Frolenkov GI, Klenerman D, Korchev YE (2009) Nanoscale live-cell imaging using hopping probe ion conductance microscopy. Nat Methods 6(4): 279–281
44. Ossola D, Dorwling-Carter L, Dermutz H, Behr P, Voros J, Zambelli T (2015) Simultaneous scanning ion conductance microscopy and atomic force microscopy with microchanneled cantilevers. Phys Rev Lett 115(23):238103
45. Schäffer TE, Ionescu-Zanetti C, Proksch R, Fritz M, Walters DA, Almqvist N, Zaremba CM, Belcher AM, Smith BL, Stucky GD, Morse DE, Hansma PK (1997) Does abalone nacre form by heteroepitaxial nucleation or by growth through mineral bridges? Chem Mater 9(8): 1731–1740
46. Korchev YE, Bashford CL, Milovanovic M, Vodyanoy I, Lab MJ (1997) Scanning ion conductance microscopy of living cells. Biophys J 73(2):653–658
47. Korchev YE, Gorelik J, Lab MJ, Sviderskaya EV, Johnston CL, Coombes CR, Vodyanoy I, Edwards CRW (2000) Cell volume measurement using scanning ion conductance microscopy. Biophys J 78(1):451–457
48. Shevchuk AI, Gorelik J, Harding SE, Lab MJ, Klenerman D, Korchev YE (2001) Simultaneous measurement of Ca2+ and cellular dynamics: combined scanning ion conductance and optical microscopy to study contracting cardiac myocytes. Biophys J 81(3):1759–1764
49. Mann SA, Hoffmann G, Hengstenberg A, Schuhmann W, Dietzel ID (2002) Pulse-mode scanning ion conductance microscopy – a method to investigate cultured hippocampal cells. J Neurosci Methods 116(2):113–117
50. Gorelik J, Zhang Y, Shevchuk AI, Frolenkov GI, Sanchez D, Lab MJ, Vodyanoy I, Edwards CR, Klenerman D, Korchev YE (2004) The use of scanning ion conductance microscopy to image A6 cells. Mol Cell Endocrinol 217(1-2):101–108
51. Ying L, Bruckbauer A, Zhou D, Gorelik J, Shevchuk A, Lab M, Korchev Y, Klenerman D (2005) The scanned nanopipette: a new tool for high resolution bioimaging and controlled deposition of biomolecules. Phys Chem Chem Phys 7(15):2859–2866
52. Shevchuk AI, Frolenkov GI, Sanchez D, James PS, Freedman N, Lab MJ, Jones R, Klenerman D, Korchev YE (2006) Imaging proteins in membranes of living cells by high-resolution scanning ion conductance microscopy. Angew Chem Int Ed Engl 45(14): 2212–2216

53. Böcker M, Anczykowski B, Wegener J, Schäffer TE (2007) Scanning ion conductance microscopy with distance-modulated shear force control. Nanotechnology 18(14):145505
54. Böcker M, Muschter S, Schmitt E, Steinem C, Schäffer TE (2009) Imaging and patterning of pore-suspending membranes with scanning ion conductance microscopy. Langmuir 25(5): 3022–3028
55. Zhang S, Cho SJ, Busuttil K, Wang C, Besenbacher F, Dong M (2012) Scanning ion conductance microscopy studies of amyloid fibrils at nanoscale. Nanoscale 4(10):3105–3110
56. Rheinlaender J, Schäffer TE (2013) Mapping the mechanical stiffness of live cells with the scanning ion conductance microscope. Soft Matter 9(12):3230–3236
57. Schäffer TE (2013) Nanomechanics of molecules and living cells with scanning ion conductance microscopy. Anal Chem 85(15):6988–6994
58. Rheinlaender J, Vogel S, Seifert J, Schachtele M, Borst O, Lang F, Gawaz M, Schäffer TE (2015) Imaging the elastic modulus of human platelets during thrombin-induced activation using scanning ion conductance microscopy. Thromb Haemost 113(2):305–311
59. Scheenen WJ, Celikel T (2015) Nanophysiology: bridging synapse ultrastructure, biology, and physiology using scanning ion conductance microscopy. Synapse 69(5):233–241
60. Rheinlaender J, Schäffer TE (2019) Mapping the creep compliance of living cells with scanning ion conductance microscopy reveals a subcellular correlation between stiffness and fluidity. Nanoscale 11(14):6982–6989
61. Simeonov S, Schäffer TE (2019) Ultrafast imaging of cardiomyocyte contractions by combining scanning ion conductance microscopy with a microelectrode array. Anal Chem 91(15): 9648–9655
62. Swiatlowska P, Sanchez-Alonso JL, Mansfield C, Scaini D, Korchev Y, Novak P, Gorelik J (2020) Short-term angiotensin II treatment regulates cardiac nanomechanics via microtubule modifications. Nanoscale 12(30):16315–16329
63. Takahashi Y, Zhou Y, Miyamoto T, Higashi H, Nakamichi N, Takeda Y, Kato Y, Korchev Y, Fukuma T (2020) High-speed SICM for the visualization of nanoscale dynamic structural changes in hippocampal neurons. Anal Chem 92(2):2159–2167
64. Tognoni E (2021) High-speed multifunctional scanning ion conductance microscopy: innovative strategies to study dynamic cellular processes. Curr Opin Electrochem 28:100738
65. Zhu C, Huang K, Siepser NP, Baker LA (2021) Scanning ion conductance microscopy. Chem Rev 121(19):11726–11768
66. Wang D, Sun L, Okuda S, Yamamoto D, Nakayama M, Oshima H, Saito H, Kouyama Y, Mimori K, Ando T, Watanabe S, Oshima M (2022) Nano-scale physical properties characteristic to metastatic intestinal cancer cells identified by high-speed scanning ion conductance microscope. Biomaterials 280:121256
67. Meyer G, Amer NM (1988) Novel optical approach to atomic force microscopy. Appl Phys Lett 53(12):1045–1047
68. Alexander S, Hellemans L, Marti O, Schneir J, Elings V, Hansma PK, Longmire M, Gurley J (1989) An atomic-resolution atomic-force microscope implemented using an optical-lever. J Appl Phys 65(1):164–167
69. Meyer G, Amer NM (1990) Optical-beam-deflection atomic force microscopy – the Nacl (001) surface. Appl Phys Lett 56(21):2100–2101
70. Prater CB, Hansma PK, Tortonese M, Quate CF (1991) Improved scanning ion-conductance microscope using microfabricated probes. Rev Sci Instrum 62(11):2634–2638
71. Neher E, Sakmann B (1976) Single-channel currents recorded from membrane of denervated frog muscle fibres. Nature 260(5554):799–802
72. Sakmann B, Neher E (1984) Patch clamp techniques for studying ionic channels in excitable membranes. Annu Rev Physiol 46(1):455–472
73. Schierbaum N, Hack M, Betz O, Schäffer TE (2018) Macro-SICM: a scanning ion conductance microscope for large-range imaging. Anal Chem 90(8):5048–5054

74. Seifert J, Rheinlaender J, Novak P, Korchev YE, Schäffer TE (2015) Comparison of atomic force microscopy and scanning ion conductance microscopy for live cell imaging. Langmuir 31(24):6807–6813
75. Rheinlaender J, Schäffer TE (2017) An accurate model for the ion current-distance behavior in scanning ion conductance microscopy allows for calibration of pipet tip geometry and tip-sample distance. Anal Chem 89(21):11875–11880
76. Thatenhorst D, Rheinlaender J, Schäffer TE, Dietzel ID, Happel P (2014) Effect of sample slope on image formation in scanning ion conductance microscopy. Anal Chem 86(19): 9838–9845
77. Martin Y, Williams CC, Wickramasinghe HK (1987) Atomic force microscope force mapping and profiling on a sub 100-Å scale. J Appl Phys 61(10):4723–4729
78. Albrecht TR, Grütter P, Horne D, Rugar D (1991) Frequency modulation detection using high-Q cantilevers for enhanced force microscope sensitivity. J Appl Phys 69(2):668–673
79. Zhong Q, Inniss D, Kjoller K, Elings VB (1993) Fractured polymer silica fiber surface studied by tapping mode atomic-force microscopy. Surf Sci 290(1-2):L688–L692
80. Hansma PK, Cleveland JP, Radmacher M, Walters DA, Hillner PE, Bezanilla M, Fritz M, Vie D, Hansma HG, Prater CB, Massie J, Fukunaga L, Gurley J, Elings V (1994) Tapping mode atomic-force microscopy in liquids. Appl Phys Lett 64(13):1738–1740
81. Radmacher M, Cleveland JP, Fritz M, Hansma HG, Hansma PK (1994) Mapping interaction forces with the atomic force microscope. Biophys J 66(6):2159–2165
82. Baselt DR, Baldeschwieler JD (1994) Imaging spectroscopy with the atomic-force microscope. J Appl Phys 76(1):33–38
83. You HX, Lau JM, Zhang S, Yu L (2000) Atomic force microscopy imaging of living cells: a preliminary study of the disruptive effect of the cantilever tip on cell morphology. Ultramicroscopy 82(1-4):297–305
84. Hansma HG, Sinsheimer RL, Groppe J, Bruice TC, Elings V, Gurley G, Bezanilla M, Mastrangelo IA, Hough PV, Hansma PK (1993) Recent advances in atomic force microscopy of DNA. Scanning 15(5):296–299
85. Pi J, Cai J (2019) Cell topography and its quantitative imaging by AFM. In: Santos NC, Carvalho FA (eds) Atomic force microscopy: methods and protocols. Springer, New York, pp 99–113
86. Cappella B, Baschieri P, Frediani C, Miccoli P, Ascoli C (1997) Improvements in AFM imaging of the spatial variation of force-distance curves: on-line images. Nanotechnology 8(2):82–87
87. Jiao Y, Schäffer TE (2004) Accurate height and volume measurements on soft samples with the atomic force microscope. Langmuir 20(23):10038–10045
88. Giessibl FJ (1997) Forces and frequency shifts in atomic-resolution dynamic-force microscopy. Phys Rev B 56(24):16010–16015
89. Abe M (2018) Frequency-modulation atomic force microscopy. In: Compendium of surface and Interface analysis. Springer, Singapore, pp 201–204
90. Martinez-Martin D, Carrasco C, Hernando-Perez M, de Pablo PJ, Gomez-Herrero J, Perez R, Mateu MG, Carrascosa JL, Kiracofe D, Melcher J, Raman A (2012) Resolving structure and mechanical properties at the nanoscale of viruses with frequency modulation atomic force microscopy. PLoS One 7(1):e30204
91. Al-Rekabi Z, Contera S (2018) Multifrequency AFM reveals lipid membrane mechanical properties and the effect of cholesterol in modulating viscoelasticity. Proc Natl Acad Sci U S A 115(11):2658–2663
92. Proksch R, Lal R, Hansma PK, Morse D, Stucky G (1996) Imaging the internal and external pore structure of membranes in fluid: TappingMode scanning ion conductance microscopy. Biophys J 71(4):2155–2157
93. Pastre D, Iwamoto H, Liu J, Szabo G, Shao Z (2001) Characterization of AC mode scanning ion-conductance microscopy. Ultramicroscopy 90(1):13–19

94. Takahashi Y, Murakami Y, Nagamine K, Shiku H, Aoyagi S, Yasukawa T, Kanzaki M, Matsue T (2010) Topographic imaging of convoluted surface of live cells by scanning ion conductance microscopy in a standing approach mode. Phys Chem Chem Phys 12(34): 10012–10017
95. Zhukov A, Richards O, Ostanin V, Korchev Y, Klenerman D (2012) A hybrid scanning mode for fast scanning ion conductance microscopy (SICM) imaging. Ultramicroscopy 121(11):1–7
96. Simeonov S, Schäffer TE (2019) High-speed scanning ion conductance microscopy for sub-second topography imaging of live cells. Nanoscale 11(17):8579–8587
97. Rheinlaender J, Schäffer TE (2009) Image formation, resolution, and height measurement in scanning ion conductance microscopy. J Appl Phys 105(9):094905
98. McKelvey K, Perry D, Byers JC, Colburn AW, Unwin PR (2014) Bias modulated scanning ion conductance microscopy. Anal Chem 86(7):3639–3646
99. Li P, Liu LQ, Yang Y, Wang YC, Li GY (2015) In-phase bias modulation mode of scanning ion conductance microscopy with capacitance compensation. IEEE Trans Ind Electron 62(10): 6508–6518
100. Li P, Liu LQ, Wang YC, Yang Y, Zhang CL, Li GY (2014) Phase modulation mode of scanning ion conductance microscopy. Appl Phys Lett 105(5):053113
101. Li P, Liu L, Yang Y, Zhou L, Wang D, Wang Y, Li G (2015) Amplitude modulation mode of scanning ion conductance microscopy. J Lab Autom 20(4):457–462
102. Rheinlaender J, Geisse NA, Proksch R, Schäffer TE (2011) Comparison of scanning ion conductance microscopy with atomic force microscopy for cell imaging. Langmuir 27(2): 697–704
103. Rodolfa KT, Bruckbauer A, Zhou D, Korchev YE, Klenerman D (2005) Two-component graded deposition of biomolecules with a double-barreled nanopipette. Angew Chem Int Ed Engl 44(42):6854–6859
104. Rodolfa KT, Bruckbauer A, Zhou D, Schevchuk AI, Korchev YE, Klenerman D (2006) Nanoscale pipetting for controlled chemistry in small arrayed water droplets using a double-barrel pipet. Nano Lett 6(2):252–257
105. Happel P, Thatenhorst D, Dietzel ID (2012) Scanning ion conductance microscopy for studying biological samples. Sensors (Basel) 12(11):14983–15008
106. Klar TA, Hell SW (1999) Subdiffraction resolution in far-field fluorescence microscopy. Opt Lett 24(14):954–956
107. Oyabu N, Custance O, Yi I, Sugawara Y, Morita S (2003) Mechanical vertical manipulation of selected single atoms by soft nanoindentation using near contact atomic force microscopy. Phys Rev Lett 90(17):176102
108. Giessibl FJ (2005) AFM's path to atomic resolution. Mater Today 8(5):32–41
109. Gan Y (2009) Atomic and subnanometer resolution in ambient conditions by atomic force microscopy. Surf Sci Rep 64(3):99–121
110. Butt HJ, Jaschke M (1995) Calculation of thermal noise in atomic-force microscopy. Nanotechnology 6(1):1–7
111. Schäffer TE (2005) Calculation of thermal noise in an atomic force microscope with a finite optical spot size. Nanotechnology 16(6):664–670
112. Bustamante C, Keller D (1995) Scanning force microscopy in biology. Phys Today 48(12): 32–38
113. Stolz M, Raiteri R, Daniels AU, VanLandingham MR, Baschong W, Aebi U (2004) Dynamic elastic modulus of porcine articular cartilage determined at two different levels of tissue organization by indentation-type atomic force microscopy. Biophys J 86(5):3269–3283
114. Dufrêne YF, Martinez-Martin D, Medalsy I, Alsteens D, Muller DJ (2013) Multiparametric imaging of biological systems by force-distance curve-based AFM. Nat Methods 10(9): 847–854
115. Radmacher M, Fritz M, Hansma PK (1995) Imaging soft samples with the atomic force microscope: gelatin in water and propanol. Biophys J 69(1):264–270

116. Alonso JL, Goldmann WH (2003) Feeling the forces: atomic force microscopy in cell biology. Life Sci 72(23):2553–2560
117. Kuznetsova TG, Starodubtseva MN, Yegorenkov NI, Chizhik SA, Zhdanov RI (2007) Atomic force microscopy probing of cell elasticity. Micron 38(8):824–833
118. Braet F, Rotsch C, Wisse E, Radmacher M (1998) Comparison of fixed and living liver endothelial cells by atomic force microscopy. Appl Phys A 66(7):S575–S578
119. Francis LW, Gonzalez D, Ryder T, Baer K, Rees M, White JO, Conlan RS, Wright CJ (2010) Optimized sample preparation for high-resolution AFM characterization of fixed human cells. J Microsc 240(2):111–121
120. Hinterdorfer P, Dufrene YF (2006) Detection and localization of single molecular recognition events using atomic force microscopy. Nat Methods 3(5):347–355
121. Bezanilla M, Drake B, Nudler E, Kashlev M, Hansma PK, Hansma HG (1994) Motion and enzymatic degradation of DNA in the atomic force microscope. Biophys J 67(6):2454–2459
122. Dubrovin EV, Schächtele M, Klinov DV, Schäffer TE (2017) Time-lapsed single biomolecule atomic force microscopy investigation on modified graphite in solution. Langmuir 33:10027–10034
123. Kasas S, Thomson NH, Smith BL, Hansma HG, Zhu X, Guthold M, Bustamante C, Kool ET, Kashlev M, Hansma PK (1997) Escherichia coli RNA polymerase activity observed using atomic force microscopy. Biochemistry 36(3):461–468
124. Jiao Y, Cherny DI, Heim G, Jovin TM, Schäffer TE (2001) Dynamic interactions of p53 with DNA in solution by time-lapse atomic force microscopy. J Mol Biol 314(2):233–243
125. Leung C, Bestembayeva A, Thorogate R, Stinson J, Pyne A, Marcovich C, Yang J, Drechsler U, Despont M, Jankowski T (2012) Atomic force microscopy with nanoscale cantilevers resolves different structural conformations of the DNA double helix. Nano Lett 12(7):3846–3850
126. Müller DJ, Sass HJ, Müller SA, Buldt G, Engel A (1999) Surface structures of native bacteriorhodopsin depend on the molecular packing arrangement in the membrane. J Mol Biol 285(5):1903–1909
127. Mulvihill E, van Pee K, Mari SA, Muller DJ, Yildiz O (2015) Directly observing the lipid-dependent self-assembly and pore-forming mechanism of the cytolytic toxin Listeriolysin O. Nano Lett 15(10):6965–6973
128. Janovjak H, Kedrov A, Cisneros DA, Sapra KT, Struckmeier J, Muller DJ (2006) Imaging and detecting molecular interactions of single transmembrane proteins. Neurobiol Aging 27(4): 546–561
129. Pfreundschuh M, Alsteens D, Hilbert M, Steinmetz MO, Muller DJ (2014) Localizing chemical groups while imaging single native proteins by high-resolution atomic force microscopy. Nano Lett 14(5):2957–2964
130. Zhang L, Yang F, Cai JY, Yang PH, Liang ZH (2014) In-situ detection of resveratrol inhibition effect on epidermal growth factor receptor of living MCF-7 cells by Atomic Force Microscopy. Biosens Bioelectron 56:271–277
131. Dupres V, Alsteens D, Pauwels K, Dufrene YF (2009) In vivo imaging of S-layer nanoarrays on Corynebacterium glutamicum. Langmuir 25(17):9653–9655
132. Pasquina-Lemonche L, Burns J, Turner RD, Kumar S, Tank R, Mullin N, Wilson JS, Chakrabarti B, Bullough PA, Foster SJ, Hobbs JK (2020) The architecture of the gram-positive bacterial cell wall. Nature 582(7811):294–297
133. Hirvonen LM, Marsh RJ, Jones GE, Cox S (2020) Combined AFM and super-resolution localisation microscopy: investigating the structure and dynamics of podosomes. Eur J Cell Biol 99(7):151106
134. Mondeja B, Valdes O, Resik S, Vizcaino A, Acosta E, Montalvan A, Paez A, Mune M, Rodriguez R, Valdes J, Gonzalez G, Sanchez D, Falcon V, Gonzalez Y, Kouri V, I.P.K.V.R. Group, Diaz A, Guzman M (2021) SARS-CoV-2: preliminary study of infected human nasopharyngeal tissue by high resolution microscopy. Virol J 18(1):149

135. Rheinlaender J, Schäffer TE (2015) Lateral resolution and image formation in scanning ion conductance microscopy. Anal Chem 87(14):7117–7124
136. Sun L, Shigyou K, Ando T, Watanabe S (2019) Thermally driven approach to fill sub-10-nm pipettes with batch production. Anal Chem 91(21):14080–14084
137. Navikas V, Leitao SM, Marion S, Davis SJ, Drake B, Fantner GE, Radenovic A (2020) High-throughput nanocapillary filling enabled by microwave radiation for scanning ion conductance microscopy Imaging. ACS Appl Nano Mater 3(8):7829–7834
138. Shigyou K, Sun L, Yajima R, Takigaura S, Tajima M, Furusho H, Kikuchi Y, Miyazawa K, Fukuma T, Taoka A, Ando T, Watanabe S (2020) Geometrical characterization of glass nanopipettes with sub-10 nm pore diameter by transmission electron microscopy. Anal Chem 92(23):15388–15393
139. Korchev YE, Milovanovic M, Bashford CL, Bennett DC, Sviderskaya EV, Vodyanoy I, Lab MJ (1997) Specialized scanning ion-conductance microscope for imaging of living cells. J Microsc 188(1):17–23
140. Walker SC, Allen S, Bell G, Roberts CJ (2015) Analysis of leaf surfaces using scanning ion conductance microscopy. J Microsc 258(2):119–126
141. Ushiki T, Nakajima M, Choi M, Cho SJ, Iwata F (2012) Scanning ion conductance microscopy for imaging biological samples in liquid: a comparative study with atomic force microscopy and scanning electron microscopy. Micron 43(12):1390–1398
142. Nakajima M, Mizutani Y, Iwata F, Ushiki T (2018) Scanning ion conductance microscopy for visualizing the three-dimensional surface topography of cells and tissues. Semin Cell Dev Biol 73:125–131
143. Zhou Y, Saito M, Miyamoto T, Novak P, Shevchuk AI, Korchev YE, Fukuma T, Takahashi Y (2018) Nanoscale Imaging of primary cilia with scanning ion conductance microscopy. Anal Chem 90(4):2891–2895
144. Gorelik J, Yang LQ, Zhang Y, Lab M, Korchev Y, Harding SE (2006) A novel Z-groove index characterizing myocardial surface structure. Cardiovasc Res 72(3):422–429
145. Voelkner C, Wendt M, Lange R, Ulbrich M, Gruening M, Staehlke S, Nebe B, Barke I, Speller S (2021) The nanomorphology of cell surfaces of adhered osteoblasts. Beilstein J Nanotechnol 12:242–256
146. Müller DJ, Dufrene YF (2011) Atomic force microscopy: a nanoscopic window on the cell surface. Trends Cell Biol 21(8):461–469
147. Braet F, de Zanger R, Seynaeve C, Baekeland M, Wisse E (2001) A comparative atomic force microscopy study on living skin fibroblasts and liver endothelial cells. J Electron Microsc (Tokyo) 50(4):283–290
148. Gorelik J, Shevchuk AI, Frolenkov GI, Diakonov IA, Lab MJ, Kros CJ, Richardson GP, Vodyanoy I, Edwards CR, Klenerman D, Korchev YE (2003) Dynamic assembly of surface structures in living cells. Proc Natl Acad Sci U S A 100(10):5819–5822
149. Klingauf J, Kavalali ET, Tsien RW (1998) Kinetics and regulation of fast endocytosis at hippocampal synapses. Nature 394(6693):581–585
150. Shibata M, Uchihashi T, Ando T, Yasuda R (2015) Long-tip high-speed atomic force microscopy for nanometer-scale imaging in live cells. Sci Rep 5(1):8724
151. Guo Y, Li D, Zhang S, Yang Y, Liu JJ, Wang X, Liu C, Milkie DE, Moore RP, Tulu US, Kiehart DP, Hu J, Lippincott-Schwartz J, Betzig E, Li D (2018) Visualizing intracellular organelle and cytoskeletal interactions at nanoscale resolution on millisecond timescales. Cell 175(5):1430–1442.e17
152. Fuller PWW (2009) An introduction to high speed photography and photonics. Imaging Sci J 57(6):293–302
153. Betzig E, Patterson GH, Sougrat R, Lindwasser OW, Olenych S, Bonifacino JS, Davidson MW, Lippincott-Schwartz J, Hess HF (2006) Imaging intracellular fluorescent proteins at nanometer resolution. Science 313(5793):1642–1645

154. Heilemann M, van de Linde S, Schuttpelz M, Kasper R, Seefeldt B, Mukherjee A, Tinnefeld P, Sauer M (2008) Subdiffraction-resolution fluorescence imaging with conventional fluorescent probes. Angew Chem Int Ed Engl 47(33):6172–6176
155. Huang B, Babcock H, Zhuang X (2010) Breaking the diffraction barrier: super-resolution imaging of cells. Cell 143(7):1047–1058
156. Jacquemet G, Carisey AF, Hamidi H, Henriques R, Leterrier C (2020) The cell biologist's guide to super-resolution microscopy. J Cell Sci 133(11):jcs240713
157. Walters DA, Cleveland JP, Thomson NH, Hansma PK, Wendman MA, Gurley G, Elings V (1996) Short cantilevers for atomic force microscopy. Rev Sci Instrum 67(10):3583–3590
158. Schäffer T, Cleveland J, Ohnesorge F, Walters D, Hansma P (1996) Studies of vibrating atomic force microscope cantilevers in liquid. J Appl Phys 80(7):3622–3627
159. Viani MB, Schäffer TE, Paloczi GT, Pietrasanta LI, Smith BL, Thompson JB, Richter M, Rief M, Gaub HE, Plaxco KW, Cleland AN, Hansma HG, Hansma PK (1999) Fast imaging and fast force spectroscopy of single biopolymers with a new atomic force microscope designed for small cantilevers. Rev Sci Instrum 70(11):4300–4303
160. Ando T, Kodera N, Takai E, Maruyama D, Saito K, Toda A (2001) A high-speed atomic force microscope for studying biological macromolecules. Proc Natl Acad Sci U S A 98(22): 12468–12472
161. Ando T, Kodera N, Naito Y, Kinoshita T, Furuta K, Toyoshima YY (2003) A high-speed atomic force microscope for studying biological macromolecules in action. ChemPhysChem 4(11):1196–1202
162. Ando T, Uchihashi T, Kodera N, Miyagi A, Nakakita R, Yamashita H, Sakashita M (2006) High-speed atomic force microscopy for studying the dynamic behavior of protein molecules at work. Jpn J Appl Phys Pt 1 45(3b):1897–1903
163. Yamashita H, Kodera N, Miyagi A, Uchihashi T, Yamamoto D, Ando T (2007) Tip-sample distance control using photothermal actuation of a small cantilever for high-speed atomic force microscopy. Rev Sci Instrum 78(8):083702
164. Ando T, Uchihashi T, Kodera N, Yamamoto D, Miyagi A, Taniguchi M, Yamashita H (2008) High-speed AFM and nano-visualization of biomolecular processes. Pflugers Arch 456(1): 211–225
165. Miyagi A, Chipot C, Rangl M, Scheuring S (2016) High-speed atomic force microscopy shows that annexin V stabilizes membranes on the second timescale. Nat Nanotechnol 11(9): 783–790
166. Lin YC, Guo YR, Miyagi A, Levring J, MacKinnon R, Scheuring S (2019) Force-induced conformational changes in PIEZO1. Nature 573(7773):230–234
167. Lim K, Kodera N, Wang H, Mohamed MS, Hazawa M, Kobayashi A, Yoshida T, Hanayama R, Yano S, Ando T, Wong RW (2020) High-speed AFM reveals molecular dynamics of human influenza A hemagglutinin and its interaction with exosomes. Nano Lett 20(9):6320–6328
168. Kodera N, Noshiro D, Dora SK, Mori T, Habchi J, Blocquel D, Gruet A, Dosnon M, Salladini E, Bignon C, Fujioka Y, Oda T, Noda NN, Sato M, Lotti M, Mizuguchi M, Longhi S, Ando T (2021) Structural and dynamics analysis of intrinsically disordered proteins by high-speed atomic force microscopy. Nat Nanotechnol 16(2):181–189
169. Braunsmann C, Schäffer TE (2010) High-speed atomic force microscopy for large scan sizes using small cantilevers. Nanotechnology 21(22):225705
170. Braunsmann C, Seifert J, Rheinlaender J, Schäffer TE (2014) High-speed force mapping on living cells with a small cantilever atomic force microscope. Rev Sci Instrum 85(7):073703
171. Casuso I, Redondo-Morata L, Rico F (2020) Biological physics by high-speed atomic force microscopy. Philos Trans A Math Phys Eng Sci 378(2186):20190604
172. Kodera N, Ando T (2022) Visualization of intrinsically disordered proteins by high-speed atomic force microscopy. Curr Opin Struct Biol 72:260–266
173. Happel P, Hoffmann G, Mann SA, Dietzel ID (2003) Monitoring cell movements and volume changes with pulse-mode scanning ion conductance microscopy. J Microsc 212(Pt 2):144–151

174. Shevchuk AI, Novak P, Taylor M, Diakonov IA, Ziyadeh-Isleem A, Bitoun M, Guicheney P, Lab MJ, Gorelik J, Merrifield CJ, Klenerman D, Korchev YE (2012) An alternative mechanism of clathrin-coated pit closure revealed by ion conductance microscopy. J Cell Biol 197(4):499–508
175. Zhuang J, Jiao Y, Li Z, Lang J, Li F (2018) A continuous control mode with improved imaging rate for scanning ion conductance microscope (SICM). Ultramicroscopy 190:66–76
176. Zhuang J, Wang Z, Li Z, Liang P, Vincent M (2018) Smart scanning ion-conductance microscopy Imaging technique using horizontal fast scanning method. Microsc Microanal 24(3):264–276
177. Ida H, Takahashi Y, Kumatani A, Shiku H, Matsue T (2017) High speed scanning ion conductance microscopy for quantitative analysis of nanoscale dynamics of microvilli. Anal Chem 89(11):6015–6020
178. Jung GE, Noh H, Shin YK, Kahng SJ, Baik KY, Kim HB, Cho NJ, Cho SJ (2015) Closed-loop ARS mode for scanning ion conductance microscopy with improved speed and stability for live cell imaging applications. Nanoscale 7(25):10989–10997
179. Watanabe S, Ando T (2017) High-speed XYZ-nanopositioner for scanning ion conductance microscopy. Appl Phys Lett 111(11):113106
180. Watanabe S, Kitazawa S, Sun L, Kodera N, Ando T (2019) Development of high-speed ion conductance microscopy. Rev Sci Instrum 90(12):123704
181. Leitao SM, Drake B, Pinjusic K, Pierrat X, Navikas V, Nievergelt AP, Brillard C, Djekic D, Radenovic A, Persat A (2021) Time-resolved scanning ion conductance microscopy for three-dimensional tracking of nanoscale cell surface dynamics. ACS Nano 15(11):17613–17622
182. Seifert J, Rheinlaender J, Lang F, Gawaz M, Schäffer TE (2017) Thrombin-induced cytoskeleton dynamics in spread human platelets observed with fast scanning ion conductance microscopy. Sci Rep 7:4810
183. Haase K, Pelling AE (2015) Investigating cell mechanics with atomic force microscopy. J R Soc Interface 12(104):20140970
184. Pegoraro AF, Janmey P, Weitz DA (2017) Mechanical properties of the cytoskeleton and cells. Cold Spring Harb Perspect Biol 9(11):a022038
185. Sanyour HJ, Li N, Rickel AP, Childs JD, Kinser CN, Hong Z (2019) Membrane cholesterol and substrate stiffness co-ordinate to induce the remodelling of the cytoskeleton and the alteration in the biomechanics of vascular smooth muscle cells. Cardiovasc Res 115(8): 1369–1380
186. Ketene AN, Roberts PC, Shea AA, Schmelz EM, Agah M (2012) Actin filaments play a primary role for structural integrity and viscoelastic response in cells. Integr Biol (Camb) 4(5): 540–549
187. Guo M, Pegoraro AF, Mao A, Zhou EH, Arany PR, Han Y, Burnette DT, Jensen MH, Kasza KE, Moore JR, Mackintosh FC, Fredberg JJ, Mooney DJ, Lippincott-Schwartz J, Weitz DA (2017) Cell volume change through water efflux impacts cell stiffness and stem cell fate. Proc Natl Acad Sci U S A 114(41):E8618–E8627
188. Roan E, Wilhelm KR, Waters CM (2015) Kymographic Imaging of the elastic modulus of epithelial cells during the onset of migration. Biophys J 109(10):2051–2057
189. Schächtele M, Kemmler J, Rheinlaender J, Schäffer TE (2021) Combined high-speed atomic force and optical microscopy shows that viscoelastic properties of melanoma cancer cells change during the cell cycle. Adv Mater Technol:2101000
190. Taubenberger AV, Baum B, Matthews HK (2020) The mechanics of mitotic cell rounding. Front Cell Dev Biol 8:687
191. Suresh S (2007) Biomechanics and biophysics of cancer cells. Acta Biomater 3(4):413–438
192. Cross SE, Jin YS, Rao J, Gimzewski JK (2007) Nanomechanical analysis of cells from cancer patients. Nat Nanotechnol 2(12):780–783
193. Quan FS, Kim KS (2016) Medical applications of the intrinsic mechanical properties of single cells. Acta Biochim Biophys Sin Shanghai 48(10):865–871

194. Plodinec M, Loparic M, Monnier CA, Obermann EC, Zanetti-Dallenbach R, Oertle P, Hyotyla JT, Aebi U, Bentires-Alj M, Lim RY (2012) The nanomechanical signature of breast cancer. Nat Nanotechnol 7(11):757–765
195. Ozkan AD, Topal AE, Dana A, Guler MO, Tekinay AB (2016) Atomic force microscopy for the investigation of molecular and cellular behavior. Micron 89:60–76
196. Wu PH, Aroush DR, Asnacios A, Chen WC, Dokukin ME, Doss BL, Durand-Smet P, Ekpenyong A, Guck J, Guz NV, Janmey PA, Lee JSH, Moore NM, Ott A, Poh YC, Ros R, Sander M, Sokolov I, Staunton JR, Wang N, Whyte G, Wirtz D (2018) A comparison of methods to assess cell mechanical properties. Nat Methods 15(7):491–498
197. Tao NJ, Lindsay SM, Lees S (1992) Measuring the microelastic properties of biological material. Biophys J 63(4):1165–1169
198. Weisenhorn AL, Khorsandi M, Kasas S, Gotzos V, Butt HJ (1993) Deformation and height anomaly of soft surfaces studied with an AFM. Nanotechnology 4(2):106–113
199. Radmacher M, Tillmann RW, Gaub HE (1993) Imaging viscoelasticity by force modulation with the atomic force microscope. Biophys J 64(3):735–742
200. Butt HJ, Cappella B, Kappl M (2005) Force measurements with the atomic force microscope: technique, interpretation and applications. Surf Sci Rep 59(1-6):1–152
201. Heredia A, Bui CC, Suter U, Young P, Schäffer TE (2007) AFM combines functional and morphological analysis of peripheral myelinated and demyelinated nerve fibers. Neuroimage 37(4):1218–1226
202. Kasas S, Stupar P, Dietler G (2018) AFM contribution to unveil pro- and eukaryotic cell mechanical properties. Semin Cell Dev Biol 73:177–187
203. Schillers H (2019) Measuring the elastic properties of living cells. Springer, New York, pp 291–313
204. Liang W, Shi H, Yang X, Wang J, Yang W, Zhang H, Liu L (2020) Recent advances in AFM-based biological characterization and applications at multiple levels. Soft Matter 16(39): 8962–8984
205. Hertz H (1882) Ueber die Berührung fester elastischer Körper. Journal für die reine und angewandte Mathematik (Crelles Journal) 1882(92):156–171
206. Sneddon IN (1965) The relation between load and penetration in the axisymmetric boussinesq problem for a punch of arbitrary profile. Int J Eng Sci 3(1):47–57
207. Kilpatrick JI, Revenko I, Rodriguez BJ (2015) Nanomechanics of cells and biomaterials studied by atomic force microscopy. Adv Healthc Mater 4(16):2456–2474
208. Targosz-Korecka M, Malek-Zietek KE, Brzezinka GD, Jaglarz M (2016) Morphological and nanomechanical changes in mechanosensitive endothelial cells induced by colloidal AFM probes. Scanning 38(6):654–664
209. Cook SM, Schäffer TE, Chynoweth KM, Wigton M, Simmonds RW, Lang KM (2006) Practical implementation of dynamic methods for measuring atomic force microscope cantilever spring constants. Nanotechnology 17(9):2135–2145
210. Zemla J, Bobrowska J, Kubiak A, Zielinski T, Pabijan J, Pogoda K, Bobrowski P, Lekka M (2020) Indenting soft samples (hydrogels and cells) with cantilevers possessing various shapes of probing tip. Eur Biophys J 49(6):485–495
211. Sanchez D, Johnson N, Li C, Novak P, Rheinlaender J, Zhang Y, Anand U, Anand P, Gorelik J, Frolenkov GI, Benham C, Lab M, Ostanin VP, Schäffer TE, Klenerman D, Korchev YE (2008) Noncontact measurement of the local mechanical properties of living cells using pressure applied via a pipette. Biophys J 95(6):3017–3027
212. Pellegrino M, Orsini P, Pellegrini M, Baschieri P, Dinelli F, Petracchi D, Tognoni E, Ascoli C (2011) Weak hydrostatic forces in far-scanning ion conductance microscopy used to guide neuronal growth cones. Neurosci Res 69(3):234–240
213. Pellegrino M, Orsini P, De Gregorio F (2009) Use of scanning ion conductance microscopy to guide and redirect neuronal growth cones. Neurosci Res 64(3):290–296

214. Pellegrino M, Orsini P, Pellegrini M, Baschieri P, Dinelli F, Petracchi D, Tognoni E, Ascoli C (2012) Integrated SICM-AFM-optical microscope to measure forces due to hydrostatic pressure applied to a pipette. Micro Nano Lett 7(4):317–320
215. Pellegrino M, Pellegrini M, Orsini P, Tognoni E, Ascoli C, Baschieri P, Dinelli F (2012) Measuring the elastic properties of living cells through the analysis of current-displacement curves in scanning ion conductance microscopy. Pflugers Arch 464(3):307–316
216. Baumann J, Sachs L, Otto O, Schoen I, Nestler P, Zaninetti C, Kenny M, Kranz R, von Eysmondt H, Rodriguez J, Schäffer TE, Nagy Z, Greinacher A, Palankar R, Bender M (2022) Reduced platelet forces underlie impaired hemostasis in mouse models of MYH9-related disease. Sci Adv 8(20):eabn2627
217. Rheinlaender J, Wirbel H, Schäffer TE (2021) Spatial correlation of cell stiffness and traction forces in cancer cells measured with combined SICM and TFM. RSC Adv 11:13951–13956
218. Rheinlaender J, Schäffer TE (2020) The effect of finite sample thickness in scanning ion conductance microscopy stiffness measurements. Appl Phys Lett 117(11):113701
219. Clarke RW, Novak P, Zhukov A, Tyler EJ, Cano-Jaimez M, Drews A, Richards O, Volynski K, Bishop C, Klenerman D (2016) Low stress ion conductance microscopy of sub-cellular stiffness. Soft Matter 12(38):7953–7958
220. Kolmogorov VS, Erofeev AS, Woodcock E, Efremov YM, Iakovlev AP, Savin NA, Alova AV, Lavrushkina SV, Kireev II AO, Prelovskaya EV, Sviderskaya D, Scaini NL, Klyachko PS, Timashev Y, Takahashi SV, Salikhov YN, Parkhomenko AG, Majouga CRW, Edwards P, Novak YEK, Gorelkin PV (2021) Mapping mechanical properties of living cells at nanoscale using intrinsic nanopipette-sample force interactions. Nanoscale 13(13):6558–6568
221. Loftus JC, Choate J, Albrecht RM (1984) Platelet activation and cytoskeletal reorganization: high voltage electron microscopic examination of intact and triton-extracted whole mounts. J Cell Biol 98(6):2019–2025
222. Shin EK, Park H, Noh JY, Lim KM, Chung JH (2017) Platelet shape changes and cytoskeleton dynamics as novel therapeutic targets for anti-thrombotic drugs. Biomol Ther (Seoul) 25(3): 223–230
223. Kuwahara M, Sugimoto M, Tsuji S, Matsui H, Mizuno T, Miyata S, Yoshioka A (2002) Platelet shape changes and adhesion under high shear flow. Arterioscler Thromb Vasc Biol 22(2):329–334
224. Lee D, Fong KP, King MR, Brass LF, Hammer DA (2012) Differential dynamics of platelet contact and spreading. Biophys J 102(3):472–482
225. Watson SP (2009) Platelet activation by extracellular matrix proteins in haemostasis and thrombosis. Curr Pharm Des 15(12):1358–1372
226. Harmon JT, Jamieson GA (1986) Platelet activation by alpha-thrombin is a receptor-mediated event. Ann N Y Acad Sci 485(1):387–395
227. Casella JF, Flanagan MD, Lin S (1981) Cytochalasin D inhibits actin polymerization and induces depolymerization of actin filaments formed during platelet shape change. Nature 293(5830):302–305
228. Verdier C (2003) Rheological properties of living materials. From cells to tissues. J Theor Med 5(2):67–91
229. Kollmannsberger P, Fabry B (2011) Linear and nonlinear rheology of living cells. Annu Rev Mat Res 41(1):75–97
230. Hoffman BD, Crocker JC (2009) Cell mechanics: dissecting the physical responses of cells to force. Annu Rev Biomed Eng 11(1):259–288
231. Mahaffy R, Shih C, MacKintosh F, Käs J (2000) Scanning probe-based frequency-dependent microrheology of polymer gels and biological cells. Phys Rev Lett 85(4):880
232. Alcaraz J, Buscemi L, Grabulosa M, Trepat X, Fabry B, Farre R, Navajas D (2003) Microrheology of human lung epithelial cells measured by atomic force microscopy. Biophys J 84(3):2071–2079

233. Hiratsuka S, Mizutani Y, Toda A, Fukushima N, Kawahara K, Tokumoto H, Okajima T (2009) Power-law stress and creep relaxations of single cells measured by colloidal probe atomic force microscopy. Jpn J Appl Phys 48(8S2):08JB17
234. Moeendarbary E, Harris AR (2014) Cell mechanics: principles, practices, and prospects. Wiley Interdiscip Rev Syst Biol Med 6(5):371–388
235. Efremov YM, Okajima T, Raman A (2020) Measuring viscoelasticity of soft biological samples using atomic force microscopy. Soft Matter 16(1):64–81
236. Hecht FM, Rheinlaender J, Schierbaum N, Goldmann WH, Fabry B, Schäffer TE (2015) Imaging viscoelastic properties of live cells by AFM: power-law rheology on the nanoscale. Soft Matter 11(23):4584–4591
237. Schierbaum N, Rheinlaender J, Schäffer TE (2017) Viscoelastic properties of normal and cancerous human breast cells are affected differently by contact to adjacent cells. Acta Biomater 55:239–248
238. Fabry B, Maksym GN, Butler JP, Glogauer M, Navajas D, Fredberg JJ (2001) Scaling the microrheology of living cells. Phys Rev Lett 87(14):148102
239. Cai P, Mizutani Y, Tsuchiya M, Maloney JM, Fabry B, Van Vliet KJ, Okajima T (2013) Quantifying cell-to-cell variation in power-law rheology. Biophys J 105(5):1093–1102
240. Geiger B (1979) A 130K protein from chicken gizzard: its localization at the termini of microfilament bundles in cultured chicken cells. Cell 18(1):193–205
241. Burridge K, Feramisco JR (1980) Microinjection and localization of a 130K protein in living fibroblasts: a relationship to actin and fibronectin. Cell 19(3):587–595
242. Schierbaum N, Rheinlaender J, Schäffer TE (2019) Combined atomic force microscopy (AFM) and traction force microscopy (TFM) reveals a correlation between viscoelastic material properties and contractile prestress of living cells. Soft Matter 15:1721–1729
243. Moreno-Flores S, Benitez R, Vivanco MD, Toca-Herrera JL (2010) Stress relaxation microscopy: imaging local stress in cells. J Biomech 43(2):349–354
244. Takahashi R, Okajima T (2015) Mapping power-law rheology of living cells using multi-frequency force modulation atomic force microscopy. Appl Phys Lett 107(17):173702
245. Schächtele M, Hänel E, Schäffer TE (2018) Resonance compensating chirp mode for mapping the rheology of live cells by high-speed atomic force microscopy. Appl Phys Lett 113:093701
246. Efremov YM, Cartagena-Rivera AX, Athamneh AIM, Suter DM, Raman A (2018) Mapping heterogeneity of cellular mechanics by multi-harmonic atomic force microscopy. Nat Protoc 13(10):2200–2216
247. Liu X, Li Y, Zhu H, Zhao Z, Zhou Y, Zaske AM, Liu L, Li M, Lu H, Liu W, Dong JF, Zhang J, Zhang Y (2015) Use of non-contact hopping probe ion conductance microscopy to investigate dynamic morphology of live platelets. Platelets 26(5):480–485
248. Seifert J, Rheinlaender J, von Eysmondt H, Schäffer TE (2022) Mechanics of migrating platelets investigated with scanning ion conductance microscopy. Nanoscale 14:8192–8199
249. Barnard H, Drake B, Randall C, Hansma PK (2013) Deep atomic force microscopy. Rev Sci Instrum 84(12):123701
250. Drake B, Randall C, Bridges D, Hansma PK (2014) A new ion sensing deep atomic force microscope. Rev Sci Instrum 85(8):083706
251. Meister A, Gabi M, Behr P, Studer P, Voros J, Niedermann P, Bitterli J, Polesel-Maris J, Liley M, Heinzelmann H, Zambelli T (2009) FluidFM: combining atomic force microscopy and nanofluidics in a universal liquid delivery system for single cell applications and beyond. Nano Lett 9(6):2501–2507
252. Dorwling-Carter L, Aramesh M, Forro C, Tiefenauer RF, Shorubalko I, Voros J, Zambelli T (2018) Simultaneous scanning ion conductance and atomic force microscopy with a nanopore: effect of the aperture edge on the ion current images. J Appl Phys 124(17):174902
253. Ossola D, Amarouch M-Y, Behr P, Vörös JN, Abriel H, Zambelli T (2015) Force-controlled patch clamp of beating cardiac cells. Nano Lett 15(3):1743–1750
254. Li M, Liu LQ, Zambelli T (2022) FluidFM for single-cell biophysics. Nano Res 15(2): 773–786

255. Alesutan I, Seifert J, Pakladok T, Rheinlaender J, Lebedeva A, Towhid ST, Stournaras C, Voelkl J, Schäffer TE, Lang F (2013) Chorein sensitivity of actin polymerization, cell shape and mechanical stiffness of vascular endothelial cells. Cell Physiol Biochem 32(3):728–742
256. Mizutani Y, Choi M-H, Cho S-J, Okajima T (2013) Nanoscale fluctuations on epithelial cell surfaces investigated by scanning ion conductance microscopy. Appl Phys Lett 102(17): 173703
257. Kim SO, Kim J, Okajima T, Cho NJ (2017) Mechanical properties of paraformaldehyde-treated individual cells investigated by atomic force microscopy and scanning ion conductance microscopy. Nano Converg 4(1):5
258. Wang K, Zhou L, Li J, Liu W, Wei Y, Guo Z, Fan C, Hu J, Li B, Wang L (2020) Label-free and three-dimensional visualization reveals the dynamics of plasma membrane-derived extra-cellular vesicles. Nano Lett 20(9):6313–6319

BIOREV (2022) 3: 73–122
https://doi.org/10.1007/11663_2021_9

Published online: 21 October 2021

Analytical Applications of Scanning Ion Conductance Microscopy: Measuring Ions and Electrons

Myung-hoon Choi, Cody W. Leasor, and Lane A. Baker

Contents

Abstract Applications of SICM extend well beyond topographic imaging. Opportunities to measure ion concentrations or electron transfer reactions with SICM probes have proven particularly powerful approaches. Electrochemical imaging applications of SICM have advanced rapidly and benefit from simplicity in probe (electrode) design, advanced methods of feedback, and developments in instrumentation for imaging. Here we consider select analytical applications of the scanning ion conductance microscope, considering first methods to measure ions, and second methods to measure electrons. Special attention is paid to the design of the probe and analytical caveats considered in such measurements.

M.-h. Choi, C. W. Leasor, and L. A. Baker (✉)
Department of Chemistry, Indiana University, Bloomington, IN, USA
e-mail: lanbaker@indiana.edu

Keywords Electrochemical · Electron transfer · Ion selective · Pipet · Probe

1 Understanding Ions with SICM

1.1 Introduction

Understanding the concentration, distribution, and transport of ions at interfaces are critical aspects to understanding the physical and chemical nature of said interfaces. A variety of SICM studies have developed routes to analytical measurements, based on the intrinsic ability of SICM instrumentation to manipulate ions. For example, reports on surface charge measurements [1], ion current rectification (ICR) effects [2–6], and sensing/mapping of topography with current [7–9] or local apparent conductance [10–14] have all been reported. Additionally, modifications of SICM probes have led to new detection possibilities, such as sensing/mapping pH [15–17], and measuring ion channel dependent signals [18, 19].

As discussed in detail in preceding chapters, for a typical SICM setup, an electrolyte-filled pipette is placed into the bath electrolyte in which the sample to be analyzed is immersed. An electric potential applied between an electrode inside the pipette and an electrode in the bath solution induces an ionic current between the two electrodes. As the pipette is moved toward a surface, an increase in access resistance (R_{acc}) develops which can be used as a feedback signal to control the pipette position. In most scanning probe techniques, the dimensions of the probe are inversely related to the resolution of the image produced. Though, for SICM, the desire to fabricate a probe to achieve high resolution with the smallest diameter possible must be balanced with optimizing resistance and generating sufficient ion current for SICM feedback and imaging. Additionally, nanoscale structures with asymmetry and surface charge, such as the nanopipettes used for SICM probes, possess unique ion transport properties that should be considered for quantitative applications of ion current measurements.

1.2 SICM Surface Charge Measurements

Nano- and micropipettes used for SICM probes are commonly fabricated from quartz or borosilicate glass capillaries. The chemistry of the glass surface is such that silanol groups on the walls of the pipette can be protonated or deprotonated depending on the pH of the solutions that surround and fill the probe. For bare quartz surfaces, silanol groups are deprotonated at moderate to high pH conditions, leading to an immobile anionic charge at the walls of the pipette [20].

When the surface is in contact with an electrolyte solution, charge on the silanol groups is compensated by formation of an electrical double layer (EDL). For small volumes of electrolyte, such as those in the tip region of a nanopipette, forming the EDL can perturb local ion concentrations (both counter and co-ions). When an electric field is applied between the electrode inside and outside the nanopipette, the presence of significant surface charge, along with an asymmetry in shape (such as the conical shape of a nanopipette) can result in selective transport of cations or anions. Selective transport can be quite significant and result in a diode-like current-voltage response (ICR), which is strongly dependent on charge screening from the electrolyte solution, with higher electrolyte concentrations suppressing ICR, and lower concentrations promoting ICR. In addition to electrolyte concentration, ICR effects can be attributed to the surface chemistry of the pipette, pH of the solution [21], composition of solvent/electrolyte [2], and probe geometry [22]. ICR has been reviewed extensively in the context of nanoscale pores, which include nanopipettes considered here [21].

In 2011, Sa and Baker [3] examined the influence of the EDL and surface charge on the operation of SICM. For this study, when the nanopipette tip is approached to a surface, the EDL at both the tip of the pipette and the surface interact, which alters observed ion currents through the pipette. Specifically, ion current rectification ratios (ICRR) were used here to describe the degree of rectification of ion current flowing through a charged nanopipette in close proximity to charged substrate surfaces [22]. The ICRR is the absolute value of the current ratio between two equivalent positive (I_{+V}) and negative (I_{-V}) potentials: [22]

$$\mathrm{ICRR} = \left| \frac{I_{-V}}{I_{+V}} \right| \tag{1}$$

Results (Fig. 1) [3] showed a dependence in *ICRR* as a function of probe-substrate distance (D_{PS}) when the pipette tip was approached to a surface with a negative charge (poly(dimethyl siloxane), Fig. 1c, e) or a positive charge (amine-modified poly(dimethylsiloxane), Fig. 1d, f). Changes in *ICRR* can be visualized clearly in differences of the current-voltage response of the nanopipette at distances far and close to the sample surface (Fig. 1c–f). Further, for surfaces employed in this study [3], the pH dependence was examined. By changing the pH of the bath solution, the relative charge of the surface could be modulated, enhancing observed effects, or removing the dependence all together.

In 2013, Baker and coworkers used a combination of experimental data and finite element models (FEM) to examine the influence of nanopipette geometry on observed ion current rectification in the absence of a surface [22]. Subsequently, an additional report, largely based on FEM, examined the influence of charge on the pipette and surface and the influence of D_{PS} on experiments like those shown in Fig. 1 [3]. Six different conditions with respect to surface charge on the pipette wall and the sample were examined. Conditions included: nanopipettes with no net surface charge interacting with a sample surface with no net charge, a positive sample surface charge, and a negative sample surface charge (Fig. 2a), and nanopipettes with a negative surface charge interacting with a sample surface with no net charge (Fig. 2d), a positive sample surface charge (Fig. 2c), and a negative sample surface charge (Fig. 2b).

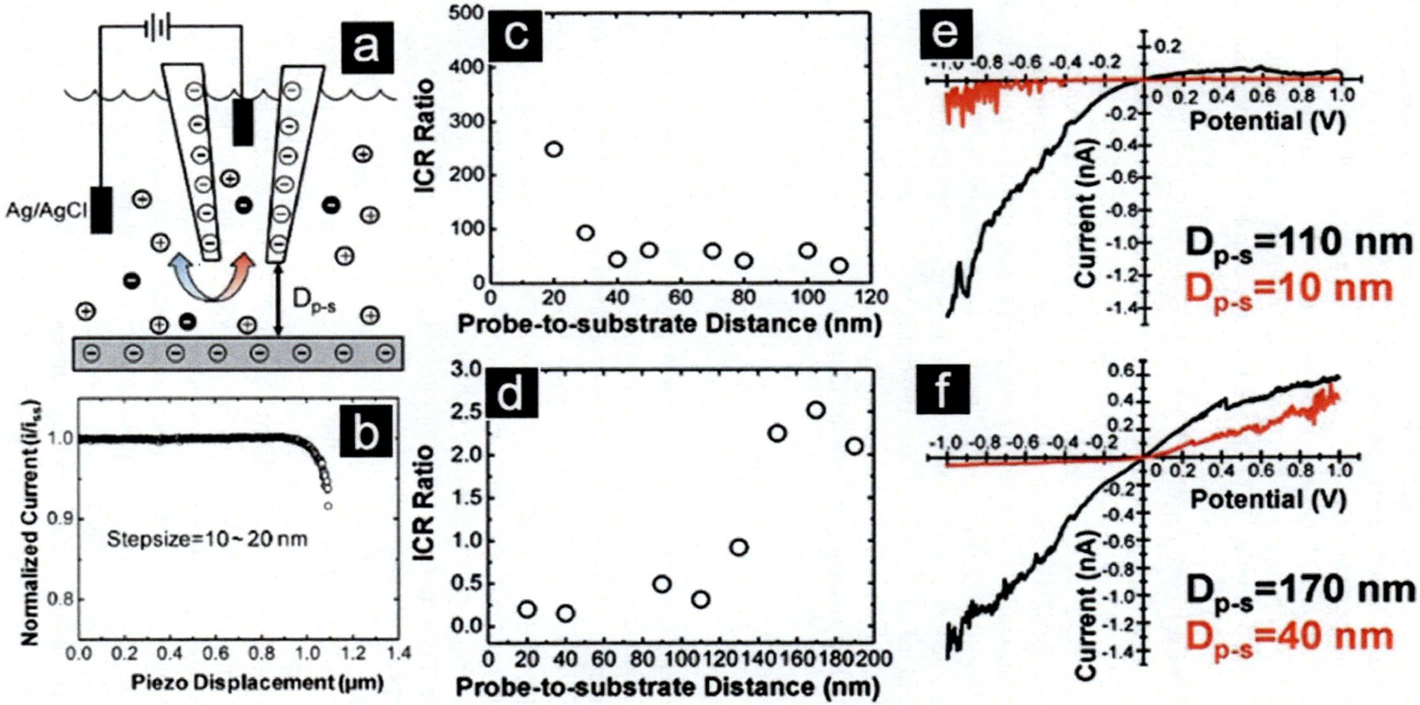

Fig. 1 (Left) Measurement of ion current through nanopipettes in proximity of a substrate. (**a**) Schematic of experiment described. A Ag/AgCl electrode is placed inside an electrolyte-filled nanopipette. The ion current between this electrode and a second Ag/AgCl electrode in the bath solution is measured as a function of both the potential difference applied and the nanopipette position relative to a surface of interest ($D_{p\text{-}s}$ is the probe-substrate distance). (**b**) Approach-curve of a nanopipette, in which normalized ion current as a function of piezo displacement is measured. (**c-d**) ICR ratio plotted as a function of nanopipette

distance from a substrate and illustrative current-voltage response. (**c**) ICR ratio as a function of probe-substrate distance ($D_{p\text{-}s}$) for a nanopipette approaching an anionic ($PDMS^{OH}$) substrate. (**e**) Current-voltage response at probe-substrate distances of 110 nm (black) and 10 nm (red). For the anionic surface, data was recorded in 50 mM KCl, 2 mM KH_2PO_4, pH 8.0. (**d**) ICR ratio as a function of probe-substrate distance ($D_{p\text{-}s}$) for a nanopipette approaching a cationic ($PDMS^{NH2}$) substrate. (**f**) Current-voltage response at probe-substrate distances of 170 nm (black) and 40 nm (red). For the cationic surface, data was recorded in 50 mM KCl, 2 mM KH_2PO_4, pH 6.8. Adapted with permission from Ref. [3]. Copyright 2011 American Chemical Society

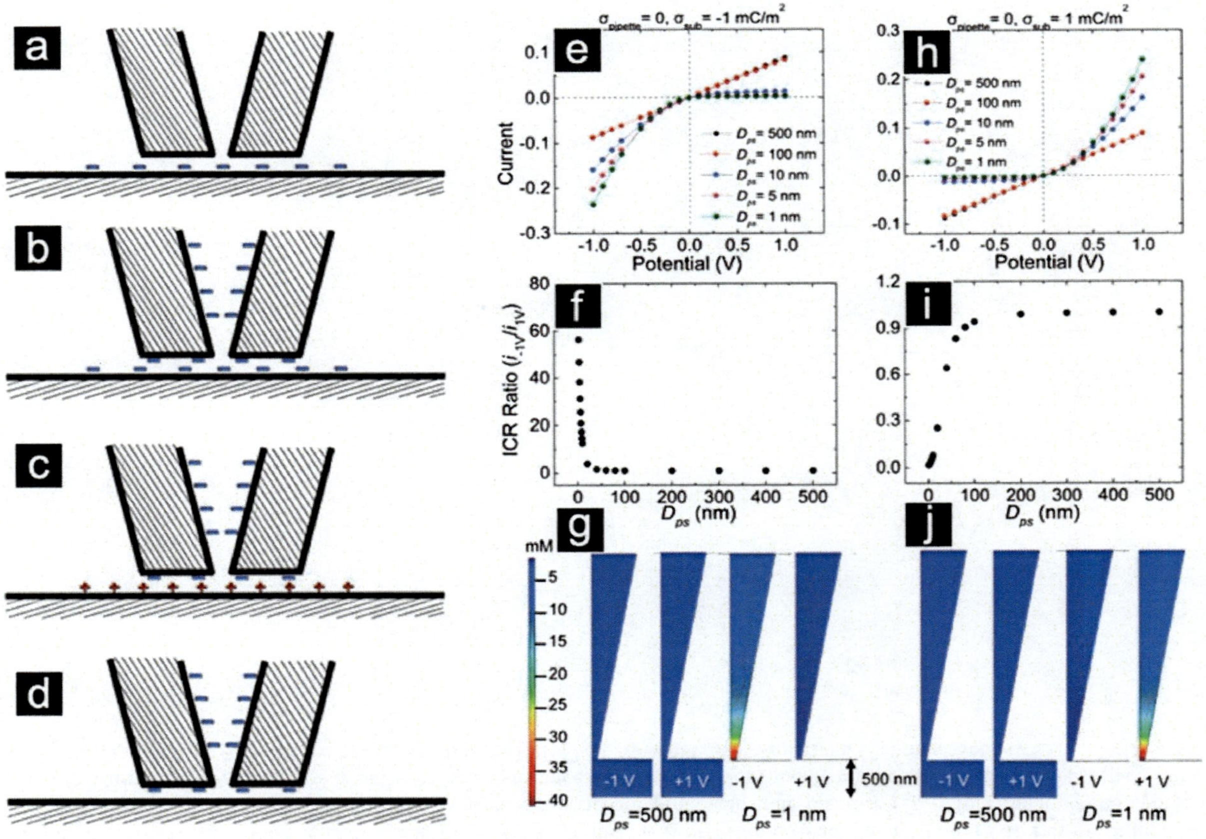

Fig. 2 (Left) (**a–d**) Schematic representation of a cross-sectional view for a nanopipette in series with a nanochannel formed between the gap of the nanopipette and substrate at close probe-to-substrate distances: (**a**) a neutral pipette with a negatively charged substrate, (**b**) negatively charged pipette and substrate, (**c**) a

negatively charged pipette with a positively charged substrate, and (**d**) a negatively charged pipette with a neutral substrate. (Right) Simulated current-voltage (I-V) curves as a neutral nanopipette ($\sigma_{pipet} = 0$ mC/m^2) approaches (**e**) a negatively ($\sigma_{sub} = -1$ mC/m^2) and (**h**) a positively charged substrate ($\sigma_{sub} = +1$ mC/m^2). The corresponding ion current rectification (ICR) ratio is plotted as a function of the distance (D_{ps}) between the probe and the (**f**) negatively and (**i**) positively charged substrate. D_{ps} strongly varies when decreasing the distance from 500 to 1 nm. Simulated total ion concentration distribution at $D_{ps} = 500$ and 1 nm for (**g**) $\sigma_{sub} = -1$ mC/m^2 and (j) $\sigma_{sub} = +1$ mC/m^2. The inner diameter (i.d.) and outer diameter (o.d.) of the probe are set as 25 and 200 nm, respectively. The ionic concentration (C_0) in the bulk solution is equal to 1 mM. The ICR ratio is defined as the ratio of the magnitude of the ion current at -1 V relative to that at +1 V. Adapted with permission from Ref. [5]. Copyright 2013 American Chemical Society

The simplest case is that with no net charge on the nanopipette walls and the sample surface. As the nanopipette approaches the sample, a reduction in ion current is observed, solely due to the access resistance R_{acc}, which is the typical case considered in SICM approach-curves. At distances far from the surface, current-voltage curves (*I-V curves*) collected from a probe with no net charge approaching to a negatively or positively charged surface showed no rectification. When the pipette is approached toward the surface, at distances <100 nm, rectified current effects, dependent on the polarity of the charge on the sample surface were observed. Modeled *I-V curves*, *ICRR* as a function of D_{PS} plots, and electrolyte concentration profiles are shown in Fig. 2e–g (for a substrate with a negative surface charge) and Fig. 2h–j (for a substrate with a positive surface charge) [5].

McKelvey et al. were the first to utilize surface charge for mapping in 2014 [23], illustrating in both experiment and models, the influence of surface charge at the pipette and sample on ions in solution in SICM measurements through phase-based AC current measurements. This report made use of distance-modulated SICM and showed the formation of selective zones of ion enrichment and depletion at the tip of the nanopipette based on the charge of the substrate. By correlating FEM models with experimental results, the study developed a semi-quantitative model for extracting surface charge values and underscored the possible influence of surface charge on SICM measurements [23].

1.3 Selective SICM Studies of Surface Charge: Living Cells and Materials

On the basis of these initial reports, subsequent reports have described surface charge interactions or surface charge mapping with SICM. Unwin and coworkers have described additional applications of mapping surface charge of polymers [24], living cells [25], hair follicles and most recently Gram-positive/negative bacterial cells [26]. Baker and coworkers described mapping surface charge of polymers, employing a differential approach to ICR measurement at interfaces [1]. Klenerman and coworkers examined fundamental surface charge interactions to understand current enhancements in SICM [27]. Investigations of surface charge at lipid bilayers have also been undertaken by Dong and coworkers [28, 29], where the distance of approach was strongly correlated with the inherent charge of the lipid species present in different domains. We discuss selected examples of these studies in more detail.

Zhu et al. [1] developed a differential measurement approach with SICM that allows visualization of apparent surface charge. Charged substrates were created by patterning a polydimethylsiloxane (PDMS) substrate with 3-aminopropylthiethoxysilane (APTES (+)) (to create a positively charged substrate) and with 3-(trihydroxysilyl)-1-propanesulfonic acid (TPS (−)) (to create a negatively charged substrate) [1]. The influence of surface charge on approach-curves was examined by approaching the TPS (−) and APTES (+) modified substrates at positive or negative pipette potentials.

Additionally, approach-curves to the TPS (−) substrate were recorded as for a range of KCl concentrations (1 M to 0.05 M). For the TPS (−) substrate, approach-curves were strongly influenced by the KCl concentration, and under selected conditions, approach-curves exhibited a current enhancement during approach, which was postulated as an area of ion enrichment due to the negatively charged substrate [1]. The current-voltage response of the nanopipette was measured at two probe-sample distances (extended and retracted). By taking the difference of the ICRR at both probe-sample distances, the polarity and magnitude of surface charge could be determined qualitatively [1].

$$\Delta \mathrm{ICRR} = \mathrm{ICRR}_{\mathrm{extended}} - \mathrm{ICRR}_{\mathrm{retracted}} \tag{2}$$

Perry et al. demonstrated analysis of surface charge on biological samples by developing FEM simulations that could be compared to experimental data. Bias modulated SICM feedback was used to image *Z. mays* root hairs and human adipocyte cells [25]. Page et al. [24] expanded this work in a similar approach that reduced imaging time by using a pulsed-potential chronoamperometry approach that self-calibrates at each pixel. Surface charge maps of a polystyrene film on a glass substrate and a PC-12 neurite on a glass substrate were collected. An optical image of the PC-12 neurite on a glass substrate is shown in Fig. 3a, and the SICM topographical (Fig. 3b) and quantified charge map (Fig. 3e) of the cell. The simulated approach-curve is displayed in Fig. 3c, which was used to develop the normalized current dependency of surface charge (Fig. 3d) and then to generate a quantitative charge map. The experimental current-time (I-t) curves for the points indicated in Fig. 3b are plotted in Fig. 3f [24].

1.4 SICM Nanopore and Ion Transport Measurements

In the original description of SICM, Hansma and coworkers imaged polymer membranes and noted the possibility of measuring ion currents at such types of structures [30]. An early study also combined nanopores in a polymer membrane and SICM by Siwy et al., to observe current flickering at single polymer nanopores in a manner reminiscent of ion channels [21].

Chen and Baker examined nanopores of polymer membranes in a variety of configurations [31]. Ion transport through a single nanopore was examined in the case of polyimide membranes with a single pore [31]. Pores imaged in this study were compared to scanning electron microscope (SEM) images that showed nominal pore sizes of 460 nm and 690 nm for the cylindrical shaped pore and the conical shaped pore, respectively. Imaging the pores with SICM allowed current and topography maps to be collected that showed the influence of the local electric fields. When compared to a topographical image (Fig. 4a), current maps, displayed as changes in current (ΔI) as a function of applied transmembrane potential (V_{TM}) −0.3 V (Fig. 4c), 0 V (Fig. 4b), and +0.3 V (Fig. 4d), showed distinct differences [31].

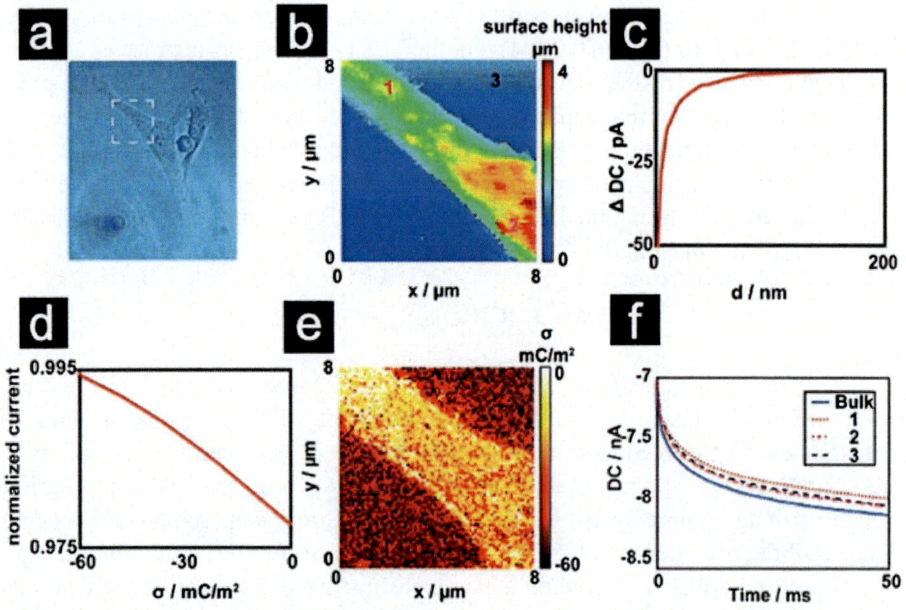

Fig. 3 Simultaneous topography and charge maps of a PC-12 neurite on a glass substrate. (**a**) Optical image of the scanned cell, the white square showing the scan area. (**b**) Topographical image of the neurite, collected concurrently with the quantified charge map (**e**). (**c**) FEM simulation of the change in DC as the probe approaches the surface at +20 mV, showing dependence of probe−substrate separation on the feedback threshold chosen. (**d**) Simulated dependence of the normalized current on the charge at the surface, used to generate the quantified charge map in (**e**). Experimental I−t curves at the points of the scan labeled in (**b**) are shown in (**f**), along with a bulk I−t curve for comparison. Adapted with permission from Ref. [24]. Copyright 2016 American Chemical Society

Another important observation within this experiment was the current trend measured as a function of V_{TM} when the pipette is directly over the pore as compared to when the pipette was displaced laterally such that the pipette was not directly over the pore. The current response was recorded as a function of horizontal displacement from nominally zero to ~9 μm away from the pore. Influence of the potential on currents measured axially from the pore was also measured and plotted as *I-V curves* when the probe was directly over the pore at various D_{PS} values that ranged from 610 nm to −2.00 μm (where negative D_{PS} values indicate the pipette moved into the pore). In total, this study demonstrated the opportunity to couple electric fields in porous materials with the SICM pipette probe and to map the electric field in three dimensions. In subsequent studies, this experiment was expanded to study pores with multiple membranes where the current from individual pores could be extracted based on the probe's horizontal location [32]. In these experiments a circuit model that treated the system as a voltage divider was developed that was successful at reproducing experimental observations.

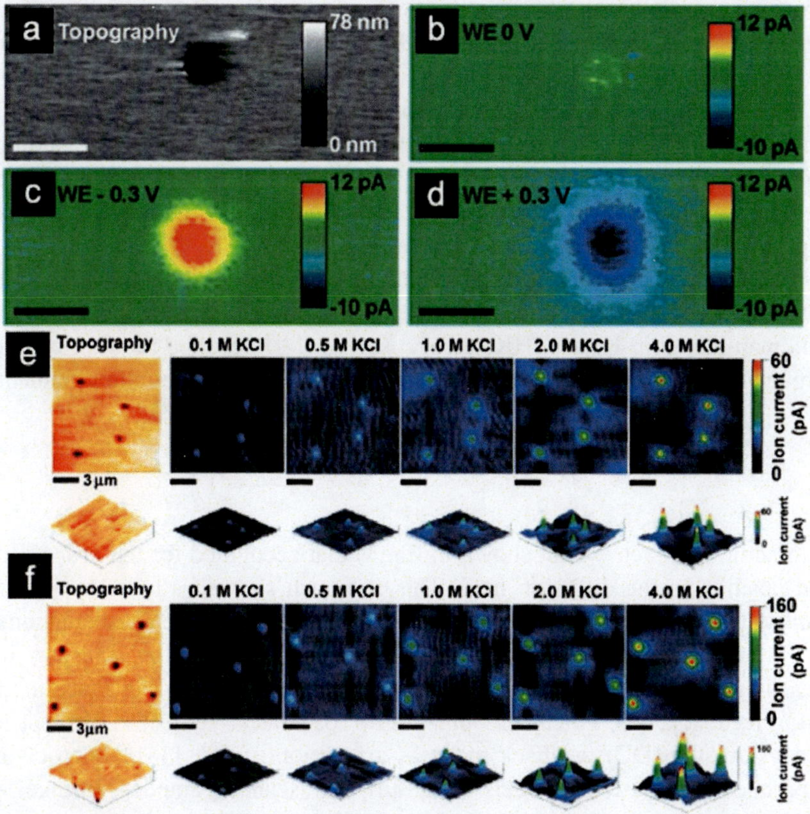

Fig. 4 (Above) Topography (**a**) and the corresponding current images (**b**, **c**, **d**) of an isolated single nanopore measured with the three-electrode SICM described here. Images correspond to (**b**) 0 V, (**c**) −0.3 V, and (**d**) +0.3 V applied to the working electrode (WE). Scale bar: 1 μm. Adapted with permission from Ref. [26]. Copyright 2011 American Chemical Society. (Below) Topographic and ion current images of (**e**) a 300 nm diameter nanopore membrane and (**f**) a 500 nm diameter nanopore membrane as a function of the KCl concentration in the lower chamber (concentration indicated above each image). For all images, the electrolyte in the upper chamber of the diffusion cell is maintained at 0.1 M KCl. Adapted with permission from Ref. [7]. Copyright 2009 American Chemical Society

Alternatively, the transport of ions through pores driven primarily by diffusion (concentration gradients), with a minimal contribution from migration induced by the SICM probe was studied [7]. A perfusion cell which had an upper and lower chamber of solution separated by a porous membrane was used. A two-electrode system where a Ag/AgCl electrode was inserted into the nanopipette (PE) and another Ag/AgCl electrode was placed in the bottom chamber as the reference electrode (RE) served to control the pipette for imaging and data acquisition. SICM images collected were compared to SEM and atomic force microscopy (AFM) images to validate topographic data collected.

The effects of concentration gradients on the transport of ions were examined by holding the concentration of KCl in the upper chamber constant (0.1 M) and varying the KCl concentration ([KCl]) in the lower chamber (0.1 M to 4.0 M). The SICM topographic and ion current maps for the 300 nm and 500 nm diameter pores are shown in Fig. 4e, f, respectively. In both Fig. 4e, f, an increase in [KCl] in the lower chamber resulted in an increase in the ion current measured near and over the pores, while no significant changes were observed in the topography images. Additionally, for these experiments, the nanopipette was biased at +0.1 V which induced an electric field gradient that can influence translocation of the ions across the pore depending on the probe-sample distance (D_{PS}) [7].

The peak ion current plotted as a function of the [KCl] in the lower chamber for both pore sizes showed a linear relationship. Data were compared to a model where Goldman-Hodgkin-Katz (GHK) theory was used to describe the flux of ions ($j_{\pm}$) through a nanopore under an applied electric field and concentration gradient [7]:

$$j_{\pm} = \frac{\mp D_{\pm}\Delta}{L}\left(\frac{C_S^L}{1 - e^{\mp\Delta}} - \frac{C_S^U}{e^{\pm\Delta} - 1}\right) \tag{3}$$

with additional factors (not shown in Eq. 3) that accounted for the probe geometry and position in the diffusion field. This approach showed a route to quantify ion transport and concentrations with SICM based on first principles calculations.

Further models assessed the role of convection-driven ion transport in such measurements by comparing results acquired with distance-modulated (AC) feedback, in which the pipette tip oscillates vertically, to that of the non-modulated (DC) feedback mode as a function of D_{PS} [33]. When comparing experimental data with models from different scanning modes, normalized ion currents measured with SICM were found to be more accurately reported with DC feedback mode. However, distance-modulated approaches provided better feedback regulation, which is beneficial for mapping ion flux from surfaces where conductance variations can interfere with probe control [33].

1.5 Potentiometric-SICM (P-SICM): SICM Apparent Local Conductance Measurements

Potentiometric SICM (P-SICM) is another version of SICM which employs a dual barrel pipette (theta pipette) to measure both the topography and apparent local conductance of a sample surface. P-SICM was first described by Chen et al. in 2013 [13]. In 2017, Zhou et al. advanced this technique's ability to measure local conductance, such that topography and local conductance mapping could be performed simultaneously [12]. In P-SICM two Ag/AgCl electrodes are back-inserted into the two barrels of the theta pipette, where one of the electrodes (PE) measures the current for SICM feedback to develop a topographical map of the sample surface. The other electrode measures the local potential (potentiometric electrode, UE) to simultaneously develop a coincident conductance map. This allows for direct comparison between topography and local conductance data to be collected. P-SICM requires that the sample of interest is permeable to ions or molecules. The experimental setup for P-SICM is shown in Fig. 5a, b [34]. There is one Ag/AgCl back-inserted electrode in each barrel of the probe (the PE and UE respectively). Another two Ag/AgCl electrodes are used, where one is in the upper chamber of the perfusion cell (the reference electrode, RE) and the second is in the lower chamber of the perfusion cell (the working electrode, WE), and finally there is a Pt counter electrode (CE) in the upper chamber. Local potential variations in the vicinity of the sample surface induced by the V_{TM} are recorded by the UE and used to calculate the local conductance (G) of the probed position through a previously defined equation: [13]

$$G = \left(\frac{E}{\rho V^e}\right) = \frac{\Delta V_{\mathrm{close}} - \Delta V_{\mathrm{far}}}{\rho V^e \Delta z} \quad (4)$$

In Eq. (4), E is the electric field, determined by dividing the difference between the potential differences recorded at two different D_{PS} values ($\Delta V_{\mathrm{close}} - \Delta V_{\mathrm{far}}$) by the vertical displacement of the pipette (Δz), ρ is the bath solution resistivity, and V^e is the potential range that is applied to the WE (otherwise understood as ΔV_{TM}) to induce a voltage deflection. The voltage deflection, plotted as a function of V_{TM} as shown in Fig. 5d, can be used to measure the potential distribution at different positions. For example, the SICM topographical map, shown in Fig. 5c, is an example of Madin-Darby canine kidney strain II (MDCKII) cells that was used to guide the pipette to the desired positions for conductance measurements (e.g., cell bodies and cell–cell junctions) [13]. With the nanometer-scale resolution offered by SICM, this technique has the ability to distinguish between submicron sample features and correlate the surface local conductance with topography.

Alternatively, instead of using a DC V_{TM} for surface charge mapping [35], P-SICM can also be used in AC mode similar to surface charge mapping that makes use of phase measurements between the AC applied potential and the position of a hopping pipette [23, 35]. In this setup an AC potential is applied to the WE in the bottom chamber so that when the pipette scans over a pore, the resistance of the pore can be determined and to distinguish heterogeneous conductive pathways such as nanopores, as shown in Fig. 6 [34]. An electron micrograph of a pore (Fig. 6a), a cartoon of the experimental setup (Fig. 6b), and the current maps of a pore under +0.9 V and −0.9 V applied to the WE, respectively (Fig. 6c), are shown. In 2015 [34], Zhou et al. described AC-PSICM and found that the frequency range of the AC potential applied to the WE was related to spatial resolution and sensitivity of the measurement. Additionally, sensitivity in the phase angle of the potential measured at the UE can be seen as a function of D_{PS} and as a function of the lateral displacement from the pore center [34]. The resistance of the pore was found to be a critical factor in optimizing the imaging frequencies employed.

In 2014, Zhou et al. assessed the instrumental limitations of P-SICM with measurements of synthetic membranes [14]. Current measurements were not as

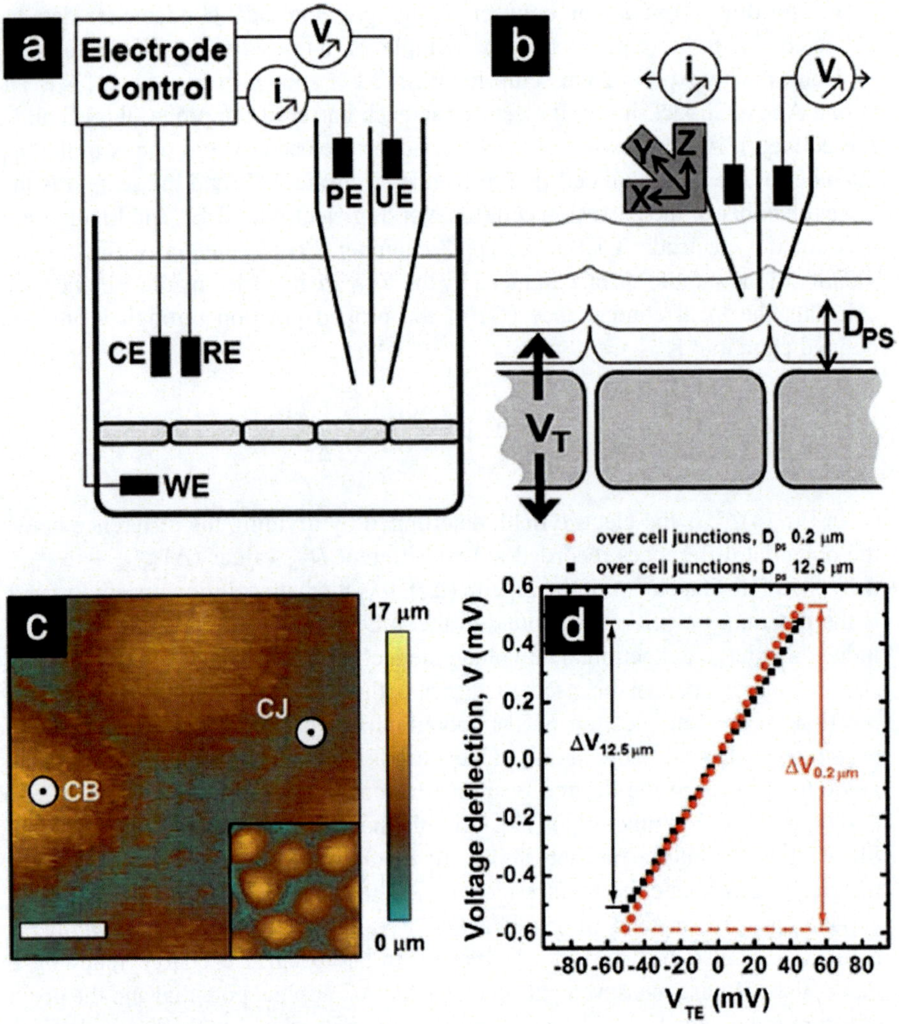

Fig. 5 (Above) Illustration of potentiometric scanning ion conductance microscopy (P-SICM). (**a**) A dual-barrel pipette is utilized to measure topographic and potential gradients of a sample of interest. Here the pipette electrode (PE) is used to control pipette position and record topographic images. The potentiometric electrode (UE) records the local potential at the pipette tip. A potential is applied across the sample between the working electrode (WE) and counter electrode (CE). All electrode potentials are referenced to a common reference electrode (RE). (**b**) Expanded view of the process. The potential applied across the sample (V_T) results in local changes in electric field at conductive pathways within the sample. Positioning of the dual-barrel probe is achieved with piezo electric positioners that control the probe–surface distance (D_{ps}) and translate the pipette for image collection. (Below) Characterization of epithelial monolayers with P-SICM. (**c**) Topographic image of the apical surface of a cell monolayer was imaged to locate the position of cell bodies and the cell junctions. Scale bar: 5 μm. The inset shows the zoom-out image of the cell monolayer under study (40 × 40 μm). Two positions which illustrate pipette placement are indicated for CB (cell body) and CJ (cell junction). The black marker in the center of the larger white marker at these positions approximates the size of the pipette tip utilized. (**d**) Potential deflection recorded with fixed position mode at two D_{ps}, far (12.5 μm) and close (0.2 μm), was carried out over the cell junctions to evaluate apparent paracellular conductance. Adapted with permission from Ref. [13]. Copyright 2013 American Chemical Society

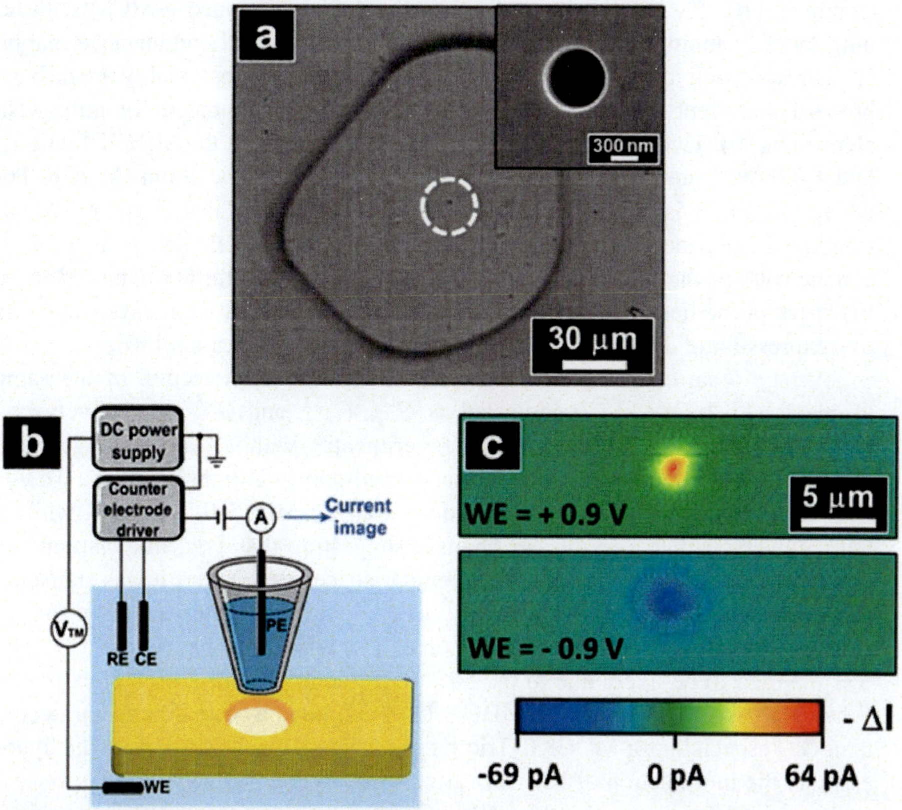

Fig. 6 (**a**) Optical image of a single nanopore membrane. Boundary of painted epoxy for single pore isolation can be seen, and the isolated nanopore is indicated with the dashed circle. Inset: scanning electron microscopy (SEM) image of the isolated single pore in (**a**). Nanopore diameter is 687 nm. (**b**) Illustration of the four-electrode SICM setup. Pipette electrode (PE) inside the nanopipette is used to control pipette position as well as to record ion current images. (**c**) SICM ion current images of the single pore shown in (**a**) obtained under −0.9 V and +0.9 V. Adapted with permission from Ref. [34]. Copyright 2015 American Chemical Society

sensitive as the potential measurements with a dual barrel probe, which was a noted experimental trade-off, as smaller probe diameters typically result in better resolution. Additionally, smaller probe diameters result in smaller D_{PS} values that can be used to measure larger differences in the local electric field as the probe can operate closer to the sample. Results showed that as the size of the pipette decreased, gains in signal-to-noise ratio were observed. Parameters for a quantitative comparison of transport properties of small permeable features were also developed further [14].

To facilitate understanding ion transport in biological samples, P-SICM was applied to a model MDCKII epithelial cell line [10, 12, 36]. An example of the experimental setup and the corresponding topographical image and the voltage deflections can be found in Fig. 5 [13] for MDCKII cells. In later studies, transport properties of tight junctions (both bicellular (bTJs) and tricellular (tTJs)) were examined [10, 12, 36]. Advances in instrumentation combined P-SICM with hopping mode scanning for simultaneous topography and local conductance mapping [12], an approach further used to explore the effects of hyperosmolality on the barrier properties of tight junctions (TJs) with the aid of an automated computer vision algorithm [10]. Two variants of MDCKII cells were studied, the MDCKII wild type (MDCKII-WT) and MDCKII with the overexpression of tricellulin (MDCK-Tric), where tricellulin is an integral membrane proteins found in tTJ central tubules. Cation size dependent studies were performed for each cell line to measure the conductance of the tTJs, bTJs, and cell bodies (CBs) that further explored the ion transport properties of tTJs, bTJs, and CBs of epithelial monolayers and how overexpression of Tric correlated with the cation size permeation [36].

Average local conductance (G) measurements for each section of the sample (Table 1, tTJ, bTJ, and CB indicated as G^{tTJ}, G^{bTJ}, and G^{CB}, respectively) were recorded using ion replacement experiments with N-methyl-D-glucamine ($NMDG^+$), a bulky cation, which is transported poorly at tight junctions. The data showed that paracellular conductance decreased for MDCKII-Tric cells, while the transcellular conductance did not change, suggesting that Tric, like claudins, primarily regulates transport across intercellular spaces. Moreover, it was statistically shown that the observed 16% decrease in G^{tTJ} was less significant than the 25% decrease in G^{bTJ} under the overexpression of Tric where the change in G^{bTJ} (ΔG^{bTJ}) was reported to be likely due to the 50% decrease in claudin-2 protein expression. When Na^+ was replaced with $NMDG^+$, G^{bTJ} exhibited a more significant decrease than G^{tTJ} especially for MDCKII-Tric samples. These results agree with the hypothesis that the introduction of $NMDG^+$ affects mainly the pathways at bTJs since the pore size of bTJs is smaller than $NMDG^+$ thus blocking the transport of this ion through bTJs. This further demonstrated the ability of P-SICM to measure and differentiate the transport properties of bTJs and tTJs. However, it was discussed that the less significant ΔG^{tTJ} observed when claudin-2 protein expression was used could be from the significant ΔG^{bTJ} which is understood as the bTJ background. Background noise limited explicit differentiation of the two paracellular pathways, relative to the tTJs that had neighboring bTJs. However, tTJ and bTJ conductances values could be determined within experimental limits [36].

Combining P-SICM with hopping mode to allow simultaneous topography and conductance mapping improved the overall system performance. Operation of the hopping mode system was validated using polymer membranes, which allowed for optimization of imaging parameters, such as hop height selection. Topographical and conductance maps were then collected from the MDCKII cells where TJs can

Table 1 Conductance measurements of cell samples for distinguishable tight junctions

$G^{Section}$ (mS/cm^2)				
Section\sample	WT (NaCl)	WT (NMDG-Cl)	Tric (NaCl)	Tric (NMDG-Cl)
tTJ	12.47 ± 1.04	11.16 ± 0.87	10.44 ± 0.67	10.49 ± 0.60
bTJ	11.08 ± 0.82	8.67 ± 0.61	8.35 ± 0.63	6.86 ± 0.49
CB	5.83 ± 0.99	5.95 ± 0.59	5.34 ± 0.62	5.32 ± 0.57

clearly be seen in the topographical map (Fig. 7d) and the correlated conductance from TJs can be observed in the conductance map (Fig. 7e) [12]. In the conductance map the TJs are observed to be more conductive than the cell bodies, where tTJs showed a general increase in conductance at or slightly around the junctions, relative to bTJs.

To further evaluate the condition of cell monolayers under prolonged exposure to transmembrane potentials, impedance spectroscopy was used to study the resistance of the cell monolayer before and after the experiment, which demonstrated that

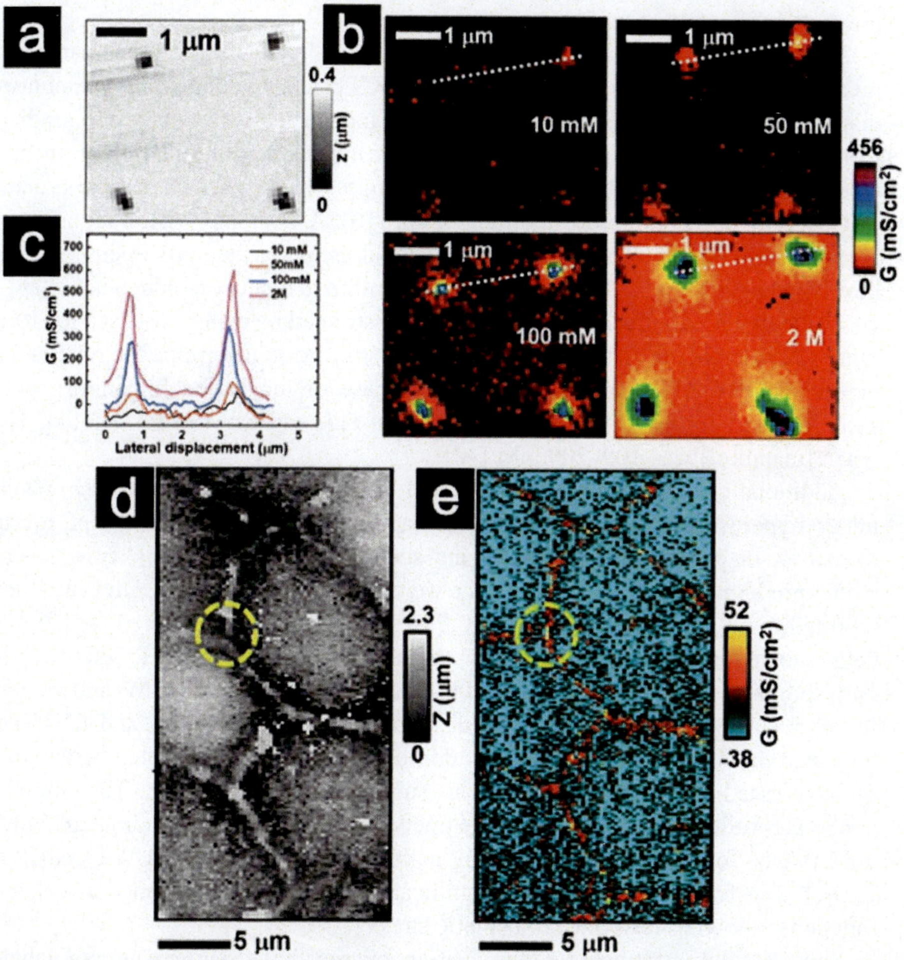

Fig. 7 (Above) Topography (**a**) and apparent conductance (**b**) images of four nanopores in a membrane. Apparent conductance images of the same area are shown as a function of KCl concentration in the bottom chamber of the perfusion cell (values indicated in images) while the top chamber concentration was maintained at 0.1 M. (**c**) Line profiles extracted from the conductance maps (dashed line). (Below) Topography (**d**) and conductance (**e**) images of a MDCKII cell monolayer in modified DMEM; V_{TM} was swept between ±100 mV. Adapted with permission from Ref. [12]. Copyright 2017 American Chemical Society

extended application of V_{TM} tightened the epithelial barrier. Moreover, immunofluorescence staining of the fixed cell monolayer and freeze-fracture electron micrographs (FF-EMs) of the cell monolayer were conducted with and without exposing the cells to V_{TM}. Immunofluorescence staining showed the expression of zonula occludens-1 (ZO-1), an important scaffold protein that anchors TJ strand proteins, was significantly increased. Additionally, the FF-EMs revealed a higher number and complexity of the TJ strands. The overexpression of ZO-1 and the increase in number and complexity of the TJ strands explained why the resistance of the cell monolayer increased after exposure to V_{TM} and therefore why bTJs were hardly visible in the conductance map, which was due to a tightening of the TJs which prevented ion permeation [12].

In a more recent P-SICM study by Huang et al. [10], the effect of hyperosmolality on epithelial barriers was explored with MDCKII cells by adding varied amounts of mannitol to the basolateral side of the cells to induce an osmotic pressure gradient. Conductance changes at junctional areas under hyperosmolality mainly occurred at tTJs, while bTJs did not display significant change, which indicated that the structure of tTJs were susceptible to osmotic pressure [10]. A computer vision algorithm to automatically recognize the pixels of junctional areas for data processing was also developed in this report. The ability of the algorithm to identify conductance changes was validated by a cell–cell junction model constructed by using a focused ion beam in a Si_3N_4 membrane. The algorithm was then utilized to determine the relative ΔG values of CBs, bTJs, and tTJs under each case of the varied [mannitol] which showed saturation of hyperosmolar effects on tTJs of 380–480 mOsmol/kg H_2O when [mannitol] was 100–200 mM [10].

Additionally in this study, the depletion of Ca^{2+} was explored for tTJ opening under hyperosmolar conditions, where Ca^{2+} is known to be important for the formation of junction forming proteins such as E-cadherins [37]. Effects from overexpression of angulin-2 (ILDR1) were also examined as ILDR1 has been reported to be found predominantly at tTJs. MDCKII cells transfected with ILDR1 cells showed a four-fold increase in resistance compared to WT cells and the corresponding mannitol treatments (100 mM, 200 mM, and 300 mM) and changes in conductance could not be easily distinguished for CBs, bTJs, and tTJs. This indicated that bTJs were tightened in addition to the tTJs by possible interaction of overexpressed ILDR1 with bTJ proteins such as ZOs and claudins. The computer vision algorithm was applied here to extract conductance values, finding ΔG values <3 mS/cm^2 for ILDR1 overexpressing cells, versus 40 mS/cm^2 in WT cells, thus ILDR1 transfection was found to significantly alter the conductance at cell–cell junctions, allowing resistance to osmotic stress [10].

Super-resolution structured illumination microscopy (SIM) images of labeled cells for ZO-1, Tric, and claudin-2 proteins for the fine characterization of both tTJs' and bTJs' structures before and after the cells were exposed to 300 mM mannitol, supported P-SICM measurements. Tric was observed intact under normal physiological conditions (before mannitol addition). After subjecting the cell monolayer to mannitol treatment, the appearance of dumbbell-shaped features was observed that were rationalized as the splitting of tTJs structure by the external forces induced by osmotic pressure. Lastly, no changes in the ZO-1 nor claudin-2 protein were observed in SIM images, which explained why there was little to no effect on transport properties across the bTJs in P-SICM data recorded [10].

1.6 Modified SICM Probe Applications

Modified nanopipettes have also been utilized as biological sensors for glucose in cancer cells [38], label-free sensors [39], calcium binding affinities [40], STING sensors [41, 42] and sensors for DNA sequences [43]. Thus, chemical modification of nanopipettes proves an intriguing approach to study chemical sensing. For example, Morris et al. [17] demonstrated pH detection by measuring potential differences at electropolymerized films on the exterior of a nanopipette, an approach described in detail below.

Previous reports described a single barrel micropipette constructed with black lipid membranes that supported insertion of α-Hemolysin (α-HL) ion channels at the pipette orifice to create an ion channel sensor integrated with SICM, where the number of channels is reported to determine the maximum current that can be measured and can be selectively chosen by adjusting the concentration of the bath electrolyte [17]. This was then used to develop line scan topographical data of PDMS that matched well with a control image of the same surface imaged with SEM. This work was followed by reports from other groups [18]. SICM has also been used in a more intriguing manner to extract membrane patches from cells, trapping ion channels at the pipette barrel in an inside-out or outside-out patch configuration. Shi et al. [44] demonstrated the use of a dual barrel or a theta pipette with one barrel used for supporting inside-out patches of TRPV1 or BK channels from HEK 293 cells. TRPV1 and BK ion channel modified pipettes were used to measure capsaicin (TRPV1) or Ca^{2+} (BK) based on channel activities [19]. Spatial resolution of the detection of agonist concentration the probe was demonstrated by laterally moving over a pore and plotting TRPV1 channel activity when near a source of agonist [19]. Other studies with similar ion channel probe modifications have been made with smart patch clamp systems for biological samples [45] and α-HL ion channels for β-cyclodextrin [18]. An additional work has used modification of the interior probe walls to develop sensors based on selective recognition [46, 47], pH [48], temperature [48], and light [48] by monitoring the response of ionic flux or potential changes.

1.7 Reporting Faradaic Electron Transfer with Ion Currents

In 2016, Momotenko et al. [49] demonstrated the ability for SICM to spatially resolve current changes due to ionic flux generated at an ultramicroelectrode (UME) induced by local electrochemical reactions, for instance the oxidation of ferrocene (neutral) to ferrocenium (cationic) species. FEM was used to determine relative concentration changes induced by electrochemical reactions at the UME. Compared to scanning electrochemical microscopy, this approach serves as an attractive alternative for correlating surface topography and surface reactivity based on local changes in conductivity that can occur when oxidation or reduction of a molecule results in a corresponding change in ion concentrations. This study showed correlated topography and surface activity with a combination of simultaneous AC and DC SICM measurements. Furthermore, this work showed chemical mapping at scan rates that approached 4 s per image with high pixel resolution.

2 Understanding Electrons with Scanning Electrochemical Microscopy (SECM)-SICM

2.1 *Introduction*

Scanning electrochemical microscopy (SECM) and SICM share many common features, including the positioning hardware. SECM differs significantly from SICM in that the current recorded is typically a result of faradaic electron transfer (change in oxidation state) of a molecule that interacts at the tip of a conductive probe. SECM was reported around the same time as SICM [50, 51], with SECM initially gaining wider adoption as a tool for chemical imaging.

For the first few decades after inception (i.e., 1980~2000), SECM typically employed electrodes with microscale dimensions composed of carbon, gold, and platinum. Ever shrinking electrode dimensions allowed successes in SECM measurement at high resolution (submicron [52] to nanometer [53]). However, tedious probe fabrication protocols are required to realize nanoscale SECM. In recent years, approaches to combine SECM and SICM in hybrid formats have been undertaken that have led to new modes of hybrid imaging. Several routes to modify nanoscale pipettes as SECM-SICM probes have been described which have proven suitable alternatives [17, 54–56] to traditional SECM imaging methods [57]. In most approaches, a multifunctional probe is used to achieve SECM-SICM, with SICM controlling probe position by ion current feedback and SECM measuring an electrochemical response.

2.2 *SECM-SICM Probe Fabrication*

2.2.1 Gold Electrodes in Single Barrel Pipettes

SECM-SICM probes were initially fabricated with gold as the SECM electrode, where gold acts as an inert, ideal non-polarizable electrode. Gold SECM-SICM probes were fabricated by first depositing a thin chromium adhesion layer, followed by deposition of a thicker gold layer on the outside of a micropipette via thermal evaporation. Typically, gold SECM-SICM probes offered good electrochemical stability with straightforward fabrication techniques.

The first study combining SECM and SICM was demonstrated by Walsh et al. in 2005 [54]. In their study, a ring-shaped electrode was fabricated by depositing gold on a microscale borosilicate pipette with a 1.9 μm outer diameter, followed by insulation with anodic electrophoretic paint. The ring electrode at the pipette tip was kept exposed by applying nitrogen gas through the pipette during the insulation process. This technique enabled SECM-SICM probe fabrication with diameters on the order of 3 μm. The open pipette inside the gold ring electrode was used to dispense small (pL) amounts of solution, rather than directly controlling the probe-substrate distance. The electrochemical response detected at the Au-ring-electrode was utilized for feedback to control probe-substrate separation.

In 2010, Comstock et al. presented a route to fabrication of smaller, crescent-shaped electrodes on glass capillaries (Fig. 8a, b) [55]. By depositing gold at only half of the outer pipette surface, they were able to successfully decrease the physical size of SECM-SICM probe tip (relative to the Walsh report), to outside diameters on the order of ~1 μm. Atomic layer deposition (ALD) of aluminum oxide (Al_2O_3) comprised provided a highly uniform and insulating coating. Focused ion beam (FIB)-milling was used to expose the crescent-shaped electrode at the probe tip. However, the Al_2O_3 layer was not deposited selectively, causing some Al_2O_3 to deposit on inner walls of the nanopipette. In the same year, Takahashi et al. also reported a ring-shaped Ti/Au electrode with an equivalent probe dimension of 1 μm outer diameter that (Fig. 8c, d) [58] combined the approaches of Walsh and Comstock, in that the Ti/Au probes were insulated by anodic electrophoretic paint, and FIB-milling was used to expose the tip. In 2012 bulk fabrication of SECM-SICM probes was demonstrated by Baker and coworkers by using vapor deposition of parylene C as an insulation layer in which several tens of Au coated nanopipettes were insulated at once [56]. Instead of exposing only the tip of the probe, a greater area of the gold electrode, up to 40 μm in height from tip-end, was left uninsulated. Employing parylene C provided uniform and pinhole free insulation for multiple SECM-SICM probes with similar size of exposed areas between each probe. Soon after, Baker and coworkers were able to use the parylene C insulated probes for local pH measurements by SECM-SICM [17], in which a pH-responsive polyaniline (PANi) was coated on the Au electrode surface on the sidewall of the probe by electropolymerization. Potentiometric methods were used to calibrate the pH-dependent response of the probe, and during imaging simultaneous topography and changes in pH were measured at single nanopores. In addition, Shi and Baker improved the resolution of the SECM-SICM by insulating Au crescent-shaped electrodes and cutting the parylene C layer at the tip by FIB-milling. Consequently, improved resolution allowed observation of redox mediators diffusing through nanopores, randomly formed on a degraded Nafion membrane [60].

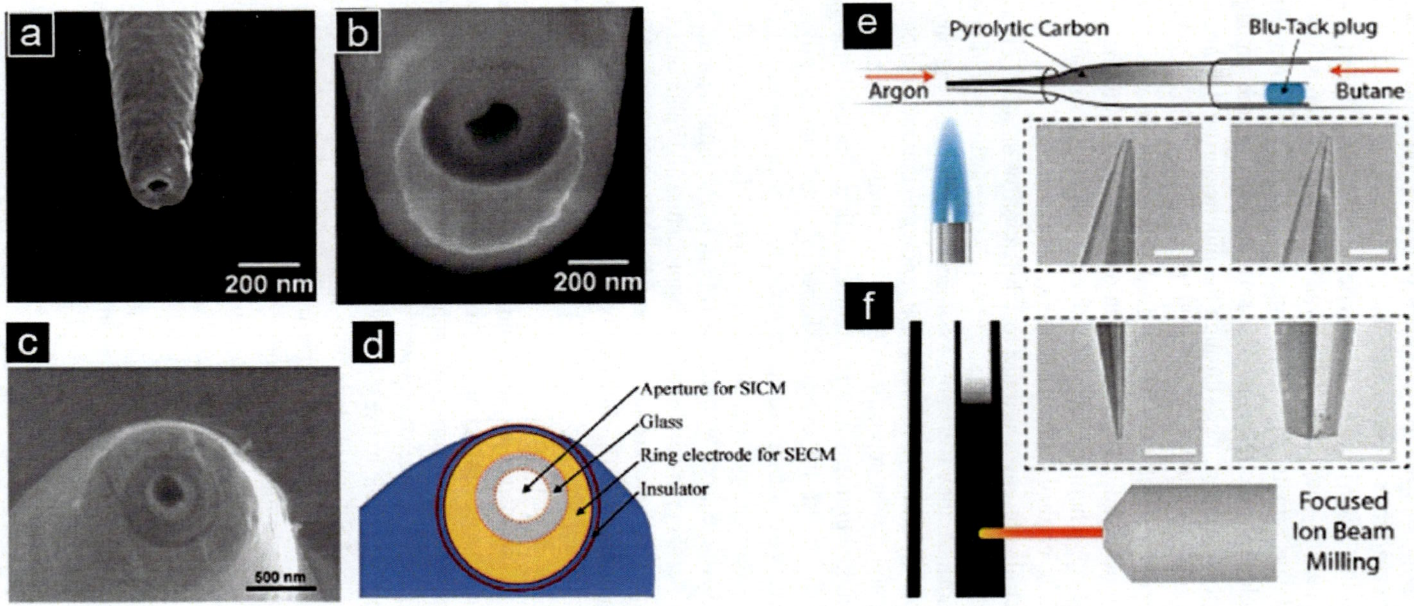

Fig. 8 Scanning electron micrographs of SECM-SICM nanopipettes throughout the fabrication process: (**a**) as-pulled borosilicate nanopipette with a 30 nm AuPd film to minimize charging, (**b**) after FIB-milling to expose the gold and open the nanopipette tip. Adapted from Ref. [55]. Copyright 2010 American Chemical Society. (**c**) SEM micrographs of the SECM-SICM probe. (**d**) Schematic illustration of probe structure. The nanopipette aperture radius was 220 nm. Inner and outer radii of the SECM Au nanoring electrode were 330 and 550 nm, respectively. The insulator was 70 nm thick. Adapted from Ref. [58]. Copyright 2010 American Chemical Society. Fabrication of dual-barrel nanoprobes for use in SICM-SECM. (**e**) Carbon was deposited in one barrel of the probe via the pyrolysis of butane (SECM) while the other was kept open (SICM). Inset transmission electron microscopy (TEM) images show an example of both complete (left) and incomplete (right) carbon deposition. Scale bar in both micrographs is 500 nm. (**f**) The probe diameter was regulated using focused ion beam (FIB) milling. Inset TEM images show a probe with scale bars of 5 μm (left) and 500 nm (right) after FIB-milling. Adapted with permission from Ref. [59]. Copyright 2017 American Chemical Society

2.2.2 Carbon Probes in Single and Dual Barrel Pipettes

Carbon materials are often used as working electrodes for electrochemical studies. Carbon is an excellent electrode material due to chemical inertness, resistance to fouling, convenience in surface modification, a large potential window for both oxidation and reduction reactions, and well-understood surface chemistry [61, 62].

In 2011, Takahashi et al. reported SECM-SICM probes incorporating carbon electrodes fabricated with theta pipettes, where one barrel contained conductive carbon and the other allowed for SICM. These carbon SECM-SICM probes were referred to as "double barrel carbon nanoprobes (DBCNPs)." Pyrolyzing hydrocarbon gas, such as butane, in one barrel of a theta pipette produced the carbon electrode simply with a reproducible 200 nm outer diameter of the entire probe [63]. Historically, the pyrolytic decomposition of butane within a single barrel pipette had been established for SECM probe fabrication in which one barrel of the theta pipette was internally coated with carbon [64–66]. In contrast, Thakar et al. reported a ring-shaped carbon electrode fabrication method using a thermal pyrolysis technique in 2013 [67]. Pyrolyzed parylene C (PPC) was formed on the outer surface of the single barrel nanopipette as a thin carbon electrode layer by chemical vapor deposition and subsequent pyrolysis, followed by additional parylene C coating for insulation. The carbon electrode at the tip was exposed by FIB-milling as well.

Carbon electrode integration in SECM-SICM theta pipettes brought significant benefits compared to the gold deposition method, where the new method (pyrolyzing hydrocarbon inside one barrel of theta pipette) provides higher spatial resolution, a less complex fabrication procedure, and more solid insulation on the sidewall of the conductive probe. Accordingly, more accurate distance control of SICM in nanoscale probes became feasible due to the smaller inner diameter of the empty barrel (ca. 100 nm). This enabled the probe operation at probe-substrate distances as close as 50 nm [63]. Advances in SECM-SICM probes have enabled research topics such as stoichiometric surface redox kinetics [63], spatially resolved electrochemical imaging for nanostructures [68], and the activity of individual nanoparticles [69]. Carbon probes also allow for ease of surface modification at the electrode. Electrochemical deposition of other metals such as iridium and platinum on carbon enabled the development of pH-SICM and hydrogen peroxide sensing applications. However, carbon disk electrodes can be irregularly shaped occasionally due to the high temperatures required during pyrolysis. Therefore, carbon tips are often milled by FIB, such as in a recent work for studies of cell membrane permeability (Fig. 8e, f) [59].

2.2.3 Platinum and Iridium Modified Carbon Electrodes for SECM-SICM Probes

Platinum modified carbon electrode SECM-SICM probes were first introduced by O'Connell and Wain in 2014 to detect the oxygen reduction reaction (ORR) at platinum nanospheres [68]. In this study, carbon electrodes were modified with platinum nanoparticles via cathodic electrodeposition in 3 mM K_2PtCl_6 and 0.1 M $HClO_4$. Platinum modified SECM-SICM probes showed outstanding ORR performance at a lower potential (−0.15 V vs SCE) than that of the carbon electrode (−0.3 V vs SCE) in oxygen-saturated 0.1 M NaOH. The radius of the carbon electrode was ca. 100 nm outer diameter, and that of the whole theta pipette radius was ca. 200 nm, which allowed single, 200 nm diameter platinum spheres to be resolved via SECM-SICM probes through SICM topographical mapping. Concurrently, an acquired electrochemical response map, collected via the SECM component, successfully showed a higher oxygen consumption rate at each platinum nanoparticle. Although, lateral particle dimensions recorded in the electrochemical response appeared to be broader than the measured topographical map's lateral particle dimensions, likely a consequence of solution diffusion.

In 2015, O'Connell et al. applied a platinized carbon SECM-SICM probe to map H_2O_2 generation at electrodeposited gold nanoparticles (Au NPs) [69]. H_2O_2 production was recorded in substrate generation-tip collection (SG-TC) mode following a technique commonly conducted with SECM [70]. A glassy carbon substrate with deposited Au NPs was biased at −0.4 V vs RHE to carry out ORR via a 2-electron reduction path, and the Pt electrode SECM-SICM probe was biased at 1 V to oxidize and detect H_2O_2. Hydrogen peroxide generation at ca. 400 nm sized Au NPs was easily distinguishable in the electrochemical map generated, allowing investigation of heterogeneous sample activity. Recently, such platinized carbon SECM-SICM probes were applied to detect the electrochemical activity in immunocytochemically stained proteins on living cells for the first time by Sen et al. [71] Other metals, besides Pt are also suitable for the electrodeposition approach. For instance, iridium deposited SECM-SICM probe fabrication has been reported by Nadappuram et al. [16] In this work, a pH-sensitive SICM method was realized by electrochemically depositing an iridium oxide layer at the carbon electrode. Demonstration of a pH-dependent response was shown by calibrating the probe in various pH solutions by measuring the open-circuit potential.

2.2.4 Routes to High Throughput Carbon Probe Fabrication

As was briefly introduced in Sect. 2.2.2, fabrication of PPC SECM-SICM probes is a promising method for bulk scale production [67]. An important advantage of this method is the simultaneous coating of parylene polymer for carbon electrode and insulating layer around borosilicate or quartz capillaries (SICM probe). Parylene coated pipettes were pyrolyzed at 900°C under inert nitrogen atmosphere forming a thin conductive carbon layer. Subsequent parylene coating for insulation was followed by FIB-milling of tip. A FIB-milling step is a disadvantage for bulk probe production due to its time-consuming nature. Recently, Wilde et al. reported the development of an automated carbon probe fabrication with a propane/butane gas mixture in a pre-pulled quartz capillary by constructing a programmable pyrolysis stage [72]. This work focused on single barrel capillaries and the automated probe fabrication stage accurately controlled pyrolysis. Features such as constant gas flow in the chamber, heating coil position control, and current control were automated such that the quality of pyrolysis was maintained between probes. Repeatability reported for this method possibly provides a route to a high yield, bulk production of SECM-SICM probes.

2.3 *Evaluation of SECM-SICM Performance*

In conventional SECM, feedback controls the vertical movement of the SECM probe based on the conductivity of the sample. Negative feedback (hindered diffusion) is applied for insulative samples and positive feedback (enhanced redox regeneration) for conductive samples [50]. Due to the change in feedback with sample conductivity, nanoscale SECM probes with dynamic feedback can be difficult to operate on samples with heterogenous conductivity, as in constant-current imaging mode, and generating both topographical and electrochemical surface information can prove difficult. However, local topographical and surface information can be collected if signals can be separated such that topography is collected at one channel and electrochemical activity is collected at a separate channel. The ionic current based feedback signal in SICM operates in a negative mode (due to hindered ion transport) as a function of distance for both insulative and conductive samples, making SICM a compelling candidate to complement SECM. A few important criteria to evaluate SECM-SICM performance with various types of hybrid SECM-SICM probes are listed below.

- Independent electrochemical/ion current response: measure approach-curves in SICM feedback mode for conductive and insulative substrates.
- SECM electrode quality: measure steady-state current by cyclic voltammetry (CV).
- SECM electrode quality: measure approach-curves in SECM feedback mode for conductive and insulative substrates.
- Lateral resolution: comparison of the probe geometry, SICM topography, and SECM electrochemical response map.

2.3.1 Characterization of SICM Approach-Curves

Ion current approach-curves collected via SECM-SICM probes were characterized and compared to theoretical simulations [55, 58, 59, 63, 68, 73]. In 2010, Takahashi et al. demonstrated that SICM approach-curves produced with an SECM-SICM probe had much higher sensitivity to tip-substrate distance when compared to SECM curves at a gold ring electrode on the same probe. Ion current recorded during the approach was consistent for both conductive and insulative substrates, indicating that electrochemical current of the SECM electrode does not interfere with the measured ion current for SICM [58]. In addition, the SICM approach-curve matched theory, as the probe-substrate distance at 1% current reduction was comparable to the inner radius of a ring-shaped SECM-SICM probe (ca. 200 nm). Results presented by Takahashi et al. directly verified that SECM-SICM can provide accurate distance control for various electrochemical imaging and detection. A similar trend using a crescent-shaped gold electrode was demonstrated by Comstock et al., exhibiting consistent approach-curves regardless of the substrate in both dc and ac ion current response (Fig. 9) [55].

2.3.2 Characterization of Steady-State Current of SECM Electrodes

To quantitate response at the SECM electrode of SECM-SICM probes, steady-state current must be determined, typically, by cyclic voltammetry (CV) of a redox mediator. Steady-state current acquired via CV at gold electrodes on SECM-SICM probes has been extensively studied with ferrocene methanol ($FcCH_2OH$) and ruthenium hexamine ($Ru(NH_3)_6^{3+}$) for crescent, ring, and other electrode geometries. Steady-state current at a ring electrode was first evaluated by Walsh [54], in which the ratio of the inner to outer radii of the gold ring was considered and compared to a disk electrode of the same size. The measured radii ratio was ~0.8, which predicts the ring electrode should produce a steady-state current 12% lower than that obtained for a disk of the same outer radius. In this approach, the current at a disk electrode can be determined by [54].

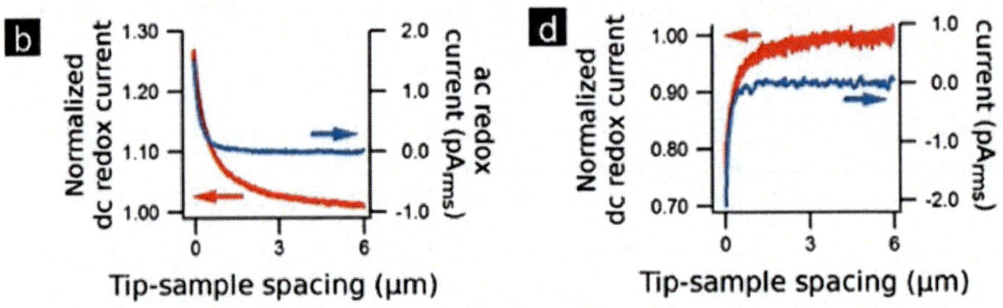

Fig. 9 Approach-curves collected by SICM-SECM, measured with gold crescent-shaped electrode in 10 mM $Ru(NH_3)_6Cl_3$, 100 mM KNO_3 solution on conductive gold surface (**a**), (**b**), and insulating Teflon surface (**c**), (**d**). Normalized dc (red) and ac (blue) ion current (**a**), (**c**), normalized dc (red) and ac (blue) redox current response (**b**), (**d**). Adapted with permission from Ref. [55]. Copyright 2010 American Chemical Society

$$i_{T,\infty} = 4nFDC_o b, \tag{5}$$

where n is the number of electrons transferred, F is the Faraday constant (96,485 C mol^{-1}), D is the diffusion coefficient of the electroactive species (e.g., $Ru(NH_3)_6^{3+}$ (5.3×10^{-6} cm^2s^{-1})), C_o is the bulk concentration of electroactive species, and b is the radius of the disk electrode (sometimes written as variable r). With Eq. (5), Walsh and Bard were able to confirm that the calculated ring dimensions matched closely with estimation based on SEM imaging. In 2010, Takahashi reported the same characterization of a smaller Au ring SECM-SICM electrode, in which the inner and outer radii of the Au SECM electrode were 330 and 550 nm, respectively [58]. However, the steady-state current acquired was approximately 6 times larger than that expected for a micro-disk electrode with the same size. The presence of defects in the aluminum oxide (Al_2O_3) insulating layer was reasoned as a possible cause of the large faradaic current, which was not observed in the previous studies by Walsh where electrophoretic paint was utilized for insulation. Although the enlarged steady-state current did not cause a serious problem in SECM imaging, significant caution was realized for the presence of defects in the insulating layer. In the crescent-shaped electrode, a well-defined equation for the steady-state current has not been defined, and the exposed electrode surface area has been approximated as an inlaid disk [55].

Electrochemical characterization of carbon electrodes on SECM-SICM probes for disk-shaped electrodes within one barrel of a theta pipette by CV of $FcCH_2OH$ was reported in 2014 [68]. In the carbon probe, precise characterization was difficult due to the recessed or deformed pipette wall during pyrolysis. For simplicity, in this study, the carbon electrode diameter was coarsely estimated within the theta pipette (diameter ca. <1 μm) assuming a disk-shaped electrode geometry, an RG (ratio of insulative surrounding radius over active electrode radius) value of 1.1, and free diffusion from the SICM barrel of the mediator by [74].

$$i_{\text{lim}} = 5.14nrFDc, \tag{6}$$

where i_{lim} is the diffusion-limited current and 5.14 is a factor accounting for an RG value equal to 1.1. One example of a carbon electrode with a 30 nm radius was estimated by calculating Eq. (6) with an experimental cyclic voltammogram in which the steady-state current (diffusion-limited) was acquired as ca. 10 pA with 1 mM FcMeOH. The calculated radius was smaller than observation by SEM (ca. >100 nm), which was hypothesized to result from adsorbed species resulting in suppressed steady-state currents.

Steady-state current measurements of a carbon-ring-shaped electrode with $Ru(NH_3)_6^{3+}$, insulated by parylene C, were reported by Thakar et al. in 2013 [67]. Carbon ring electrodes, pyrolyzed on a quartz rod and a single barrel quartz pipette were evaluated by experimental measurement and by [75].

$$I_{ss} = nFDC^* \left[\frac{\pi^2(a+b)}{\ln\left[32\left(\frac{a}{b-a}\right)\right] + \exp\left(\frac{\pi^2}{4}\right)} \right], \tag{7}$$

where I_{ss} is the steady-state current, C^* is the concentration of electroactive species, "b" is defined as the outer radius, and "a" is the inner radius for the carbon ring electrode. In this approach, the carbon ring with quartz rod (a: 350 nm and b: 800 nm) showed a close agreement between calculated I_{ss} response of 365 pA and experimental of 375 pA, while the carbon ring with quartz pipette (a: 295 nm, and b: 485 nm) revealed a higher experimental response of 270 pA compared to that of the calculated value of 200 pA (Fig. 10).

2.3.3 Characterization of Approach-Curve by SECM Electrode

SECM approach-curves can also be acquired to evaluate the electrode of an SECM-SICM probe. Both positive and negative approach-curves were collected by Bard

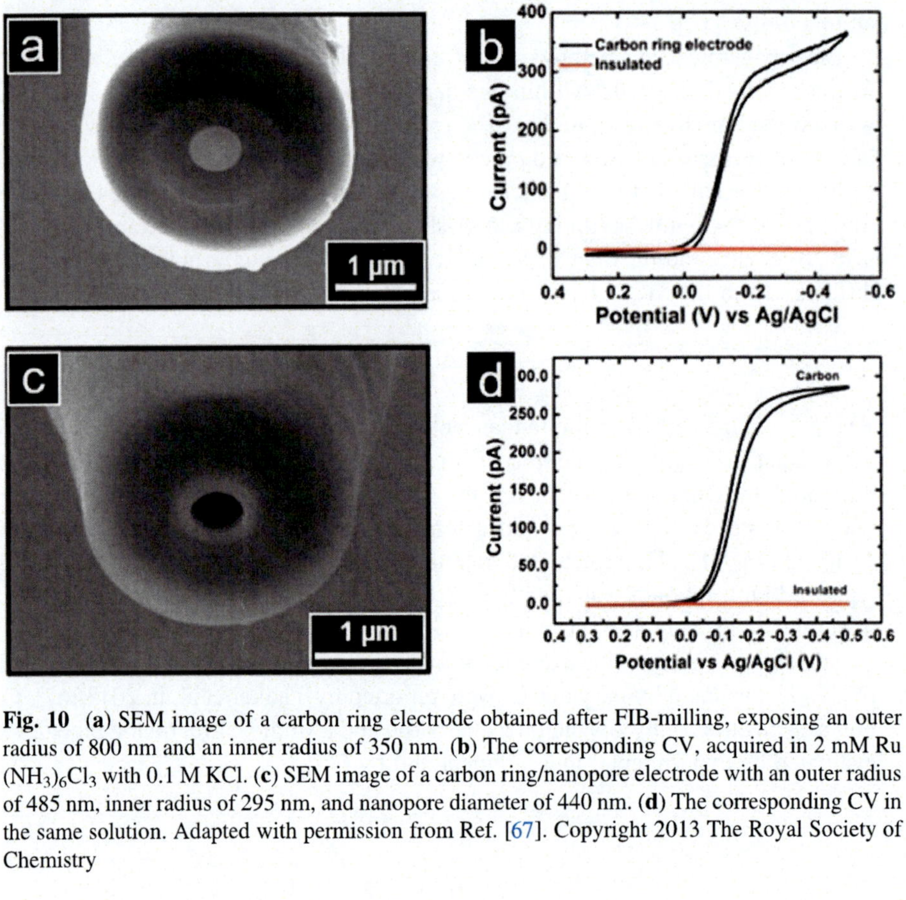

Fig. 10 (**a**) SEM image of a carbon ring electrode obtained after FIB-milling, exposing an outer radius of 800 nm and an inner radius of 350 nm. (**b**) The corresponding CV, acquired in 2 mM Ru $(NH_3)_6Cl_3$ with 0.1 M KCl. (**c**) SEM image of a carbon ring/nanopore electrode with an outer radius of 485 nm, inner radius of 295 nm, and nanopore diameter of 440 nm. (**d**) The corresponding CV in the same solution. Adapted with permission from Ref. [67]. Copyright 2013 The Royal Society of Chemistry

and coworkers for a gold ring electrode in a solution of $Ru(NH_3)_6^{3+}$ [54]. Comparison of theoretical and experimental curves provided insight into SECM sensitivity and the ring electrode geometry, where the shape of the SECM approach-curves is highly sensitive to the outer radius of the ring electrode. This comparison allowed one to understand the a/b ratio ("a" and "b" are the inner and outer radius, respectively), and the ratio of the entire (including insulating layer) electrode radius to the outer ring radius, r_g/b ("r_g" is the total radius of the electrode) [54]. In this manner, the thickness of the insulating layer around the ring electrode could be estimated. Approach-curves in both experimental and simulated data agreed well and showed a/b and r_g/b ratios to be 1.5 and 0.8, respectively. The estimated outer ring radii of 1.4 μm (positive feedback) and of 1.65 μm (negative feedback) in theoretical curves were close to those measured in SEM analysis and in CV. Au nanoring electrodes for SECM-SICM probes with 330 and 550 nm of inner and outer radii, respectively, were evaluated with $FcCH_2OH$ and a 70 nm insulating layer (anodic electrophoretic paint) by Takahashi in 2010 [58]. SECM approach-curves were performed on both a Pt electrode and a glass surface. Trends of the approach-curves were similar to theory, but not evaluated directly.

Carbon disk electrodes of an SECM-SICM probe (DBCNP) with 200 nm total outer diameter (including SICM probe) and 90 nm of disk electrode diameter were examined with $FcCH_2OH$ in 2011 [63]. The approach-curve was acquired on a Pt interdigitated array electrode. The experimental negative approach-curve fit reasonably to theoretical calculations with RG 1.5 and electrode radius 120 nm, indicating the disk-shaped carbon electrode had a planar geometry. In positive approach-curves, pure positive feedback with an unbiased Pt conductive surface was expected and the experimental result provided information of finite regeneration (surface redox) kinetics which have been reported in other studies with nanoscale tips [76, 77]. Use of ion current feedback to control probe-substrate distances as close as 50 nm without direct contact enabled calculation of heterogeneous rate constant (k) to be 0.80 cm s^{-1} [63]. This is a key aspect of the SECM-SICM technique because k is the only adjustable parameter when fitting, when the tip-substrate distance is known with high accuracy [63].

SECM-SICM has significantly contributed to improvements in determining quantitative surface kinetics and fluxes. In 2017, the regeneration rate of unbiased conductors was evaluated further by the same group [73]. Various heterogeneous rate constants from 0 to 0.5 cm s^{-1} were acquired by fitting theoretical approach-curves to experimental data recorded.

2.3.4 Evaluation of Lateral Resolution of Topography and Electrochemical Response Maps

Lateral resolution in SICM topography and SECM electrochemical response maps is of critical importance when considering SECM-SICM as an imaging tool. The addition of the SECM probe to SICM pipette significantly increases the size of the hybrid probe for SECM-SICM, often by more than two times. In general, the larger geometry of the probe naturally lowers lateral resolution of the SICM generated topographical image. However, high-resolution applications require rigorous efforts to fabricate optimal probes, which could advance areas of research that are difficult to investigate at small scales. Important applications such as surface patterning [78], cell biology [79], and drug delivery [80] demand higher resolution with precise control of small volume delivery to desired positions. Factors which have the greatest impact on the lateral resolution and measured signal in both topographical and electrochemical mapping with SECM-SICM are listed as below:

- Outer diameter of the tip-end of SECM-SICM probe.
- Electrode geometry in ring shape (symmetric), crescent and disk shape (asymmetric).
- Insulation quality from any defects and pinholes.
- Imaging modes of SICM.
- Local delivery of redox mediator or chemical species for desiring reactions.

Walsh et al. examined the first trial of SECM-SICM imaging that was coined "dispensing/feedback SECM imaging." [54] The lateral resolution of the electrochemical response map was measured with a Au ring electrode (diameter ca. 2.0 μm) by recording the oxidation of ferrocene methanol (FcMeOH) at a Pt electrode (diameter ca. 100 μm) embedded in a glass substrate and simultaneous recording to monitor the product (ferrocenium) at the probe tip. Flow of redox mediator from the inner barrel of the pipette was dispensed on the order of 1 nLs^{-1} by applying a constant pressure to the pipette. Two electrochemical response maps were constructed simultaneously by recording faradaic current from the Pt substrate electrode (held at 0.4 v vs Ag/AgCl for oxidation) and from the Au ring electrode at the tip (held at 0.0 v vs Ag/AgCl for reduction). The probe-substrate distance (D_{ps}) was kept at 2.4 μm during the scan. Notably, in this study, the pipette of SICM was utilized solely for dispensing, rather than recording ionic current for distance control. Although the lateral resolution of topography was not examined and the dimension of SECM-SICM probe was relatively large (micrometers in scale), the two electrochemical response maps from the tip and substrate recordings were recorded and contained similar circular features without significant offset, which demonstrated the advantage of dispensing the redox mediators of interest from the SICM pipette to pursue high lateral resolution electrochemical response mapping.

An evaluation of lateral resolution of SECM-SICM imaging acquired by ion current feedback via SICM was demonstrated by Comstock et al. with a Au crescent electrode (length ca. 400 nm, width ca. 200 nm) [55]. Topographical imaging was conducted in ac mode, with a modulated probe in the vertical direction at an amplitude of 50 nm and a frequency 570 Hz on a gold trench electrode (Fig. 11a). In this manner, both ionic and faradaic currents were acquired in both dc and ac signals. SECM-SICM in this feedback mode visualized the sub-micrometer sized trench structures (width 400 nm) in the topography map (Fig. 11b) and acquired dc and ac redox current maps (Fig. 11c, d). The redox mediator, 10 mM $Ru(NH_3)_6{}^{3+}$, was included in external bath solution. Remarkably, the ac redox current map showed an improvement (FWHM of the trench 580 nm) in lateral resolution,

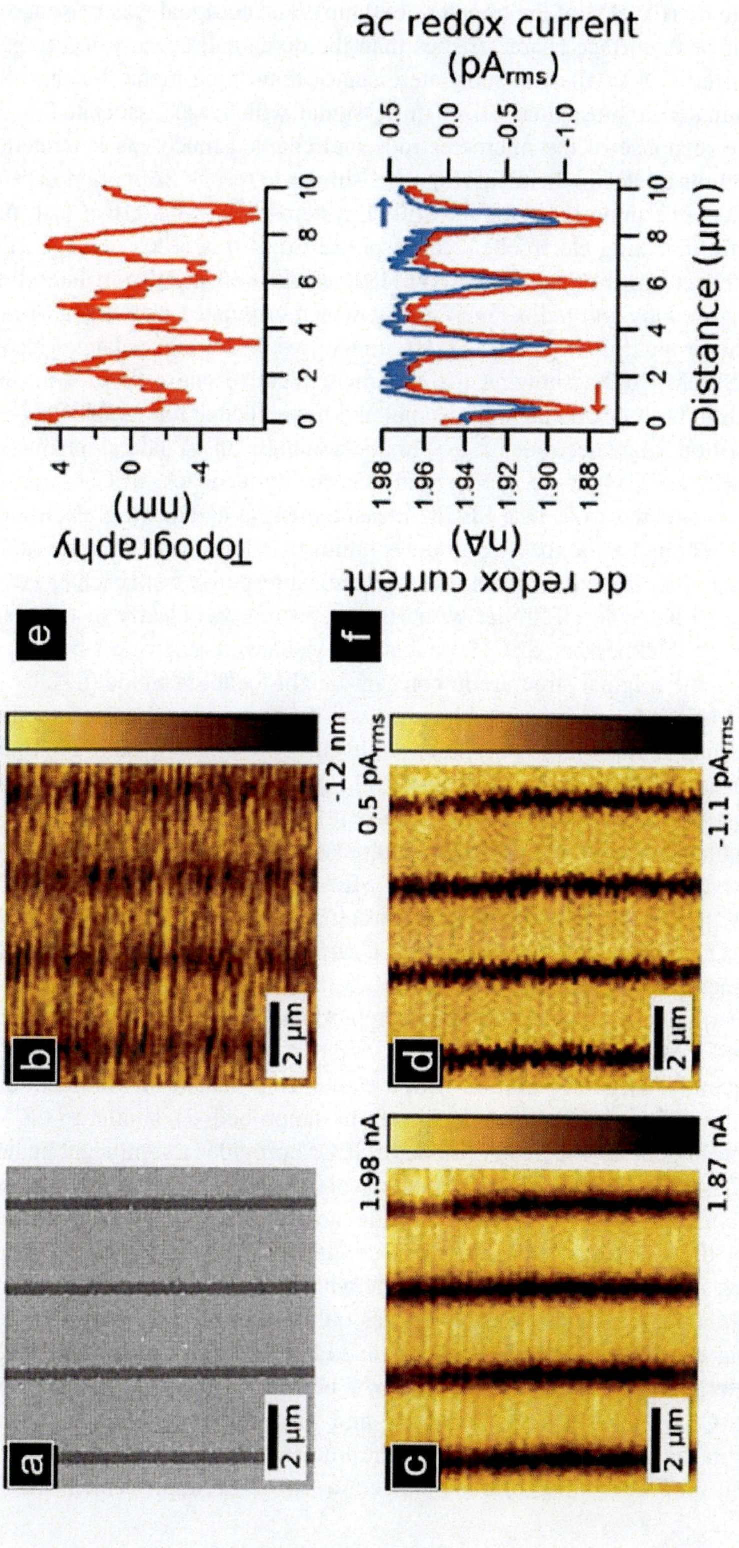

Fig. 11 **(a)** Scanning electron micrographs of trench sample of 400 nm width, milled by FIB into a gold film on a glass substrate. SICM-SECM images, acquired in 10 mM $Ru(NH_3)_6Cl_3$ + 100 mM KNO_3. **(b)** SICM topography, **(c)** SICM-SECM dc redox current, and **(d)** SICM-SECM ac redox current. Line profiles from feedback mode SICM-SECM imaging of 400 nm wide FIB-milled, glass-exposed trenches. **(e)** Topography and **(f)** dc (red) and ac (blue) redox current. Adapted with permission from Ref. [55]. Copyright 2010 American Chemical Society

compared to dc (FWHM of the trench as 680 nm). The ac signal was reasoned to be more sensitive to surface characteristics than the dc signal, because imaging in ac mode occurred at a smaller tip-substrate distance than in dc mode. Further, the ac signal eliminated a substantial offset in dc signal which was associated with the steady-state response of the microelectrode and enabled much easier detection of small variations in electrochemical response, similar to results from previous studies with atomic force microscopy-based SECM systems [81]. This effect also proves beneficial for detecting electrochemical response from stray redox signals originating from defects in the insulative layer. Images showed a 230 nm lateral offset between topography and redox signal maps, which originated from the asymmetric shape of the probe. Subsequently, SG/TC mode revealed a higher lateral resolving power of SECM-SICM, showing the electrochemical response map of a 180 nm wide trench in both dc and ac redox signal and recognizable topography.

At a carbon ring electrode, a systematic comparison of lateral resolution in SECM, SECM-SICM in ac mode, and dispensing/feedback SECM in SG/TC mode was conducted [67]. In a SECM measurement, a carbon ring electrode was fabricated on a quartz rod, having an outer radius of 800 nm and an inner radius of 350 nm (total diameter ca. 3.0 μm). Electrochemical response from each electrode in an array of 10 μm gold electrodes with 10 μm spacing was clearly distinguishable and the array thickness in SECM image was visualized as twice as thick when compared to the original structure in constant-height feedback mode [67].

In SECM-SICM, a carbon ring electrode was fabricated on a quartz nanopipette, having an outer radius of 485 nm, inner radius of 295 nm, and a nanopore diameter of 440 nm (total diameter ca. 2.5 μm) [67]. Operated in ac mode with 130 nm tip-substrate distance maintained by ion current feedback of SICM, topography and electrochemical response maps were acquired from a 5 μm gold electrode array concurrently. Both SICM topography and SECM images showed a width of each array element comparable to the original structure. This result clearly demonstrated the improvement of lateral resolution of SECM-SICM compared to just SECM with the same carbon ring electrode geometry. Small features and defects (diameter ca. 3 μm) on the electrode array were reproducibly observed in topographical maps and SECM images.

Subsequently, SECM resolution with a carbon ring SECM/SICM electrode was examined by dispensing redox molecules from the probe [67], similar to the Walsh experiment. In general, the SG/TC mode of SECM provides a significant benefit for steady-state measurements in electrochemical response acquisition because the oxidized chemical species at the electrode near the substrate is spontaneously reduced at the tip [82]. The lateral resolution improvement of SECM image in SG/TC was shown by filling the nanopipette with the redox mediator and compared to results when the mediator, $FcTMA^+$, was only present in the bath solution. The carbon ring electrode/nanopipette was scanned over a 10 μm platinum band electrode while applying a potential of 250 mV between a Ag/AgCl quasi-reference electrode (QRE) inside the nanopipette and an additional Ag/AgCl electrode immersed in the bath solution. Notably, the line scan recorded in SG/TC imaging mode with $FcTMA^+$ filled probe showed a two-fold improvement in lateral resolution compared to an electrolyte only filled probe, reporting an FWHM of 37.8 μm and 77.6 μm, respectively. These results verified that locally delivered $FcTMA^+$ ions dispensed from the aperture of the nanopipette were beneficial to achieve higher lateral resolution by limiting the local concentration of redox mediator.

In the case of DBCNP's employing a carbon disk electrode with 28 nm radius, lateral resolution in topography and electrochemical response were evaluated with nanopores (radii ca. 100 nm) in polyethylene terephthalate (PET) membranes [63]. In this experiment, both sides of membrane were filled with phosphate-buffered saline (PBS) including 1.0 mM $FcCH_2OH$ redox mediator. A constant tip-substrate distance of 30 nm was controlled by SICM feedback. Membrane topography and the corresponding electrochemical response of $FcCH_2OH$ oxidation at nanopores were imaged. The decreased oxidation current at the nanopore indicated that the diffusion of redox mediator was inhibited by the pore due to the probe–pore interaction. Topographical images of a Pt interdigitated array and living sensory neurons were also acquired with electrochemical maps. This study validated the lateral resolution of SECM-SICM with DBCNP (SICM nanopipette radius ca. 50 nm) for topographic imaging of soft living PC12 cells, with comparison to topography images obtained with a single barrel SICM probe (radius of 50 nm). Notably, the resolution quality in both topographical maps was comparable, showing clearly similar dendritic structures with diameter <200 nm.

Carbon disk electrodes (diameter ca. 200 nm), pyrolyzed in one barrel of a theta pipette reported the highest lateral resolution in both topographical and SECM images, recorded at gold nanodisk arrays (diameter ca. 150 nm, 50 nm height) on silicon, operated with SICM feedback mode [68]. Both SICM topography and electrochemical response maps of $FcCH_2OH$ faithfully visualized the disk patterns, although a small rounding effect was observed at the top of the disk. Further, asymmetric faradaic current acquisition was observed during platinum band imaging in SG/TC mode due to the asymmetric carbon disk electrode formation at one side of the theta pipette. During the SG/TC mode imaging, the SICM signal (ion current) was highly stable, while the SECM signal often degraded by a few picoamperes.

2.4 *Applications and Studies with SECM-SICM*

Fabrication and evaluation of SECM-SICM probes allows for the implementation of their various applications in local electrochemical and bioelectrochemical studies with higher accuracy in electrochemical response measured compared to typical SECM. Selected examples of such studies are described below.

2.4.1 Local Electrochemistry Studies

SECM-SICM provides an amenable route to explore local electrochemical studies for nanoscale objects such as nanoparticles and nanowires. Providing precise measurement of catalytic activity of individual single nanoparticles allows for a meaningful statistical analysis to be made, which can determine the range of heterogeneity of catalytic activity [68]. Artificially synthesized nanoparticles of various shape, size, and material have demonstrated enhanced catalytic activities for different electrochemical reactions such as the CO_2 reduction reaction [83–85], the hydrogen evolution reaction [86–88], and the ORR [89–91]. Conventionally, SECM is a typical tool to characterize such nanoparticles in electrolyte solution at sub-micrometer dimensions, although SECM has shortcomings in feedback accuracy and issues with lateral resolution mapping. Significant efforts have been made to overcome these shortcomings by developing SECM-AFM [92–94], shear force-SECM [95–97], intermittent-contact SECM [77, 98, 99], and alternating current-SECM [100–102]. Even with these efforts, improvements in lateral resolution, topography accuracy, consistency of tip fabrication, and ease of instrumental implementation are desirable. For instance, SECM-AFM provides the highest resolution among these techniques but requires complicated probe fabrication processes which are costly for both the manufacturer and end users.

Recent studies have reported pyrolyzed carbon electrode within one barrel of a theta pipette in SECM-SICM effectively achieved high lateral resolution in both topography and electrochemical response, and accurate feedback control on the order of tens of nanometers [63, 68]. The small sized carbon electrode is easily fabricated and also controllable by adjusting the diameter of the pulled dual barrel capillary. Formulating a carbon electrode on the inside of one barrel of a theta nanopipette in this method naturally provides a pinhole-free insulation layer made of borosilicate or quartz. This fabrication method has made a significant contribution to the ease of SECM-SICM fabrication by eliminating the FIB-milling process, which was essential to expose the conductive electrode from the insulation layer and the biggest disadvantage for cost-efficiency and bulk fabrication. Although it should be noted in many cases spill-over of pyrolysis materials onto the outer walls of the pipette can be observed. To overcome this issue, FIB-milling may still be required. Such carbon electrodes are highly convenient to functionalize with other materials by electrodeposition for studying specific electrochemical reactions [16, 68, 69, 103].

ORR [70, 104] is known to have multiple reaction pathways. During the 2-electron pathway, intermediate products such as HO_2^- in alkaline and H_2O_2 in acidic electrolytes are generated as shown in Table 2 [70, 104]. The 4-electron pathway is thermodynamically favored, but seldomly occurs as one elementary step due to the complexity of the reaction. Instead, H_2O_2 or HO_2^- is often generated during the ORR as the final product or an intermediate depending on the solution where the reaction takes place. The metal oxide route [70], which can occur at the surface of a metal electrode, is one favored mechanism for O_2 reduction. ORR in the

Table 2 ORR pathways on electrodes in aqueous electrolytes and at metal surface

Electrolyte	Reaction pathway	Reaction potential (V vs. NHE)
Alkaline electrolyte	$4e^-$pathway	
	$O_2 + 2H_2O + 4e^- \rightarrow 4OH^-$	+0.401
	Peroxide pathway	
	$O_2 + H_2O + 2e^- \rightarrow HO_2^- + OH^-$	−0.065
	$HO_2^- + H_2O + 2e^- \rightarrow 3OH^-$	+ 0.867
Acidic electrolyte	$4e^-$pathway	
	$O_2 + 4H^+ + 4e^- \rightarrow 2H_2O$	+1.229
	Peroxide pathway	
	$O_2 + 2H^+ + 2e^- \rightarrow H_2O_2$	+ 0.67
	$H_2O_2 + 2H^+ + 2e^- \rightarrow 2H_2O$	+1.77
Metal surface	$2M + O_2 \rightarrow 2MO$ $MO + 2H^+ + 2e^- \rightarrow M + H_2O$	

metal oxide route can also occur in series or in parallel and generate variation in H_2O_2 production and number of electrons transferred (*n*).

A platinum modified carbon probe in a theta pipette was utilized to detect oxygen consumption in a redox competition mode of SECM-SICM by O'Connell in 2014 [68]. This study demonstrated a nanoscale method to examine electrocatalytic processes at individual Pt nanoparticles and showed the value of high-resolution imaging for concurrent topography and electrochemical response, enabling quantitative measurements for electro-kinetics of single particles by SECM-SICM (Fig. 12). The ORR activity at the DBCNP was evaluated before and after platinization in an O_2 saturated solution. A significant improvement in onset potential of ORR (Pt at −0.15 V and carbon at −0.25 V vs SCE) and faradaic current (180 pA for Pt, 40 pA for carbon at −0.6 V vs SCE) were observed, indicating that a highly sensitivity ORR detection at the Pt electrode was achieved in the SECM-SICM (Fig. 12b). In the next step, a Pt-modified DBCNP was scanned over nanospheres, which were electrodeposited onto a glassy carbon substrate, to resolve local ORR activities of individual Pt nanospheres. A simplified redox competition mode of SECM was performed, in which oxygen is reduced at both the Pt sphere (biased at −0.3 V vs Ag/AgCl) and Pt-modified tip (biased at −0.6 V vs Ag/AgCl). Individual nanospheres (200~300 nm in diameter) were well resolved in SICM topography by measuring with an ~100 nm electrode radius probe although the lateral diameters of nanospheres appeared to be approximately 500 nm due to tip convolution (Fig. 12c). In the electrochemical response map of SECM, the oxygen consumption at/around each nanosphere was distinctively observed, demonstrating a higher faradaic current reading (ca. −18 pA) at the glassy carbon area where nanospheres were not present and a lower signal (ca. −14 pA) at the positions of each nanosphere (Fig. 12d). However, quantification of oxygen reduction at individual nanospheres was more difficult because of a lower lateral resolution in electrochemical response map compared to topography.

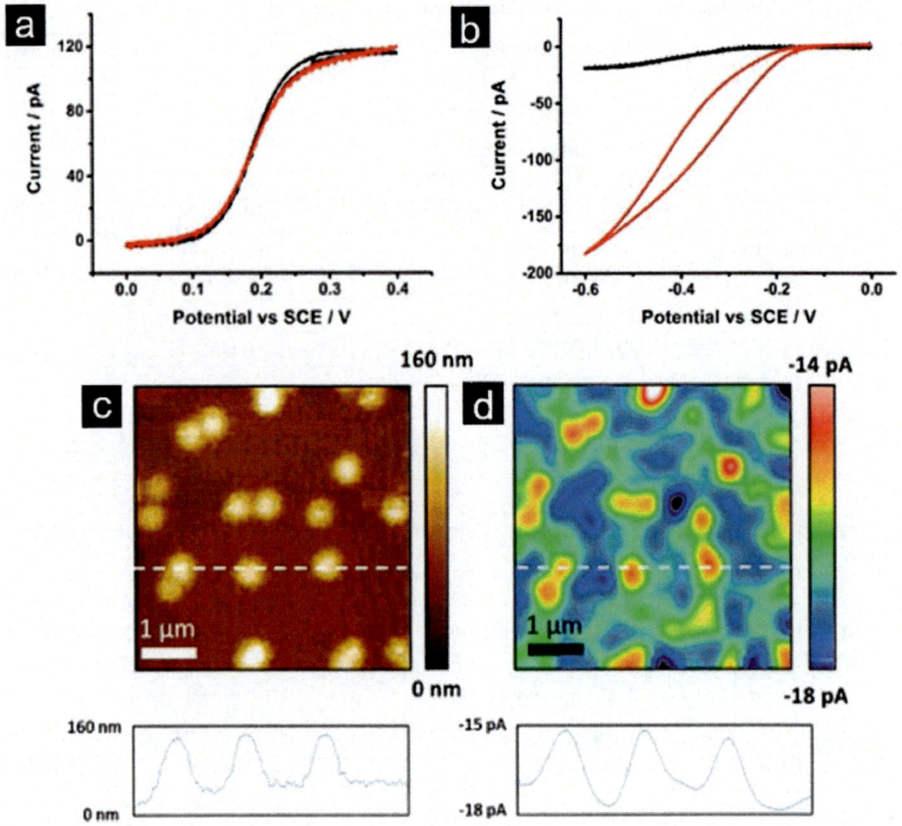

Fig. 12 Platinum modification of carbon probes: (**a**) CV in 1 mM FcMeOH/0.1 M KNO_3 before (black) and after (red) platinum deposition and (**b**) oxygen reduction CV in oxygen-saturated 0.1 M NaOH for the same probe before (black) and after (red) platinum deposition. Images of oxygen reduction at platinum nanospheres using platinized probes in 0.1 M NaOH: (**c**) SICM topography, (**d**) SECM image of competitive oxygen reduction signal during which the substrate is held at −0.3 V vs Ag/AgCl and the tip at −0.6 V vs Ag/AgCl. Cross sections are indicated by the dotted line. Adapted with permission from Ref. [68]. Copyright 2014 American Chemical Society

The same group showed that localized hydrogen peroxide generation at Au nanoparticles (Au NPs) which resulted from the 2-electron process of ORR could be measured by Pt-modified carbon SECM-SICM probes [69]. Au NPs were electrodeposited onto a glassy carbon substrate using a single potential pulse in H_2AuCl_4 electrolyte, producing particles predominately 400 nm in diameter. The crystallographic faces in the electrodeposited Au NPs were 41% Au (111), 45% Au (110), and 14% Au(100) which were detected by using Pb^{2+} under potential deposition voltammetry. Au(111) and Au(110) planes are known to favor the 2-electron ORR to hydrogen peroxide while Au(100) favors the 4-electron reaction process. Pt-modified DBCNP SECM-SICM probes were used to measure surface-generated peroxide from individual Au NPs on GC substrates. Topographical information acquired by SICM feedback revealed well-dispersed Au NPs (ca. 430 nm diameter). The GC substrate was biased at 0.4~0.9 V vs RHE to generate ORR and 1 V (vs RHE) was applied at the Pt-modified carbon electrode of SECM-SICM. At a substrate bias of 0.5 V, peroxide generation was detected at many Au NPs and an increase in collection current was observed by increasing the cathodic overpotential. Subsequently, a magnified inspection revealed that 4 Au NPs of a comparable size showed different magnitudes of peroxide collection, highlighting non-uniformity of electrochemical activity of Au NPs detected by this high-resolution electrochemical imaging technique.

Local detection of electrochemical signals by SECM-SICM was demonstrated in the study of degraded Nafion 212® membranes, revealing heterogeneity in the transport of redox mediators [60]. In this study, Shi and Baker validated concurrent measurement of SICM-SECM with a crescent-shaped Au electrode insulated by parylene C. First, the approach was validated by measuring the positions of nanopores (diameters ca. 400 nm) on a polyimide (PI) membrane which were well matched to the corresponding faradaic current response in an electrochemical map measured by oxidation of $Ru(NH_3)_6^{3+}$ diffusing through pores. Next, the ion permeability detection technique with SECM-SICM was utilized to interrogate heterogeneity in degraded N212 membranes (Fig. 13), where the Nafion membrane was degraded by Fenton's reagent [105] which produces oxidative degradation in a manner analogous to the primary cause for chemical degradation of membrane in fuel cells, namely hydroxyl and peroxyl radical decomposition of the polymer backbone [106]. After 6 h degradation, some of the defects on the membrane surface began to show faradaic current responses, which became more distinctive with prolonged degradation of the membrane after 12 h. The results suggested that leakage of degraded membranes is initiated at surface defects, and that heterogeneous permeability exists in the degraded membrane in which only approximately 50% of surface defects showed faradaic signal under conditions tested.

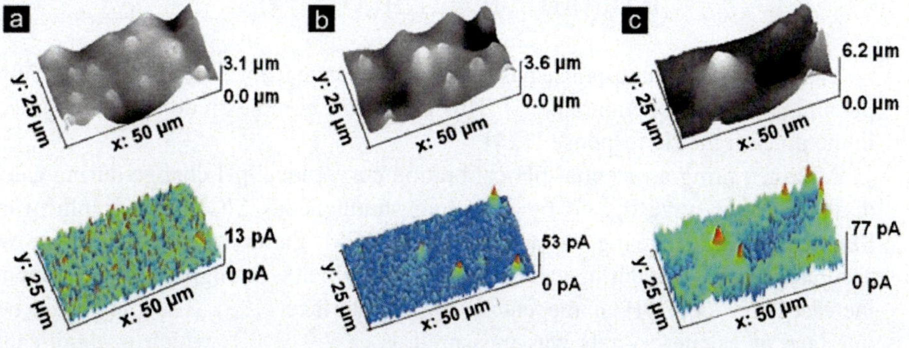

Fig. 13 SICM-SECM topographic (top) and faradaic current (bottom) images of N212 membranes degraded for (**a**) 3 h, (**b**) 6 h, (**c**) 12 h. Arrows were added to indicate bubbles/tears with distinct faradaic current responses for better correlation to topographic images. Adapted with permission from Ref. [60]. Copyright 2015 The Royal Society of Chemistry

2.4.2 Local pH Measurement

Local pH measurement is an intriguing application of SECM-SICM that is important for both cell biology and electrochemistry. In cells, physiological function is reliant on pH gradients between plasma membrane at the cellular [107] and subcellular level [108]. Tumorigenesis [109], an abnormal cellular process, is also known to be related to pH change. Numerous electrochemical phenomena such as corrosion [110, 111], metal deposition [112, 113], and acid-base dissolution [114] also often lead to local pH change. A number of techniques have been developed to detect local pH changes. Optical techniques such as wide field fluorescence microscopy [115] and laser scanning confocal microscopy [116] have been reported. Additionally, pH-sensitive fluorescent molecular probes [117], gold nanoshells [118], filed-effect nanowire transistors [119], conducting polymer-coated carbon fiber electrodes [120], cantilevers [121], and parylene-insulated iridium nanoelectrodes [122] have been reported. SECM with small electrodes for electrochemical pH measurement provide high-resolution local pH measurements at interfaces and this technique has advanced recently from the micro to the nanoscale [74, 123].

In SECM-SICM, two techniques have been demonstrated, where one utilized iridium oxide base pH electrodes [16] and the other used a pH-responsive PANi-coated electrode [17]. The iridium oxide based pH electrode was fabricated by electrodepositing iridium on the tip surface of DBCNP. Iridium oxide film formed at the electrode, according to [124]

$$\left[\mathrm{Ir(COO)_2(OH)_4}\right]^{2-}_{(aq)} + 2\mathrm{OH^-} \rightarrow \left[\mathrm{Ir(OH)_2 \cdot 2H_2O}\right]_{(s)} + 2\mathrm{CO_2} + 2e^- \quad (8)$$

Iridium oxide films have a sensitive potentiometric response to pH change and are known to exhibit a super-Nernstian pH response based on the reaction

$$\begin{aligned} &2\left[\mathrm{IrO_2(OH)_{2-x}(2+x)\cdot H_2O}\right]^{(2-x)}_{(aq)} + (3-2x)\mathrm{H^+} + 2e^- \\ &\rightleftharpoons \left[\mathrm{Ir_2O_3(OH)_3 \cdot 3H_2O}\right] + 3\mathrm{H_2O} \end{aligned} \quad (9)$$

The potentiometric response has slopes in the range from −59 mV to −90 mV per pH unit at room temperature [125]. This is due to a complex redox reaction process in the potentiometric response [124].

After acquiring a potential-pH calibration curve, local pH change during calcite dissolution was investigated by obtaining simultaneous SICM topographical and SECM pH maps. Having the advantage of SICM, the crystal height change was accurately monitored during calcite dissolution, which is strongly pH-dependent and increases the local pH at the calcite–water interface [126]. The pH at the near interface of calcite crystals was measured as ca 7.9 ± 0.1, which is significantly higher than the pH of 6.85 in bulk. Successive scans enabled the quantification of the corresponding dissolution rate of 2.6 (±0.2) × 10^{-9} mol cm^{-2} s^{-1} at pH 6.85.

Another type of pH electrode in SECM-SICM is PANi-coated electrode, which was electropolymerized on the surface of a Au electrode in 0.1 M aniline/ 1.0 M H_2SO_4 solution [120, 127]. PANi coating methods proved beneficial for fabrication of solid-state pH nanosensors with tip dimensions ranging ca. from 100 to 500 nm in diameter [120], and a wide pH response range (2–12.5). Additionally, the probe working life time improved by up to 3 weeks. The pH-sensitive response of PANi originates from protonation-deprotonation of nitrogen atoms in the polyaniline units [120]. The EMP response of these pH electrodes was calibrated by performing a series of cyclic voltammetry sweeps in a range of pH-buffered solutions for multiple iterations at room temperature (24 °C). Near-Nernstian slopes were detected (−56 mV/pH) in pH ranges of 2.5–12.0. To perform local pH measurements, a five-electrode system was employed over a polyimide membrane that possessed many single nanopores. Migration and diffusion processes were studied in the setup where solutions with different pH were filled in the upper and lower chambers with the porous polyimide membrane (ca. 900 nm diameter pores) placed between the chambers. Local pH differences in the vicinity of a pore were obvious when high transmembrane potentials were applied (above 0.7 V) with a high pH gradient (upper chamber pH 12, bottom chamber pH 2.5 in 0.1 KCl). This result indicates that local pH differences due to migration can be detected as a map of EMF response. Interestingly, such local pH differences were difficult to observe without application of a transmembrane potential.

2.4.3 Bio-Electrochemical Studies

Electrochemical studies of biological systems, such as living cells and tissues, at high resolution have been a key to understanding many important molecular level phenomena. SECM has been employed in this role to detect enzymatic activity, cellular activity, cell membrane permeability, and the presence of electroactive metabolites.

Takahashi et al. demonstrated SECM-SICM for bioelectrochemical studies and showed the simultaneous acquisition of topography and electrochemical maps of various biological molecules [58]. In this study, a Au ring electrode in SECM-SICM measured horseradish peroxidase (HRP) and glucose oxidase (GOD) enzymes immobilized on a substrate. Additionally, cardiac myocytes were also examined for permeability studies. In the case of GOD detection, $FcCH_2OH$ is oxidized to $[FcCH_2OH]^+$ at the ring electrode first, and immobilized GOD on the substrate oxidizes glucose. Concurrently, $[FcCH_2OH]^+$ is reduced and diffuses back to the ring electrode so that cycling of redox mediator occurs to amplify the electrochemical response. Consequently, in both measurements, the profile of an enzyme spot map of SECM was in good agreement to the corresponding topography (Fig. 14).

In the permeability study, faradaic current for oxygen, ferrocyanide, and $FcCH_2OH$ were measured at a cardiac myocyte [58]. In previous studies performed by SECM, several technical challenges were raised [128, 129]. First, the possible presence of heterogeneous permeability in specific regions; second, electrode

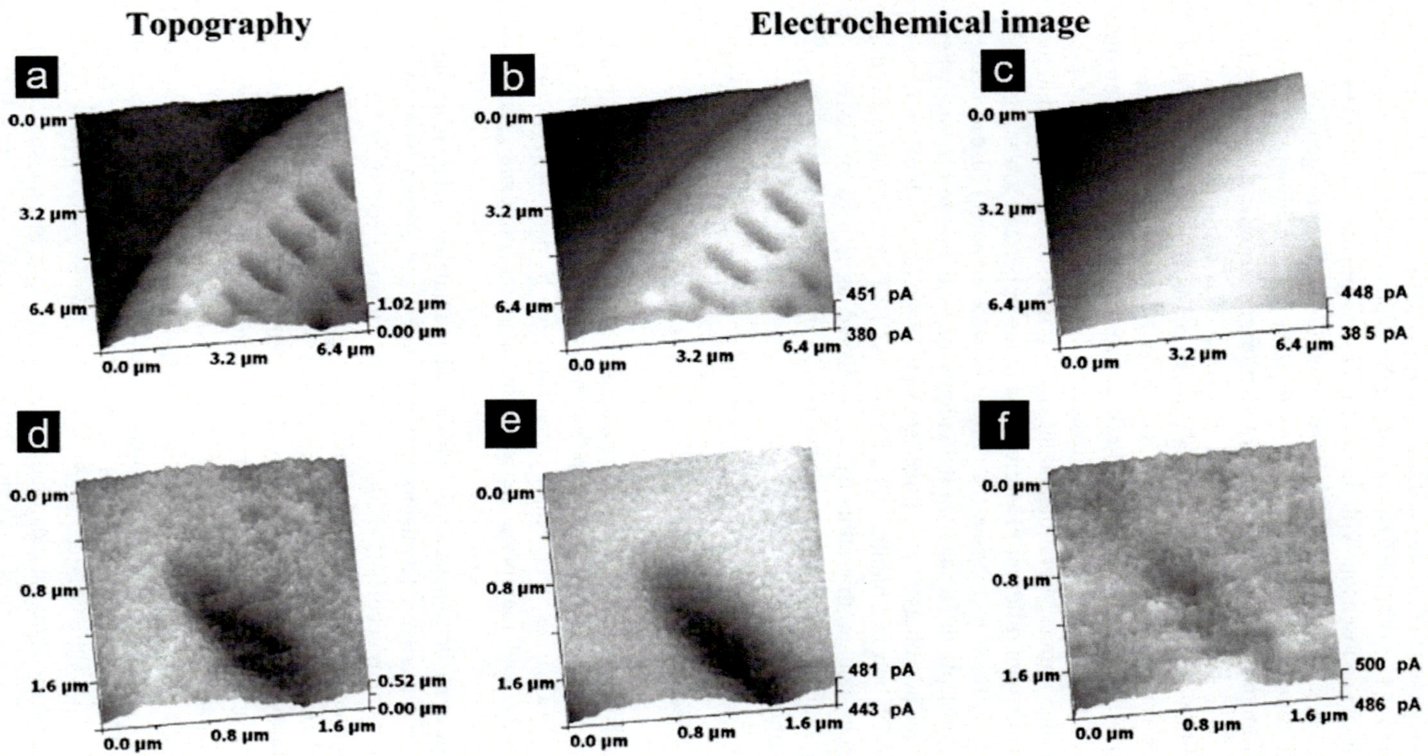

Fig. 14 (**a**, **d**) Topographic and (**b**, **c**, **e**, and **f**) electrochemical images of a GOD immobilized substrate adding 20 mM glucose in 0.50 mM $FcCH_2OH$ + 0.1 M KCl. The SECM nanoring and SICM nanopipette electrodes were held at 500 and 200 mV vs Ag/AgCl, respectively. Upper and lower images were captured with 8 μm × 8 μm and 2 μm × 2 μm, respectively. The probe-sample distances were held at 100 nm (**b**, **e**) and 600 nm (**c**, **f**), respectively. Adapted with permission from Ref. [58]. Copyright 2010 American Chemical Society

fouling, caused by biomolecules adsorption on electrode surface; and third, a variation in approach-curve measurements, originating from topographic heterogeneity in the cellular surface. In SECM-SICM with hopping mode, an approach-curve can be acquired at each measurement point so that the permeability can be examined on the living cell membrane. Although the details of sarcoma structure were not visualized clearly due to the lack of lateral resolution, the shape of the cardiac myocyte was imaged clearly. The concurrently obtained faradaic current map showed the localized permeation properties for each redox probe. Oxygen reduction current decreased when the probe-substrate distance decreased, indicating that the cell membrane exhibited high permeability for oxygen (Fig. 15a). On the contrary, $[Fe(CN)_6]^{4-/3-}$ oxidation currents were measured at the same level over the cell and substrate, indicating that the diffusion of $[Fe(CN)_6]^{4-/3-}$ was blocked because of hydrophilicity of the redox mediator (Fig. 15b).

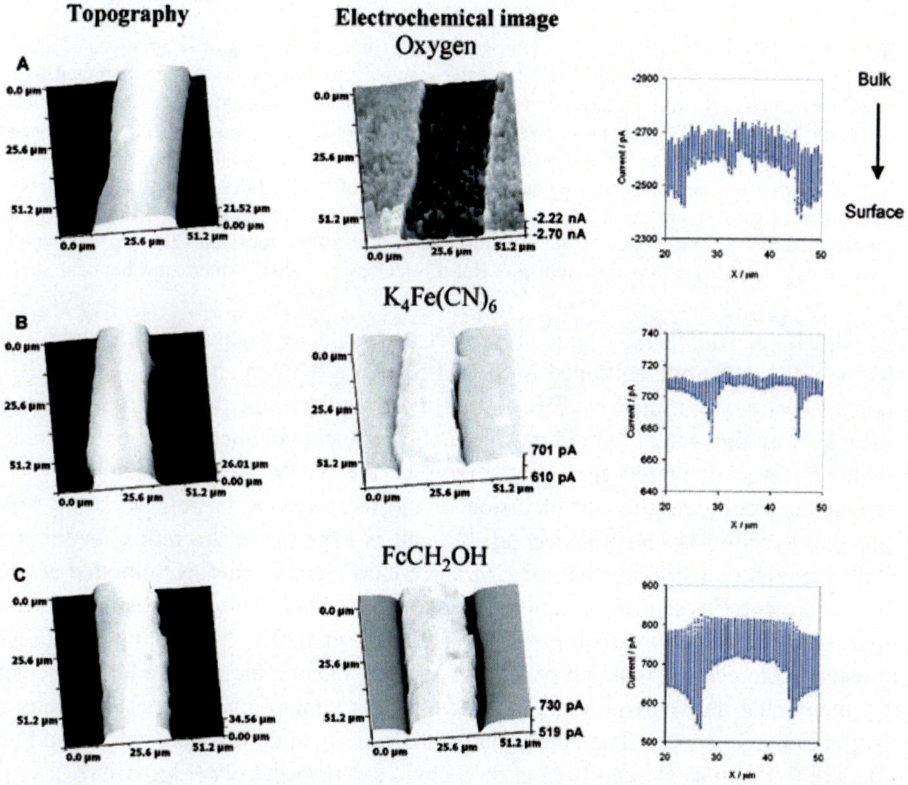

Fig. 15 Topographic (left) and electrochemical (center) images of a rat cardiac myocyte and electrochemical signal profiles of repeated approach-curves at each measurement point (right). Electrochemical images were based on (**a**) reduction of oxygen at −500 mV vs Ag/AgCl, (**b**) oxidation of 0.50 mM $K_4Fe(CN)_6$ at 500 mV vs Ag/AgCl and (**c**) oxidation of 0.50 mM $FcCH_2OH$ at 500 mV vs Ag/AgCl. Scan ranges were 64 μm × 64 μm. The approach distances were set to 10 μm. Arrows and dashed lines denote the direction of the approach and boundaries of the cell and substrate, respectively. Adapted with permission from Ref. [58]. Copyright 2010 American Chemical Society

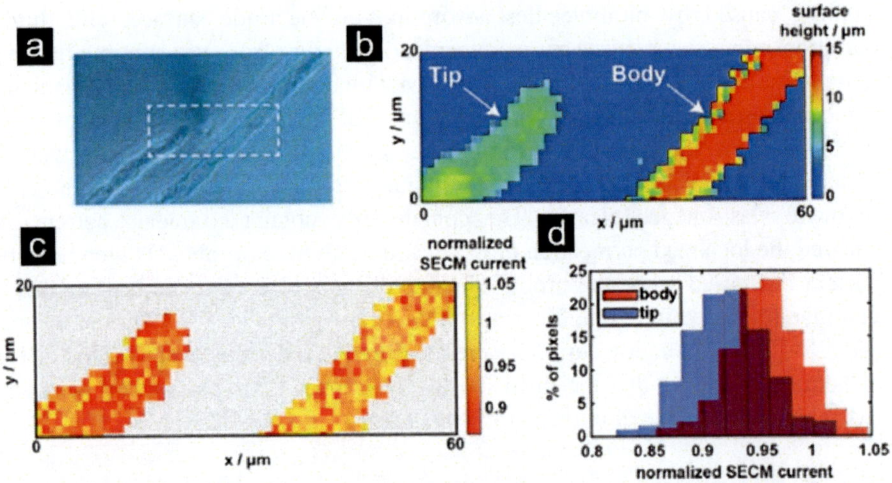

Fig. 16 SICM-SECM topographical and $[Ru(NH_3)_6]^{3+}$ uptake mapping of two regions of a single *Z. mays* root hair cell. (**a**) Optical image of the scanned root hair cell; scan area denoted by the dashed rectangle. (**b**) Substrate topography extracted from the z-position at the point of closest approach from the SICM channel. (**c**) Normalized SECM current map showing a clear difference in uptake between the root hair cell body (higher current, lower uptake) and the root hair cell tip (lower current, higher uptake). "Normalized current" is the ratio of the $[Ru(NH_3)_6]^{3+}$ reduction current at the point of closest approach to the same reduction current in bulk. (**d**) Histograms of the normalized SECM current across the two different regions of the root hair cell, "tip" and "body" (see part b). Adapted with permission from Ref. [59]. Copyright 2017 American Chemical Society

Uptake of $Ru(NH_3)_6{}^{3+}$ at *Z. mays* root hair cells were studied with a pyrolytic carbon disk electrode equipped in SECM-SICM by Unwin and coworkers in 2017 [59]. The tip was milled by FIB so that the irregularity of the electrode shape of SECM was prevented, producing a roughly 500 nm diameter probe. The redox mediator was delivered from one barrel of the SICM probe. This has several advantages compared to the inclusion of the redox probe in bulk solution. First, uptake detection is feasible; second, the cell is exposed to the redox mediator in minimum amounts; third, both SECM and SICM channel can be calibrated at each measurement. Consequently, a faster uptake rate of $[Ru(NH_3)_6]^{3+}$ at the root hair tip than at the cell body was observed in an SECM map. The higher uptake rate was postulated to occur as a result of several possible factors, including a higher density of membrane transport proteins, a typically looser membrane, or heterogeneity of surface charge density. The uptake rate at the cell tip and body were calculated to be $0.27 \pm 0.05\ cm^{-1}\ s^{-1}$ and $0.22 \pm 0.05\ cm^{-1}\ s^{-1}$ respectively (Fig. 16).

3 Future Directions

This chapter has provided an overview of advances in bringing quantitative measurement of both ions and electrons to SICM. Select examples that highlight applications and the present state of the art have also been presented. Efforts to bring quantitative methods to SICM have benefitted greatly from a deeper history of quantitative modeling and experiment with SECM. One of the most significant advantages of SICM relative to SECM is the opportunity to produce probes of small (~100 nm) dimension with little effort (compared to SECM). Incorporating SECM measurements into SICM has opened new avenues for electrochemical measurement as well. Significant experimental space remains in developing quantitative tools with SICM based approaches, both through nanopipette modification and in instrumental (hardware and software) advances. As these methods become more quantitative and commonplace, true nanoscale electrochemical imaging promises to open new vistas in electrochemical studies of biological and materials systems. As such, quantitation with SICM proves an important tool to address pressing challenges in understanding biological processes and chemical reactions important for new energy strategies.

Acknowledgement We acknowledge support from the National Science Foundation (CHE-1808133; CBET-1726642) and the National Institutes of Health (NINDS 5R01NS105888) to L.A.B.

References

1. Zhu C, Zhou L, Choi M, Baker LA (2018) Mapping surface charge of individual microdomains with scanning ion conductance microscopy. ChemElectroChem 5:2986–2990
2. Plett T, Shi W, Zeng Y, Mann W, Vlassiouk I, Baker LA, Siwy ZS (2015) Rectification of nanopores in aprotic solvents-transport properties of nanopores with surface dipoles. Nanoscale 7:19080–19091
3. Sa N, Baker LA (2011) Rectification of nanopores at surfaces. J Am Chem Soc 133:10398–10401
4. Deng XL, Takami T, Son JW, Kang EJ, Kawai T, Park BH (2014) Effect of concentration gradient on ionic current rectification in polyethyleneimine modified glass nano-pipettes. Sci Rep 4:1–7
5. Sa N, Lan WJ, Shi W, Baker L (2013) Rectification of ion current in nanopipettes by external substrates. ACS Nano 7:11272–11282
6. Liu S, Dong Y, Zhao W, Xie X, Ji T, Yin X, Liu Y, Liang Z, Momotenko D, Liang D, Girault HH, Shao Y (2012) Studies of ionic current rectification using polyethyleneimines coated glass nanopipettes. Anal Chem 84:5565–5573
7. Chen C-C, Derylo MA, Baker LA (2009) Measurement of ion currents through porous membranes with SICM. Anal Chem 81:4742–4751
8. Chen CC, Zhou Y, Baker LA (2012) Scanning ion conductance microscopy. Annu Rev Anal Chem 5:207–228
9. Choi M, Baker LA (2018) Biphasic-scanning ion conductance microscopy. Anal Chem 90:11797–11801
10. Huang K, Zhou L, Alanis K, Hou J, Baker LA (2020) Imaging effects of hyperosmolality on individual tricellular junctions. Chem Sci 11:1307–1315
11. Zhou L, Zhou Y, Baker L (2014) Measuring ions with scanning ion conductance microscopy. Electrochem Soc Interface 23:47–52
12. Zhou L, Gong Y, Hou J, Baker LA (2017) Quantitative visualization of nanoscale ion transport. Anal Chem 89:13603–13609

13. Chen CC, Zhou Y, Morris CA, Hou J, Baker LA (2013) Scanning ion conductance microscopy measurement of paracellular channel conductance in tight junctions. Anal Chem 85:3621–3628
14. Zhou Y, Chen CC, Weber AE, Zhou L, Baker LA (2014) Potentiometric-scanning ion conductance microscopy. Langmuir 30:5669–5675
15. Zhang Y, Takahashi Y, Hong SP, Liu F, Bednarska J, Goff PS, Novak P, Shevchuk A, Gopal S, Barozzi I, Magnani L, Sakai H, Suguru Y, Fujii T, Erofeev A, Gorelkin P, Majouga A, Weiss DJ, Edwards C, Ivanov AP, Klenerman D, Sviderskaya EV, Edel JB, Korchev Y (2019) High-resolution label-free 3D mapping of extracellular pH of single living cells. Nat Commun 10:5610
16. Nadappuram BP, McKelvey K, Al Botros R, Colburn AW, Unwin PR (2013) Fabrication and characterization of dual function nanoscale pH-scanning ion conductance microscopy (SICM) probes for high resolution pH mapping. Anal Chem 85:8070–8074
17. Morris CA, Chen C-C, Ito T, Baker LA (2013) Local pH measurement with scanning ion conductance microscopy. J Electrochem Soc 160:H430–H435
18. Macazo FC, White RJ (2016) Bioinspired protein channel-based scanning ion conductance microscopy (bio-SICM) for simultaneous conductance and specific molecular imaging. J Am Chem Soc 138:2793–2801
19. Shi W, Zeng Y, Zhu C, Xiao Y, Cummins TR, Hou J, Baker LA (2018) Characterization of membrane patch-ion channel probes for scanning ion conductance microscopy. Small 14:1702945
20. Shi W, Sa N, Thakar R, Baker LA (2015) Nanopipette delivery: influence of surface charge. Analyst 140:4835–4842
21. Siwy S, Gu C, Spohr HA, Baur D, Wolf-Reber A, Spohr R, Apel P, Korchev YE (2002) Rectification and voltage gating of ion currents in a nanofabricated pore. Eurphys Lett 60:349–355
22. Sa N, Baker LA (2013) Experiment and simulation of ion transport through nanopipettes of well-defined conical geometry. J Electrochem Soc 160:H376–H381
23. McKelvey K, Kinnear SL, Perry D, Momotenko D, Unwin PR (2014) Surface charge mapping with a nanopipette. J Am Chem Soc 136:13735–13744
24. Page A, Perry D, Young P, Mitchell D, Frenguelli BG, Unwin PR (2016) Fast nanoscale surface charge mapping with pulsed-potential scanning ion conductance microscopy. Anal Chem 88:10854–10859
25. Perry D, Paulose Nadappuram B, Momotenko D, Voyias PD, Page A, Tripathi G, Frenguelli BG, Unwin PR (2016) Surface charge visualization at viable living cells. J Am Chem Soc 138:3152–3160
26. Cremin K, Jones BA, Teahan J, Meloni GN, Perry D, Zerfass C, Asally M, Soyer OS, Unwin PR (2020) Scanning ion conductance microscopy reveals differences in the ionic environments of gram-positive and negative bacteria. Anal Chem 92(24):16024–16032
27. Clarke RW, Zhukov A, Richards O, Johnson N, Ostanin V, Klenerman D (2013) Pipette-surface interaction: current enhancement and intrinsic force. J Am Chem Soc 135:322–329
28. Klausen LH, Fuhs T, Dong M (2016) Mapping surface charge density of lipid bilayers by quantitative surface conductivity microscopy. Nat Commun 7:12447
29. Fuhs T, Klausen LH, Sønderskov SM, Han X, Dong M (2018) Direct measurement of surface charge distribution in phase separating supported lipid bilayers. Nanoscale 10:4538–4540
30. Hansma PK, Drake B, Marti O, Gould SA, Prater CB (1989) The scanning ion-conductance microscope. Science 243:641–643
31. Chen C-C, Zhou Y, Baker LA (2011) Single-nanopore investigations with ion conductance microscopy. ACS Nano 5:8404–8411
32. Zhou Y, Chen C-C, Baker LA (2012) Heterogeneity of multiple-pore membranes investigated with ion conductance microscopy. Anal Chem 84:3003–3009
33. Chen CC, Baker LA (2011) Effects of pipette modulation and imaging distances on ion currents measured with scanning ion conductance microscopy (SICM). Analyst 136:90–97

34. Zhou L, Zhou Y, Shi W, Baker LA (2015) Alternating current potentiometric scanning ion conductance microscopy (AC-PSICM). J Phys Chem C 119:14392–14399
35. Perry D, Botros RA, Momotenko D, Kinnear SL, Unwin PR (2015) Simultaneous nanoscale surface charge and topographical mapping. ACS Nano 9:7266–7276
36. Zhou L, Gong Y, Sunq A, Hou J, Baker LA (2016) Capturing rare conductance in epithelia with potentiometric-scanning ion conductance microscopy. Anal Chem 88:9630–9637
37. Lacaz-Vieira F (1997) Calcium site specificity early Ca^{2+} related tight junction events. J Gen Physiol 110:727–740
38. Nascimento RAS, Özel RE, Mak WH, Mulato M, Singaram B, Pourmand N (2016) Single cell "glucose nanosensor" verifies elevated glucose levels in individual cancer cells. Nano Lett 16:1194–1200
39. Umehara S, Karhanek M, Davis RW, Pourmand N (2009) Label-free biosensing with functionalized nanopipette probes. Proc Natl Acad Sci 106:4611–4616
40. Vilozny B, Actis P, Seger RA, Vallmajo-Martin Q, Pourmand N (2011) Reversible cation response with a protein-modified nanopipette. Anal Chem 83:6121–6126
41. Actis P, Jejelowo O, Pourmand N (2010) Ultrasensitive mycotoxin detection by STING sensors. Biosens Bioelectron 26:333–337
42. Actis P, Rogers A, Nivala J, Vilozny B, Seger RA, Jejelowo O, Pourmand N (2011) Reversible thrombin detection by aptamer functionalized STING sensors. Biosens Bioelectron 26:4503–4507
43. Fu Y, Tokuhisa H, Baker LA (2009) Nanopore DNA sensors based on dendrimer-modified nanopipettes. Chem Commun 32:4877–4879
44. Shi W, Zeng Y, Zhou L, Xiao Y, Cummins TR, Baker LA (2016) Membrane patches as ion channel probes for scanning ion conductance microscopy. Faraday Discuss 193:81–97
45. Gorelik J et al (2002) Ion channels in small cells and subcellular structures can be studied with a smart patch-clamp system. Biophys J 83:3296–3303
46. Actis P, Vilozny B, Seger RA, Li X, Jejelowo O, Rinaudo M, Pourmand N (2011) Voltage-controlled betal binding on polyelectrolyte-functionalized nanopores. Langmuir 27:6528–6533
47. Sa N, Fu Y, Baker LA (2010) Reversible cobalt ion binding to imidazole-modified nanopipettes. Anal Chem 82:9963–9966
48. Wen L, Hou X, Tian Y, Nie FQ, Song Y, Zhai J, Jiang L (2010) Bioinspired smart gating of nanochannels toward photoelectric-conversion systems. Adv Mater 22:1021–1024
49. Momotenko D, McKelvey K, Kang M, Meloni GN, Unwin PR (2016) Simultaneous interfacial reactivity and topography mapping with scanning ion conductance microscopy. Anal Chem 88:2838–2846
50. Kwak J-Y, Bard AJ (1989) Scanning electrochemical microscopy. Theory of the feedback mode. Anal Chem 61:1221–1227
51. Engstrom RC, Weber M, Wunder DJ, Burgess R, Winquist S (1986) Measurements within the diffusion layer using a microelectrode probe. Anal Chem 58:844–848
52. Takahashi Y, Hirano Y, Yasukawa T, Shiku H, Yamada H, Matsue T (2006) Topographic, electrochemical, and optical images captured using standing approach mode scanning electrochemical/optical microscopy. Langmuir 22:10299–10306
53. Ma W, Hu K, Chen Q, Zhou M, Mirkin MV, Bard AJ (2017) Electrochemical size measurement and characterization of electrodeposited platinum nanoparticles at nanometer resolution with scanning electrochemical microscopy. Nano Lett 17:4354–4358
54. Walsh DA, Fernández JL, Mauzeroll J, Bard AJ (2005) Scanning electrochemical microscopy.
55. Fabrication and characterization of micropipet probes. Anal Chem 77:5182–5188
55. Comstock DJ, Elam JW, Pellin MJ, Hersam MC (2010) Integrated ultramicroelectrode–nanopipet probe for concurrent scanning electrochemical microscopy and scanning ion conductance microscopy. Anal Chem 82:1270–1276

56. Morris CA, Chen C-C, Baker LA (2012) Transport of redox probes through single pores measured by scanning electrochemical-scanning ion conductance microscopy (SECM-SICM). Analyst 137:2933–2938
57. Wang Y, Noel JM, Velmurugan J, Nogala W, Mirkin MV, Lu C, Guille Collignon M, Lemaitre F, Amatore C (2012) Nanoelectrodes for determination of reactive oxygen and nitrogen species inside murine macrophages. Proc Natl Acad Sci 109:11534–11539
58. Takahashi Y, Shevchuk AI, Novak P, Murakami Y, Shiku H, Korchev YE, Matsue T (2010) Simultaneous noncontact topography and electrochemical imaging by SECM/SICM featuring ion current feedback regulation. J Am Chem Soc 132:10118–10126
59. Page A, Kang M, Armitstead A, Perry D, Unwin PR (2017) Quantitative visualization of molecular delivery and uptake at living cells with self-referencing scanning ion conductance microscopy-scanning electrochemical microscopy. Anal Chem 89:3021–3028
60. Shi W, Baker LA (2015) Imaging heterogeneity and transport of degraded nafion membranes. RSC Adv 5:99284–99290
61. Chen P, McCreery RL (1996) Control of electron transfer kinetics at glassy carbon electrodes by specific surface modification. Anal Chem 68:3958–3965
62. McCreery RL (2008) Advanced carbon electrode materials for molecular electrochemistry. Chem Rev 108:2646–2687
63. Takahashi Y, Shevchuk AI, Novak P, Zhang Y, Ebejer N, Macpherson JV, Unwin PR, Pollard AJ, Roy D, Clifford CA, Shiku H, Matsue T, Klenerman D, Korchev YE (2011) Multifunctional nanoprobes for nanoscale chemical imaging and localized chemical delivery at surfaces and interfaces. Angew Chem Int Edit 50:9638–9642
64. Kim YT, Scarnulis DM, Ewing AG (1986) Carbon-ring electrodes with 1-μm tip diameter. Anal Chem 58:1782–1786
65. Wong DKY, Xu LYF (1995) Voltammetric studies of carbon disk electrodes with submicrometer-sized structural diameters. Anal Chem 67:4086–4090
66. McNally M, Wong DKY (2001) An in vivo probe based on mechanically strong but structurally small carbon electrodes with an appreciable surface area. Anal Chem 73:4793–4800
67. Thakar R, Weber AE, Morris CA, Baker LA (2013) Multifunctional carbon nanoelectrodes fabricated by focused ion beam milling. Analyst 138:5973–5982
68. O'Connell MA, Wain AJ (2014) Mapping electroactivity at individual catalytic nanostructures using high-resolution scanning electrochemical–scanning ion conductance microcopy. Anal Chem 86:12100–12107
69. O'Connell MA, Lewis JR, Wain AJ (2015) Electrochemical imaging of hydrogen peroxide generation at individual gold nanoparticles. Chem Commun 51:10314–10317
70. Sanchez-Sanchez CM, Bard AJ (2009) Hydrogen peroxide production in the oxygen reduction reaction at different electrocatalysts as quantified by scanning electrochemical microscopy. Anal Chem 81:8094–8100
71. Şen M, Takahashi Y, Matsumae Y, Horiguchi Y, Kumatani A, Ino K, Shiku H, Matsue T (2015) Improving the electrochemical imaging sensitivity of scanning electrochemical microscopy-scanning ion conductance microscopy by using electrochemical Pt deposition. Anal Chem 87:3484–3489
72. Wilde P, Quast T, Aiyappa HB, Chen YT, Botz A, Tarnev T, Marquitan M, Feldhege S, Lindner A, Andronescu C, Schuhmann W (2018) Towards reproducible fabrication of nanometre-sized carbon electrodes: optimisation of automated nanoelectrode fabrication by means of transmission electron microscopy. ChemElectroChem 5:3083–3088
73. Takahashi Y, Ida H, Matsumae Y, Komaki H, Zhou Y, Kumatani A, Kanzaki M, Shiku H, Matsue T (2017) 3D electrochemical and ion current imaging using scanning electrochemical–scanning ion conductance microscopy. Phys Chem Chem Phys 19:26728–26733
74. Bard AJ, Mirkin MV (2012) Scanning electrochemical microscopy. CRC Press
75. Szabo A (1987) Thoery of the current at microelectrodes application to ring electrodes. J Phys Chem 91:3108–3111

76. Laforge FO, Velmurugan J, Wang Y, Mirkin MV (2009) Nanoscale imaging of surface topography and reactivity with the scanning electrochemical microscope. Anal Chem 81:3143–3150
77. McKelvey K, Edwards MA, Unwin PR (2010) Intermittent contact-scanning electrochemical microscopy (IC-SECM): a new approach for tip positioning and simultaneous imaging of interfacial topography and activity. Anal Chem 82:6334–6337
78. Fernández JL, Walsh DA, Bard AJ (2005) Thermodynamic guidelines for the design of bimetallic catalysts for oxygen electroreduction and rapid screening by scanning electrochemical microscopy. M−Co (M: Pd, Ag, Au). J Am Chem Soc 127:357–365
79. Lv J, Qian R-C, Hu Y-X, Liu S-C, Cao Y, Zheng Y-J, Long Y-T (2016) A precise pointing nanopipette for single-cell imaging via electroosmotic injection. Chem Commun 52:13909–13911
80. Sun M, Kaplan SV, Gehringer RC, Limbocker RA, Johnson MA (2014) Localized drug application and sub-second voltammetric dopamine release measurements in a brain slice perfusion device. Anal Chem 86:4151–4156
81. Davoodi A, Farzadi A, Pan J, Leygraf C, Zhu Y (2008) Developing an AFM-based SECM system; instrumental setup, SECM simulation, characterization, and calibration. J Electrochem Soc 155:C474–C485
82. Sanchez-Sanchez CM, Rodrıguez-Lopez J, Bard AJ (2008) Scanning electrochemical microscopy. 60. Quantitative calibration of the SECM substrate generation/tip collection mode and its use for the study of the oxygen reduction mechanism. Anal Chem 80:3254–3260
83. Zhu W, Michalsky R, Metin O, Lv H, Guo S, Wright CJ, Sun X, Peterson AA, Sun S (2013) Monodisperse Au nanoparticles for selective electrocatalytic reduction of CO_2 to CO. J Am Chem Soc 135:16833–16836
84. Rogers C, Perkins WS, Veber G, Williams TE, Cloke RR, Fischer FR (2017) Synergistic enhancement of electrocatalytic CO_2 reduction with gold nanoparticles embedded in functional graphene nanoribbon composite electrodes. J Am Chem Soc 139:4052–4061
85. Jin R, Ye X, Fan J, Jiang D, Chen H-Y (2019) In situ imaging of photocatalytic activity at titanium dioxide nanotubes using scanning ion conductance microscopy. Anal Chem 91:2605–2609
86. Choi M-H, Siepser NP, Jeong S, Wang Y, Jagdale G, Ye X, Baker LA (2020) Probing single-particle electrocatalytic activity at facet-controlled gold nanocrystals. Nano Lett 20:1233–1239
87. Zhou M, Bao S, Bard AJ (2019) Probing size and substrate effects on the hydrogen evolution reaction by single isolated Pt atoms, atomic clusters, and nanoparticles. J Am Chem Soc 141:7327–7332
88. Zhang D, Chen J, Deng P, Wang X, Li Y, Wen T, Li Y, Xiang Q, Liao Y (2019) Hydrogen evolution promotion of Au-nanoparticles-decorated TiO_2 nanotube arrays prepared by dip-loading approach. J Am Ceram Soc 102:5873–5880
89. Garlyyev B, Kratzl K, Ruck M, Michalicka J, Fichtner J, Macak JM, Kratky T, Gunther S, Cokoja M, Bandarenka AS, Gagliardi A, Fischer RA (2019) Optimizing the size of platinum nanoparticles for enhanced mass activity in the electrochemical oxygen reduction reaction. Angew Chem Int Ed 58:9596–9600
90. Li Z, Ma Z, Wen Y, Ren Y, Wei Z, Xing X, Sun H, Zhang YW, Song W (2018) Copper nanoflower assembled by sub-2 nm rough nanowires for efficient oxygen reduction reaction: high stability and poison resistance and density functional calculations. ACS Appl Mater Interfaces 10:26233–26240
91. Fichtner J, Garlyyev B, Watzele S, El-Sayed HA, Schwammlein JN, Li WJ, Maillard FM, Dubau L, Michalicka J, Macak JM, Holleitner A, Bandarenka AS (2019) Top-down synthesis of nanostructured platinum-lanthanide alloy oxygen reduction reaction catalysts: Pt xPr/C as an example. ACS Appl Mater Interfaces 11:5129–5135

92. Velmurugan J, Agrawal A, An S, Choudhary E, Szalai VA (2017) Fabrication of scanning electrochemical microscopy-atomic force microscopy probes to image surface topography and reactivity at the nanoscale. Anal Chem 89:2687–2691
93. Park HS, Jang JH (2016) Applications of scanning electrochemical microscopy (SECM) coupled to atomic force microscopy with sub-micrometer spatial resolution to the development and discovery of electrocatalysts. J Electrochem Sci Technol 7:316–326
94. Knittel P, Mizaikoff B, Kranz C (2016) Simultaneous nanomechanical and electrochemical mapping: combining peak force tapping atomic force microscopy with scanning electrochemical microscopy. Anal Chem 88:6174–6178
95. Nebel M, Erichsen T, Schuhmann W (2014) Constant-distance mode SECM as a tool to visualize local electrocatalytic activity of oxygen reduction catalysts. Beilstein J Nanotechnol 5:141–151
96. Kurulugama RT, Wipf DO, Takacs SA, Pongmayteegul S, Garris PA, Baur JE (2005) Scanning electrochemical microscopy of model neurons: constant distance imaging. Anal Chem 77:1111–1117
97. Haraguchi D, Yasunaga K, Yamada H (2016) A potassium-sensitive micro/sub-micro-electrode for shear force-based scanning electrochemical microscopy. Trans Mat Res Soc Jpn 41:89–92
98. Lazenby RA, McKelvey K, Peruffo M, Baghdadi M, Unwin PR (2013) Nanoscale intermittent contact-scanning electrochemical microscopy. J Solid State Electr 17:2979–2987
99. Tomlinson LI, Patten HV, Green BL, Iacobini J, Meadows KE, McKelvey K, Unwin PR, Newton ME, Macpherson JV (2016) Intermittent-contact scanning electrochemical microscopy (IC-SECM) as a quantitative probe of defects in single crystal boron doped diamond electrodes. Electroanalysis 28:2297–2302
100. Etienne M, Schulte A, Schuhmann W (2004) High resolution constant-distance mode alternating current scanning electrochemical microscopy (AC-SECM). Electrochem Commun 6:288–293
101. Eckhard K, Schuhmann W (2008) Alternating current techniques in scanning electrochemical microscopy (AC-SECM). Analyst 133:1486–1497
102. Eckhard K, Etienne M, Schulte A, Schuhmann W (2007) Constant-distance mode AC-SECM for the visualisation of corrosion pits. Electrochem Commun 9:1793–1797
103. Nadappuram BP, McKelvey K, Byers JC, Güell AG, Colburn AW, Lazenby RA, Unwin PR (2015) Quad-barrel multifunctional electrochemical and ion conductance probe for voltammetric analysis and imaging. Anal Chem 87:3566–3573
104. Yeager E (1984) Electrocatalysts for O_2 reduction. Electrochemica Acta 29:1527–1537
105. Walling C (1975) Fentons-reagent-revisited. Acc Chem Res 8:125–131
106. Curtin DE, Lousenberg RD, Henry TJ, Tangeman PC, Tisack ME (2004) Advanced materials for improved PEMFC performance and life. J Power Sources 131:41–48
107. Graham JM, Higgins JA (1993) Biomembrane protocols. Humana Press
108. Casey JR, Grinstein S, Orlowski J (2010) Sensors and regulators of intracellular pH. Nat Rev Mol Cell Biol 11:50–61
109. Gerweck LE, Seetharaman K (1996) Cellular pH gradient in tumor versus normal tissue potential exploitation for the treatment of cancer. Cancer Res 56:1194–1198
110. Tanabe H, Togashi K, Misawa T (1998) In situ pH measurements during localised corrosion of type 316LN stainless steel using scanning electrochemical microscopy. J Mater Sci Lett 17:551–553
111. Zhao M-C, Liu M, Song G-L, Atrens A (2008) Influence of pH and chloride ion concentration on the corrosion of Mg alloy ZE41. Corros Sci 50:3168–3178
112. Zhang J, Lin C, Feng Z, Tian Z (1998) Mechanistic studies of electrodeposition for bioceramic coatings of calcium phosphates by an in situ pH-microsensor technique. J Electroanal Chem 452:235–240
113. Natter H, Hempelmann R (1996) Nanocrystalline copper by pulsed electrodeposition: the effects of organic additives, bath temperature, and pH. J Phys Chem 100:19525–19532

114. Unwin PR, Compton RG (1990) Effect of carboxylic acids on the dissolution of calcite in aqueous solutions. J Chem Soc Faraday Trans 86:1517–1525
115. Miesenbock G, Angelis DA, Rothman JE (1998) Visualizing secretion and synaptic transmission with pH-sensitive green fluorescent proteins. Nature 394:192–195
116. Rudd NC, Cannan S, Bitziou E, Ciani I, Whitworth AL, Unwin PR (2005) Fluorescence confocal laser scanning microscopy as a probe of pH gradients in electrode reactions and surface activity. Anal Chem 77:6205–6217
117. Thomas JA, Buchsbaum RN, Zimniak A, Racker E (1979) Intracellular pH measurements in ehrlich ascites tumor cells utilizing spectroscopic probes generated in situ. Biochemist 18:2210–2218
118. Bishnoi SW, Rozell CJ, Levin CS, Gheith MK, Johnson BR, Johnson DH, Halas NJ (2006) All-optical nanoscale pH meter. Nano Lett 6:1687–1692
119. Ang PK, Chen W, Wee ATS, Loh KP (2008) Solution-gated epitaxial graphene as pH sensor. J Am Chem Soc 130:14392–14393
120. Zhang X, Ogorevc B, Wang J (2002) Solid-state pH nanoelectrode based on polyaniline thin film electrodeposited onto ion-beam etched carbon fiber. Anal Chim Acta 452:1–10
121. Bashir R, Hilt JZ, Elibol O, Gupta A, Peppas NA (2002) Micromechanical cantilever as an ultrasensitive pH microsensor. Appl Phys Lett 81:3091–3093
122. Ndobo-Epoy J-P, Lesniewska E, Guicquero J-P (2007) Nano-pH sensor for the study of reactive materials. Anal Chem 79:7560–7564
123. Horrocks BR, Mirkin MV, Pierce DT, Bard AJ (1993) Scanning electrochemical microscopy. 19. Ion-selective potentiometric microscopy. Anal Chem 65:1213–1224
124. Bitziou E, O'Hare D, Patel BA (2008) Simultaneous detection of pH changes and histamine release from oxyntic glands in isolated stomach. Anal Chem 80:8733–8740
125. Elsen HA, Monson CF, Majda M (2009) Effects of electrodeposition conditions and protocol on the properties of iridium oxide pH sensor electrodes. J Electrochem Soc 156:F1–F6
126. Stumm W, Morgan JJ (2012) Aquatic chemistry: chemical equilibria and rates in natural waters, vol 126. Wiley
127. Genies EM, Tsintavis C (1985) Redox mechanism and electrochemical behaviour of polyaniline deposits. J Electroanal Chem 195:109–128
128. Yasukawa T, Uchida I, Matsue T (1998) Permeation of redox species through a cell membrane of a single, living algal protoplast studied by microamperometry. Biochim Biophys Acta 1369:152–158
129. Gonsalves M, Barker AL, Macpherson JV, Unwin PR, O'Hare D, Winlove CP (2000) Scanning electrochemical microscopy as a local probe of oxygen permeability in cartilage. Biophys J 78:1578–1588

BIOREV (2022) 3: 123–138
https://doi.org/10.1007/11663_2021_8

Published online: 12 March 2022

Nanoscale Electrophysiology Using Scanning Ion Conductance Microscopy

Pavel Novak, Andrew Shevchuk, and Yuri Korchev

Contents

Abstract Scanning ion conductance microscopy shares many aspects of its operation with an electrophysiological technique known as the "patch-clamp", yet the marriage of the two techniques, referred to as "smart patch-clamp", remains perceived as challenging both in the electrophysiological community and in the SICM community. Here we set out a number of ways how an existing setup, whether it is an SICM or a patch-clamp setup, can be turned into a "smart patch-clamp" setup, and outline strategies for achieving required levels of noise and interference. We describe typical scenarios where we believe the high-resolution spatial awareness of the "smart patch-clamp" is needed as opposed to the low-resolution of conventional patch-clamp. This is followed by a detailed description of the smart patch-clamp procedure and latest addition of controlled nanopipette breaking, offering

P. Novak (✉)
National University of Science and Technology, MISiS, Moscow, Russia

Imperial College London, London, UK
e-mail: p.novak@imperial.ac.uk

A. Shevchuk and Y. Korchev
Imperial College London, London, UK

improvement of the throughput, and partially counteracting the drop in probability of seeing channel activity with increasing spatial resolution. Finally, we discuss the issue of access resistance and conclude that the combination of SICM and patch-clamp technique offers yet unmatched fidelity when it comes to recording true ion currents and membrane voltages with nanoscale resolution.

Keywords Access resistance · Electrophysiology · Gigaseal · Ion channels · Patch-clamp technique · Pipette breaking · Scanning ion conductance microscopy · Smart patch-clamp · Spatially resolved ion channel recordings · Structure-function relationship

1 Introduction

The fact that the access resistance of glass pipettes filled with electrolyte increases as the pipette approaches the membrane of a cell has been known to electrophysiologists well before the invention of scanning ion conductance microscopy. Many electrophysiologists who performed "patch-clamp" measurements [1] using manual x-y-z micromanipulators remember the rule of thumb which says keep approaching towards the cell until the amplitude of the pipette current pulse in response to a voltage pulse stimulus reduces to about one half. In the terminology of SICM, this essentially means to stop the approach at a 50% setpoint. At this point, the application of suction to the interior of the pipette would pull in the patch of the membrane in contact with the pipette, and with a bit of a luck, form a tight seal with the glass walls of the pipette. The patch of membrane encircled by the opening of the pipette tip would become mechanically and electrically isolated from the surrounding bath solution or, as the name of the technique suggests, "clamped" to the pipette tip. The resistance of the seal between the glass wall of the pipette and the membrane can reach tens of gigaohms, a state which electrophysiologists refer to as "gigaseal". At this level of electrical isolation from the surrounding environment, the ion current flowing from the pipette directly into the bath solution (known among electrophysiologists as the "leak current") is reduced below the level at which currents flow through individual ion channels in the membrane patch. Therefore, individual ion channels, which are normally obscured by the leak current, become detectable. The invention of the patch-clamp technique revolutionised the field of electrophysiology and brought it from the tissue and cellular level down to the molecular level. It is still considered the gold standard when it comes to accurate measurement of membrane currents and potentials. With so many similarities between the patch-clamp technique and scanning ion conductance microscopy, one would have thought that somebody in the electrophysiological community would have been the first to come up with the idea to use the changes in the access resistance of the pipette approaching a cell membrane as a feedback signal for a new type of scanning probe microscopy. Perhaps it was the enormous potential of the patch-clamp technique for new discoveries in neuroscience, cardiovascular sciences, and drug discovery that prevented the community from taking a step back and think out of the "electrophysiological box". It had to wait 8 more years after the invention of patch-clamp for a team of physicists, not electrophysiologists, led by Paul Hansma, to mount a nanopipette onto a 3D nanopositioning stage and establish scanning ion conductance microscopy in their seminal paper [2]. And then it would take almost another decade for scanning ion conductance microscopy to make a full circle back to electrophysiology and turn the conventional patch-clamp technique into a spatially resolved "smart" patch-clamp technique [3, 4] as it is now regularly referred to in the SICM community.

2 The Smart Patch-Clamp Technique

2.1 The Setup

Thanks to the similarities between patch-clamp and SICM one can actually build a SICM setup simply by upgrading certain parts of an existing patch-clamp setup (Fig. 1). Vice versa, most SICM setups can be used for patch-clamp after a careful review of grounding and screening issues to achieve noise levels required for low-noise single-channel recordings. The core difference between the conventional and the smart patch-clamp technique is in the positioning of the patch pipette to the place where the ion channel recording is to be made. In smart-patch, the position is selected based on the SICM topographical image of the cell membrane acquired using the same pipette with nanoscale resolution. This is compared to conventional patch-clamp where the pipette is guided on the basis of a bright-field, diffraction-limited optical image. It is worth to note here that the cell plasma membrane is barely visible in liquid due to its very low refractive index. Another major difference is that in the traditional SICM setup the nanopipette is mounted vertically above the sample, while conventional patch-clamp prefers to have the pipette under an angle of 30–60° with respect to the sample plane. The angled pipette mount in patch-clamp setups makes it easier to identify where and how far from surface the pipette tip is. This enables navigating the pipette towards the target and also allows using upright optical microscopes (with the objective above the sample) – a necessity when recording in tissues such as brain slices. In smart patch-clamp, the SICM feedback controller takes full control over navigating the nanopipette to the target point selected by the user, hence there is no need to mount the nanopipette at an angle.

The easiest way to convert an existing patch-clamp setup into a smart one is to mount nanopipette in a standard patch-clamp pipette holder onto an x-y-z piezo-actuator cube mounted onto the existing motorised patch-clamp micromanipulator (Fig. 1a, b) as it was done in the first implementation of the smart patch-clamp technique [3, 4]. The 3D piezo-actuator needs to be powered by high-voltage piezo amplifiers and a scanner controller, all commercially available from a number of

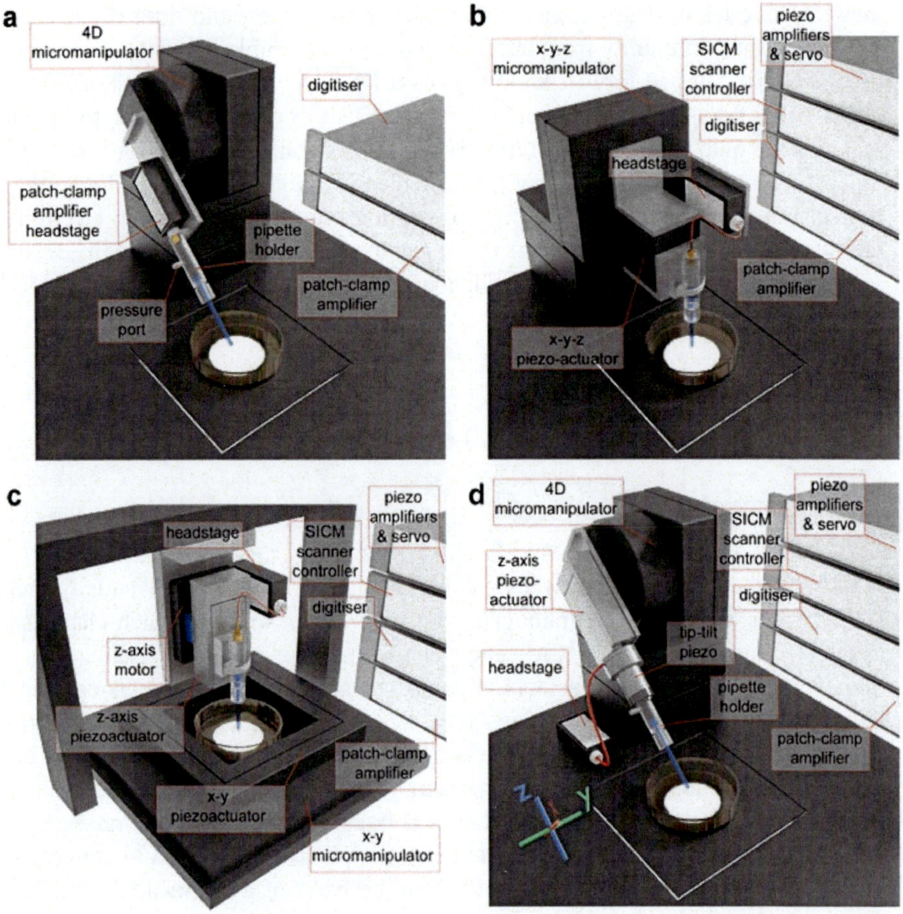

Fig. 1 Smart versus conventional patch-clamp setup. (**a**) Typical conventional patch-clamp setup with motorised 4D micromanipulator capable of moving in x-y-z and angled approach along the angled pipette axis. (**b**) Simplest implementation of a smart patch-clamp setup by upgrading an existing conventional patch-clamp in (**a**). (**c**) Typical setup of a contemporary SICM system operating in the hopping mode capable of performing smart patch-clamp recordings. (**d**) Angle-scan setup allowing the use of an upright microscope and smart patch-clamp of vertically oriented cellular structures. The orientation of the x-y-z scanning coordinate system is indicated in the bottom left corner. Please note that wired connections between some of the components of the setup were omitted for clarity

companies operating in the field of SICM. All the other components of a typical patch-clamp setup, such as an antivibration table, screening Faraday cage, low-noise patch-clamp amplifier, and the digitiser are in principle the same as the ones used in a typical SICM setup. It needs to be noted that the easiest way of converting a patch-clamp setup into SICM is not necessarily the best one. Since coining the term smart patch-clamp, the SICM technique has gone through a number of iterations, and the preferable assembly of the scanning head typically uses separate piezo-actuators for x-y sample scanning and z-axis vertical pipette movement as described in [5] and illustrated in Fig. 1c. The decoupling between the x-y and z-axis actuators provides better image quality and speed than the coupled x-y-z cube, but makes the upgrade of existing patch-clamp setups slightly more complicated and expensive. Decoupled SICM scan heads that can be mounted on top of inverted microscope are commercially available for those not keen on converting their patch-clamp setups themselves.

Although the majority of cell-based electrophysiological studies will be best accommodated by a conventional SICM setup with a vertically oriented nanopipette, an angled approach may be still beneficial in tissues, non-transparent samples, or when recording from vertically oriented parts of cells not accessible to a vertically oriented pipette. For these rather exceptional situations, "an angled-scan" SICM setup can be used (Fig. 1d) [6]. It is, in principle, a pipette-scan system (as opposed to a sample scan system such as the one in Fig. 1c) with the x-y-z coordinate system rotated with respect to the optical microscope platform (Fig. 1d). Scanning in the x-y plane is performed by a special tip-tilt piezo-actuator mounted on a separate z-axis piezo-actuator. The tip-tilt piezo requires three driving signals in contrast to a standard x-y actuator which requires a slightly different computation inside the scanner controller compared to conventional scanning systems in Fig. 1b, d. Despite a slightly more complicated assembly, the angle-scan system is capable of producing images of complex structures such as inner ear hair cells with a resolution comparable to standard SICM setups. In addition, it allows performing patch-clamp recordings from structures such as intercalated discs in cardiac myocytes normally not accessible to a vertically oriented nanopipette [6].

2.2 *Noise and Interference*

Anyone who ever tried to perform low-noise patch-clamp recordings knows it is not a trivial task. The preparation usually involves revisiting grounding and power line connections to reduce the noise and electrical interference inherently present in the recording setup to acceptable levels. Harmonic components derived from 50 Hz power line frequency are a common issue. In a properly grounded and screened SICM setup the RMS current noise within 5 kHz bandwidth measured with a 100 MΩ nanopipette immersed in the bath solution (but far from sample surface) and a patch-clamp amplifier gain of 1 mV/pA using resistive feedback (500 MΩ) should not be higher than 1–1.2 pA_{rms}. Both SICM imaging and patch-clamp recording will benefit from a high signal-to-noise ratio, however SICM is usually capable of achieving good image quality even with noise levels that are not good enough for single-channel recordings. Achieving low-noise levels with a smart patch-clamp setup requires a bit more effort since more electrical equipment is added into the setup compared to conventional patch-clamp. However, the same basic rules for reduction of noise and interference apply in case of SICM as in patch-clamp recordings. All metal components in the vicinity of the nanopipette should be grounded to a single point connected to signal ground of the patch-clamp amplifier headstage. Grounding wiring should ideally form a star topology with the central node being the signal ground point of the amplifier headstage. Power or chassis ground should be kept separated from signal ground if possible. This is usually the standard with electrophysiological equipment, however not so much with nanopositioning equipment. Commercially available high-voltage amplifiers and servocontrollers needed for running piezo-actuators mostly have only the power line ground internally connected to the chassis, which is in turn connected to chassis of the piezo-actuators. One may need to experiment with grounding of piezo-actuators but in most cases the lowest noise is achieved when the chassis of all piezo-actuators is grounded to one point – the signal ground of the headstage. This should be the only point where the ground of the nanopositioning components and the signal ground meet. Alternatively, the high-voltage piezo amplifiers and servo controllers could be powered via an isolating transformer to separate the piezo chassis ground from the power line ground. Another option for avoiding or minimising connections between power line ground and signal ground is to ground the screening of certain cable connections only on one side of the connection. Trial and error is often the only way to find out which ground connections need to be broken and which need to remain. Further guidance regarding the issues of proper grounding, screening and ground loops can be found in [7, 8].

2.3 *The Need for Smart Patch-Clamp*

Before we dive deeper into the details of the smart patch-clamp technique and its derivatives it needs to be said that conventional patch-clamp setups are still perfectly suitable for many research questions in biophysics and cellular electrophysiology. Smart patch-clamp combines high spatial resolution with high sensitivity to ionic currents and therefore would be mostly effective in those questions of cell biology where certain functionality is clustered in submicron-sized membrane structures. Examples of structures where smart patch-clamp has already provided new insights or proven to be necessary are t-tubules and intercalated disks in cardiac myocytes, small synaptic terminals in neuronal cells, microvilli and primary cilia in epithelial cells (Fig. 2a) [3, 4, 6, 9–15]. It may seem that at least some of these structures could be labelled fluorescently and targeted using optical navigation by someone with years of experience in conventional patch-clamp technique. However, a smart patch-clamp study in hippocampal neurons showed that fluorescently labelled synaptic terminals could be hidden underneath other unlabelled structures (dendrites), which opens the possibility for mistakenly assigning a recording from other structures to a synaptic terminal if performed purely based on optical navigation [12]. Even when performing cell-attached patch-camp recording on a relatively large and easy-to-target neuronal soma there is a risk of accidentally recording from structures belonging to other neurons crossing the targeted neuronal soma (Fig. 2a). Intercalated disks of cardiomyocytes are another example of a relatively large structure that could be targeted using conventional angled patch-clamp. The angle-scan smart patch-clamp, however, allows identification of smaller sub-domains which may be

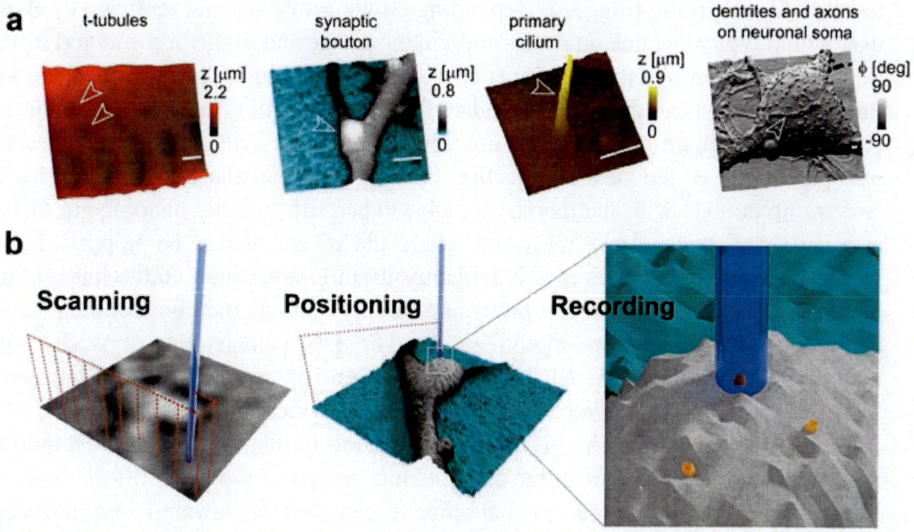

Fig. 2 The need for smart patch-clamp. (**a**) Examples of structures with particular electrophysiological functionality which are difficult, if not impossible, to target using the conventional patch-clamp technique. (**b**) Principle of the smart patch-clamp technique. An initial high-resolution scan in an area first identified using low-resolution optical methods (left) is followed by positioning the nanopipette on the structure based on high-resolution topography (middle), and finally a tight seal with the membrane is made to perform ion channel recordings (right). Scale bars 1 μm (**a**)

present (Fig. 1d, [6]), but cannot be identified using a conventional patch-clamp system. One of the most difficult structures to target – the primary cilium, a typically just 200 nm wide finger like structure, has been recorded using purely optical navigation [16], however the procedure involved suction of the ciliary tip into the pipette from a distance resulting in an unknown degree of bending of the whole structure, which is by some considered as mechanosensitive. Notably, only recordings only from the tip of the cilium were acquired using the conventional method. In contrast, the smart patch-clamp approach allows to obtain recordings from other parts of cilia as well, thanks to much higher spatial resolution [15]. Most importantly, something that is not immediately obvious from scientific papers but that many in the community around smart patch-clamp are well aware of is that the unwritten requirement for years of experience in the patch-clamp technique needed for targeting these small structures has been reduced to only weeks of experience thanks to the simplicity of the smart-patch protocol.

2.4 Standard Smart Patch-Clamp Protocol

In the traditional implementation of the smart patch-clamp, a high-resolution topographical image of the area is first obtained using SICM with typically a 100 nm-diameter or similar nanopipette. The topographical image is then used to select a location for recording (Fig. 2b). Depending on the actual scanner software used, the user would typically click on the location in the image and trigger a command from a pop-up menu to move the nanopipette to that location. Only the x-y coordinates of the location are used, the SICM distance feedback-control takes care of finding the correct z-coordinate while preventing contact with the cell membrane. Once the nanopipette is at the desired location, the distance feedback-control of SICM preventing contact with the membrane is switched off with the nanopipette locked at a certain distance above the point where the pipette would be stopped during imaging (in hopping mode this is typically the hop amplitude above the setpoint height). From this moment the procedure largely follows the conventional patch-clamp technique with few modifications. The DC voltage bias applied to the nanopipette needed for the SICM feedback-control is switched off and instead a square wave pulse of around 10 mV amplitude is typically applied to measure the pipette access resistance. Any DC offsets observed in the pipette current should be now zeroed, typically using the patch-clamp amplifier's pipette offset controls. Using the scanner software, the nanopipette can then be lowered in small steps (a distance equal to the inner diameter of the pipette is a good start) towards the cell membrane while monitoring the access resistance. However, instead of the 50% change when using angled micron-sized pipettes in conventional patch-clamp, with vertically positioned 100 nm-wide nanopipettes only a fractional increase in access resistance (<10%) is typically needed before suction can be applied to achieve a gigaseal. In practise, it is better to use the distance the nanopipette travelled below the usual imaging setpoint height to decide when one can try to start forming a gigaseal. With a 100 nm diameter nanopipette this would be around 300–1,000 nm below the imaging setpoint height, using a suction pressure up to around 10 kPa, both depending on the cell type and cell structure being targeted. As with conventional patch-clamp, few attempts may be needed to fine-tune the distances and suction pressures needed to obtain gigaseal when performing smart patch-clamp for the first time on a given cell type. One should start at the lower end of the range for both distance and pressure and gradually increase them if not successful. While this all may seem a bit random, particularly for those coming from "exact" fields of science, the success rate of getting a gigaseal with smart patch-clamp is typically relatively high, between 59–96% in cardiomyocytes [4], 67% in small synaptic boutons [12], and drops below 50% only in difficult cases such as the middle part of primary cilia (37.5%) [15]. At the lower end of the success rate spectrum are locations where the angle between the pipette axis and the membrane patch is significantly different from 90°, such as the opening of a t-tubule in a cardiomyocyte (59% [4]) or the above-mentioned middle part of a primary cilium [15]. In smart patch-clamp one has the option to select a membrane patch with more favourable spatial orientation with regard to the nanopipette axis using the information from the topographical image and thus one can consistently achieve a high success rate of patch-clamp recordings, something a user of conventional patch-clamp is unable to do.

To reach a high success rate, the initial topography image needs to be obtained with the nanopipette as far away from the cell membrane as possible to get a good image (in hopping mode SICM [5] this would correspond to a setpoint of around 0.3–0.4%) and as quickly as possible (under 15 min scan duration). In general, prolonged or repeated scanning reduces the chances of obtaining a gigaseal with the same nanopipette. This is believed to be caused by floating biological debris built-up on the glass which does not significantly impair imaging but seems to prevent the membrane from tightly adhering to the glass. Using freshly pulled nanopipettes (on the day of the patch-clamp experiment) typically results in a higher success rate of obtaining a gigaseal, while older nanopipettes seem to be less likely to get blocked during SICM imaging. This inverse relationship between the age of the nanopipette tip and the success rate of obtaining a gigaseal can be partially reverted using a trick referred to as pipette breaking.

2.5 *Smart Patch-Clamp with Controlled Nanopipette Breaking*

The importance of a clean, fresh glass surface for a high quality gigaseal has been known since the invention of the patch-clamp technique. Immediately after pipette fabrication the glass at the pipette tip is clean but often gets contaminated when crossing the water–air interface, where debris typically accumulates. Breaking the pipette tip after immersion into the bath by downward movement against the glass bottom of the recording chamber exposes new, clean glass surface with a reportedly higher chance of spontaneously forming a gigaseal [17]. At the same time, each time the pipette tip breaks, the diameter of the tip opening increases due to the conical shape of the pipette, resulting in a reduction of the pipette access resistance [17]. This old trick introduced first with a conventional patch-clamp setup has been "upgraded" to nanoscale resolution levels with the smart-patch technique targeting small synaptic boutons in hippocampal neurons [12].

The main motivation behind the reintroduction of pipette breaking into the smart patch-clamp technique was to address the fact that it was practically impossible to achieve whole-cell configuration using sharp 100 nm-diameter nanopipettes. The reasons for these limitations are not exactly known but are possibly due to the fact that the area of the membrane patch enclosed within the nanopipette opening is approximately two orders of amplitude smaller compared to conventional patch-clamp pipettes with diameter of around 1 μm. This would require suction pulses with 100 times higher pressure to achieve the same force acting on the membrane patch when trying to break into the whole-cell configuration, which in practise would be difficult to achieve without causing mechanical disturbance to nanopipette position. Many attempts to break into the whole-cell configuration with a 100 nm-wide nanopipette ended up in loss of the gigaseal, which is in agreement with the hypothesis of mechanical disturbance from the high-pressure suction pulses. In the small number of cases where the membrane patch may have broken into the whole-cell configuration, the configuration would turn out to be extremely short lived and to frequently revert back to the previous cell-attached configuration. Spontaneous closing of whole-cell configuration is well known to electrophysiologists, and it seems to be happening, not unexpectedly, at a much higher rate with the smaller, nanoscale-sized membrane patches. However, as we already mentioned, the use of nanoscale-sized nanopipettes is the main reason behind the significantly improved spatial resolution of the smart patch-clamp technique and necessary to distinguish accessible synaptic boutons from the inaccessible ones [12]. Whole-cell configuration, on the other hand, is the only way to gain access to the ion channels in the parts of synaptic membrane hidden from the nanopipette – such as the membrane facing the synaptic cleft. The answer to this dilemma lies in introducing nanopipette breaking after the initial high-resolution topographical scan with the nanopipette is obtained, and just before the formation of a gigaseal (Fig. 3). Since the initial scan is performed with a sharp nanopipette, the high spatial resolution of the method is preserved. The subsequent nanopipette breaking against a glass bottom of the recording chamber both widens the opening of the pipette and exposes fresh glass. Thanks to the vertical orientation of the nanopipette, the x-y position of the nanopipette tip remains the same, only its diameter and z-coordinate change, as proved by a subsequent rescan of the same area with the widened pipette [12]. Hence, the topographical image obtained with the nanopipette before breaking is still valid and can be used to position the now wider nanopipette with the same resolution as the original nanopipette. This would not be possible with the angled pipette used in conventional patch-clamp setups, where all three x-y-z coordinates of the tip change during pipette breaking. The success rate of obtaining a gigaseal with a broken nanopipette remains high (57% on synaptic boutons [12], 70% on cardiomyocytes [18]). Chances of obtaining a whole-cell configuration with a nanopipette widened from approximately 100 to 400 nm (Fig. 3b) increased from virtually zero to 41% [12]. Even in the cell-attached configuration, which can be easily achieved with a sharp nanopipette, the widening of the nanopipette tip to approximately 350 nm brought the benefit of increased probability of detecting voltage gated calcium channel (VGCC) activity in cardiomyocytes from negligible to 28% (Fig. 3c) [18]. Further widening to 600 nm resulted in 100% chance of observing VGCC activity and the possibility to see more than one channel in the membrane patch (Fig. 3c) [18]. This effectively increases throughput of the smart patch-clamp technique without sacrificing the high spatial resolution and as a result, nanopipette breaking after the initial topographical scan became practically a standard in smart patch-clamp studies in cardiomyocytes.

Unlike the original procedure introduced by Böhle and Benndorf [17], nanopipette breaking in the smart patch-clamp technique is performed in a highly controlled, semi-automated fashion utilising the distance feedback-control of SICM working in the hopping mode (Fig. 3a). Instead of increasing the setpoint which would simply bring nanopipette closer to the glass bottom, the procedure rather increases the approach rate. The approach rate in standard hoping mode is limited by the value given by the radius of a nanopipette per z-piezo response time and is typically around 50 nm/ms for 100 nm pipette [5]. Due to the finite response time of the z-piezo, the nanopipette tip typically "overshoots" the setpoint height, in the

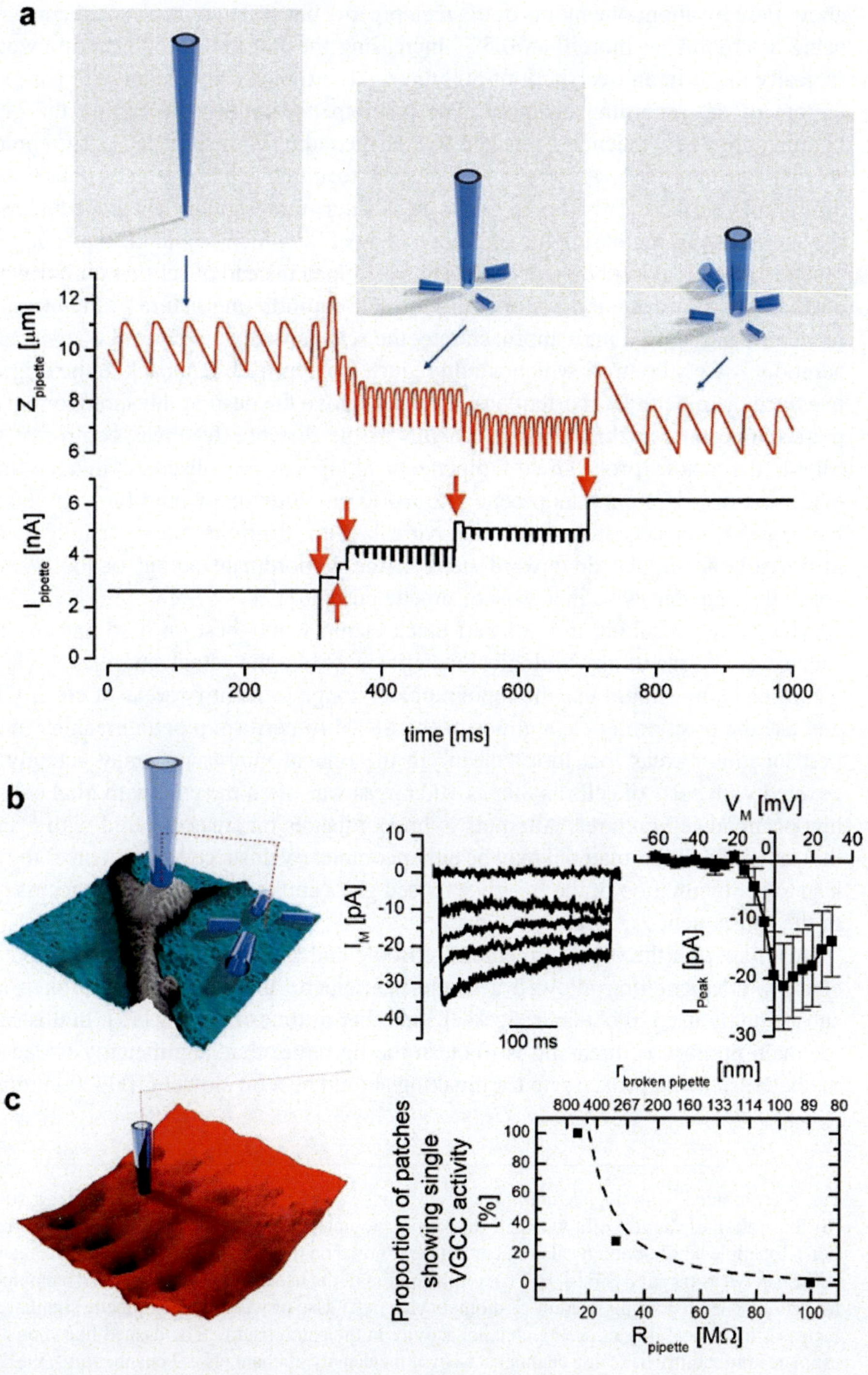

Fig. 3 Smart patch-clamp with controlled nanopipette breaking. (**a**) Example of pipette z-axis and current traces recorded during controlled nanopipette breaking. The red arrows indicate the stepwise breaking of the pipette tip. Insets illustrate the widening of the nanopipette each time a piece of the tip is clipped. Note the breaking happens during the interval between 300 ms and 700 ms where the fall rate (rate at which the pipette approaches the surface) is increased. The breaking procedure

above case by about one radius of the nanopipette, but still avoids physical contact if using a setpoint no more than 0.5%. Increasing the fall rate to 500 nm/ms would typically result in an overshoot of 500 nm right into physical contact with the glass bottom of the recording chamber. The nanopipette would keep hitting the glass bottom until a new balance is reached – when the radius of the pipette is large enough so that the approach rate does not cause it to come into physical contact with the sample surface. This turned out to be a safer, more controlled procedure than the alternative of increasing the setpoint to tens of %, which would keep pushing the pipette to the breaking point and needs to be stopped instead of settling on its own to a local equilibrium state. In addition to the self-equilibrating nature of the breaking procedure based on a high approach rate, the scanner software control does include additional safety controls which would reduce the approach rate back to the original one once the pipette DC current has increased above the desired threshold, or once a pre-set time interval has elapsed. Thanks to the distance feedback-control of the SICM, the whole procedure of pipette breaking can be observed in real time. Widening of a 100 nm nanopipette into a 400 nm wide one would typically take a few smaller breaking steps, each removing 1–2 μm from the pipette length clearly identifiable as sudden downward steps in the z-coordinate output of the scanner controller and step-wise increases of pipette current [12].

The pipette breaking in the smart patch-clamp works best on hard flat material such as glass, free of any biological material. Just like the actual smart patch-clamp recordings, one should use the topographical image to identify areas of clean glass and use the positioning capabilities of the SICM to perform pipette breaking at the best location. Areas that look "clean" in the optical microscope may actually be covered with parts of cells as thin as 100 nm as was often the case with glial cells in hippocampal cell cultures. Attempts to break pipettes on areas covered with even a thin layer of cellular material may be either completely unsuccessful (no breaking) or lead to contamination of the freshly exposed glass and thus to a reduced success rate of the subsequent gigaseal.

Just like with the original method by Böhle and Benndorf [17], the smart patch-clamp implementation allowed a single nanopipette to be repeatedly broken and subsequently used to obtain a gigaseal a number of times in a row [12]. In this case, the main purpose of breaking is to clean the tip rather than significantly widen the diameter, so the approach rate for breaking should be a bit slower (100–200 nm/ms)

Fig. 3 (continued) was stopped automatically when the pipette current increased by at least 100%. (**b**) Illustration of the use of a widened (broken) nanopipette for recording whole-bouton currents (left). Example whole-bouton calcium currents are shown on the right, together with the average IV curve (shown as mean ± S.E.M, $n = 6$) (**c**) Illustration of the use of a widened (broken) nanopipette for cell-attached recording in adult cardiomyocytes (left). Use of widened nanopipette significantly increases the probability of seeing channel activity in the patch (right). The dashed line shows the estimated probability of seeing channel activity if the density of channels is 2 channel/μm^2, the solid squares are experimental results as reported in [18]. The top axis shows estimated radius of a broken pipette corresponding to the resistance indicated on the bottom axis as reported previously [12] and the breaking should be stopped as soon as 10% or similar current increase is detected. If the nanopipette remains completely blocked after a previous gigaseal attempt, one may need to use a special SICM distance feedback-control mode in which the nanopipette keeps approaching until any current is detected again, before attempting to break the nanopipette again with the help of SICM feedback-control.

3 Access Resistance in the Smart Patch-Clamp Technique

Experienced patch-clamp operators might be wondering when we are going to discuss the issue of access resistance which is at least one order of magnitude larger for the nanopipettes than for the micropipettes typically used in the conventional patch-clamp technique. The access resistance R_A, sometimes referred to as series resistance R_S [19], represents the overall electrical resistance of the conductive pathway between the silver-chloride electrode in the nanopipette and either the membrane patch (in the cell-attached and excised patch configurations) or the entire membrane observed through access into the intracellular space (in the whole-cell configuration) (Fig. 4).

In the cell-attached mode it is mostly comprised of the resistance of the nanopipette itself while in the whole-cell configuration it is a combination of pipette resistance and the resistance of the access into the intracellular space. Its value affects the "speed" with which the membrane voltage can be controlled and the proportion of the command voltage effectively applied to the cellular membrane. The higher the access resistance, the slower the speed at which the membrane voltage reaches the desired value and the bigger the proportion of the voltage drop on the access resistance. In conventional whole-cell patch-clamp the access resistance is typically in the range of 2–20 MΩ, while in the smart patch-clamp with a nanopipette widened to $\approx$ 400 nm diameter (pipette resistance $\approx$ 40 MΩ) the average access resistance was around 150 MΩ [12]. With a typical whole-cell membrane capacitance C_M in range of 10–100 pF [19], a 150 MΩ access resistance R_A would result in a charging time constant $\tau_M = R_A \times C_M = 1.5$–15 ms, which would indeed be worryingly slow. However, one needs to remember that the main purpose of the smart patch-clamp technique is to target small sub-cellular structures. Recording from an average hippocampal synaptic bouton with an estimated membrane capacitance of around 30 fF and access resistance of 150 MΩ would result in a time constant of approximately 5 μs [12]. This is at least as fast, if not faster, than conventional patch-clamp whole-cell recording with 10 MΩ access resistance and 10–100 pF whole-cell membrane capacitance. The second worry on the mind of every patch-clamp operator is the voltage drop on the access resistance, which causes the membrane voltage to deviate from the command voltage depending on the current passing through the membrane. With current values in conventional whole-cell recording reaching up to few nA, the voltage drop on the access resistance of 10 MΩ can easily reach few tens of mV, and hundreds of mV on an access resistance of 150 MΩ. However, the membrane area of structures targeted in whole-

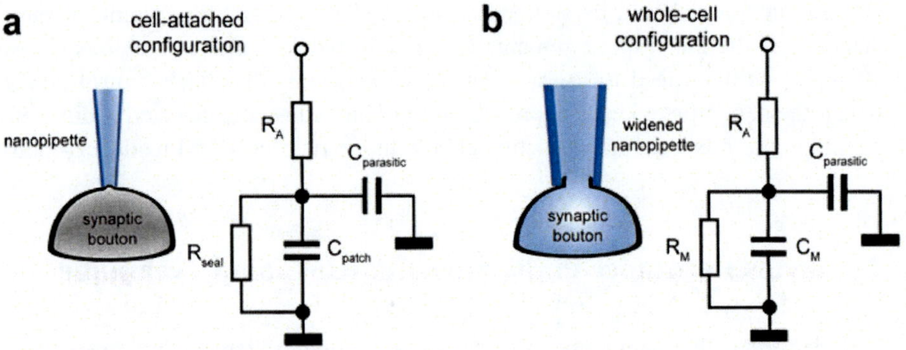

Fig. 4 Equivalent circuits of cell-attached (**a**) and whole-cell (**b**) recording configurations. The cartoons illustrate the two configurations in case of recording from small synaptic bouton. Parasitic capacitance $C_{parasitic}$ represents the capacitance of the nanopipette glass wall, C_{patch} represents the capacitance of the membrane patch in the cell-attached configuration, R_{seal} stands for the resistance of the seal between the pipette wall and the membrane, C_M capacitance of the whole-cell membrane, and R_M represents membrane resistance. In the cell-attached mode the resistance across the small membrane patch is much larger than the resistance of the seal R_{seal} and can be omitted. In the whole-cell mode, the membrane resistance is usually much lower than the seal resistance, hence the seal resistance is ignored for simplicity

cell smart patch-clamp experiments is at least one order of magnitude smaller, resulting in much smaller membrane currents (typically not more than 100 pA [12]), bringing the expected voltage drop on the access resistance into a similar range as in conventional patch-clamp recording. Hence, both negative effects of a larger access resistance in smart patch-clamp on the time constant and the voltage error are partially if not entirely compensated by the decrease in the surface area of the structure targeted by smart patch-clamp. This does not mean that the errors due to the access resistance can be ignored in smart patch-clamp, instead, one may conclude that it does not require any more attention than it already receives in the conventional patch-clamp technique. The same approaches used to mitigate these errors in conventional patch-clamp, such as the use of series resistance compensation built-in in commercially available patch-clamp amplifiers can be in principle used to deal with the issue in smart patch-clamp.

4 Conclusions

Smart patch-clamp combines high electrical sensitivity of the patch-clamp technique with the high spatial resolution of the scanning ion conductance microscopy and offers a unique tool for studying structure-function relationship in a wide range of biological samples. It brings the decades-old technology of patch-clamp into the age of nanotechnology. Alternative strategies of recording cellular potentials are being continuously developed, however none of them has yet reached the high fidelity of patch-clamp when it comes to recording true membrane potential and true currents flowing through ionic channels. Evidence suggests that many structures of interest are virtually impossible to record by directly using other available techniques. We believe that the smart patch-clamp technique, although itself almost two decades old, still has a great potential to shine new light into hidden corners of cellular physiology.

References

1. Hamill OP, Marty A, Neher E, Sakmann B, Sigworth FJ (1981) Improved patch-clamp techniques for high-resolution current recording from cells and cell-free membrane patches. Pflugers Arch 391(2):85–100
2. Hansma PK, Drake B, Gould SAC, Prater CB, Marti O, Gould SAC et al (1989) The scanning ion-conductance microscope. Science 243(4891):641–643
3. Gorelik J, Gu Y, Spohr HA, Shevchuk AI, Lab MJ, Harding SE et al (2002) Ion channels in small cells and subcellular structures can be studied with a smart patch-clamp system. Biophys J 83(6):3296–3303
4. Gu Y, Gorelik J, Spohr HHA, Shevchuk A, Lab MJ, Harding SE et al (2002) High-resolution scanning patch-clamp: new insights into cell function. FASEB J 16(7):748–750
5. Novak P, Li C, Shevchuk AI, Stepanyan R, Caldwell M, Hughes S et al (2009) Nanoscale live-cell imaging using hopping probe ion conductance microscopy. Nat Methods 6(4):279–281
6. Shevchuk A, Tokar S, Gopal S, Sanchez-Alonso JL, Tarasov AI, Vélez-Ortega AC et al (2016) Angular approach scanning ion conductance microscopy. Biophys J 110(10):2252–2265
7. Sakmann B, Neher E (eds) (1995) Single-channel recording.2nd edn. Boston, Springer. https://doi.org/10.1007/978-1-4419-1229-9
8. Molecular Devices (2020) The Axon guide. Electrophysiology and biophysics laboratory techniques. 4th edn. https://www.moleculardevices.com/sites/default/files/en/assets/ebook/dd/cns/the-axon-guide.pdf
9. Gorelik J, Zhang Y, Sánchez D, Shevchuk A, Frolenkov G, Lab M et al (2005) Aldosterone acts via an ATP autocrine/paracrine system: the Edelman ATP hypothesis revisited. Proc Natl Acad Sci U S A 102(42):15000–15005
10. Duclohier H (2005) Neuronal sodium channels in ventricular heart cells are localized near T-tubules openings. Biochem Biophys Res Commun 334(4):1135–1140
11. Bhargava A, Lin X, Novak P, Mehta K, Korchev Y, Delmar M et al (2013) Super-resolution scanning patch clamp reveals clustering of functional ion channels in adult ventricular myocyte. Circ Res. https://doi.org/10.1161/circresaha.111.300445
12. Novak P, Gorelik J, Vivekananda U, Shevchuk AI, Ermolyuk YS, Bailey RJ et al (2013) Nanoscale-targeted patch-clamp recordings of functional presynaptic ion channels. Neuron 79 (6):1067–1077
13. Glukhov AV, Balycheva M, Sanchez-Alonso JL, Ilkan Z, Alvarez-Laviada A, Bhogal N et al (2015) Direct evidence for microdomain-specific localization and remodeling of functional L-type calcium channels in rat and human atrial myocytes. Circulation 132(25):2372–2384
14. Vivekananda U, Novak P, Bello OD, Korchev YE, Krishnakumar SS, Volynski KE et al (2017) Kv1.1 channelopathy abolishes presynaptic spike width modulation by subthreshold somatic depolarization. Proc Natl Acad Sci 114(9):2395–2400
15. Torres-Pérez JV, Naeem H, Thompson CL, Knight MM, Novak P (2020) Nanoscale mapping reveals functional differences in ion channels populating the membrane of primary cilia. Cell Physiol Biochem 54(1):15–26

16. Decaen PG, Delling M, Vien TN, Clapham DE (2013) Direct recording and molecular identification of the calcium channel of primary cilia. Nature 504(7479):315–318
17. Böhle T, Benndorf K (1994) Facilitated giga-seal formation with a just originated glass surface. Pflügers Arch Eur J Physiol 427(5–6):487–491
18. Bhargava A, Lin X, Novak P, Mehta K, Korchev Y, Delmar M et al (2013) Super-resolution scanning patch clamp reveals clustering of functional ion channels in adult ventricular myocyte. Circ Res 112(8):1112–1120
19. Marty A, Neher E (1995) Tight-seal whole-cell recording. In: Sakmann B, Neher E (eds) Single-channel recording2nd edn. Boston, Springer, pp 31–52

BIOREV (2022) 3: 139–158
https://doi.org/10.1007/11663_2021_10

Published online: 23 February 2022

Correlating Cardiac Structure to Function Using Nanoscale Resolution Scanning Ion Conductance Microscopy

Anamika Bhargava and Julia Gorelik

Contents

Abstract Nanoscale resolution scanning ion conductance microscopy (SICM) is now at the forefront of imaging topography of biological cells even if they have complex and delicate surface structures. SICM offers several advantages over conventional high resolution and optical imaging techniques including super-resolution techniques. SICM can be performed on live cells without extensive sample preparation, and therefore it has also been successfully used to study cell signalling in combination with other biophysical and imaging techniques. Cardiac biology has greatly benefitted from the use of SICM to obtain high resolution topography images of cardiomyocytes and other cardiac cells. With the use of

A. Bhargava (✉)
Department of Biotechnology, Indian Institute of Technology Hyderabad (IITH), Kandi, Sangareddy, Telangana, India
e-mail: abhargava@bt.iith.ac.in

J. Gorelik (✉)
National Heart and Lung Institute, Faculty of Medicine, Imperial College London, London, UK
e-mail: j.gorelik@imperial.ac.uk

SICM combination techniques such as super-resolution scanning patch-clamp, SICM-FRET, SICM-LCM, etc., several underlying cell signalling pathways could be associated with cell surface structures reiterating the importance of structure as it relates to function. Combination techniques like super-resolution scanning patch-clamp and SICM-FRET for the first time provided direct evidence of the role of cellular microdomains in maintaining receptor function. It provided direct evidence for the concept that a receptor or a protein may become non-functional or become dysfunctional if it is relocated to a new microdomain from its native microdomain as usually happens in pathological cases.

Keywords Cardiomyocytes · Microdomains · Scanning ion conductance microscopy · SICM-FRET · Super-resolution scanning patch-clamp

1 Introduction

Scanning ion conductance microscopy (SICM) is a type of scanning probe microscopy which was developed in response to the ever growing need of high resolution non-optical methods to probe biological samples [1]. Biological surfaces are delicate and can have complex structures. Techniques to image biological cells at high resolution without disturbing their environment and/or structure are limited. Even advanced optical imaging cannot provide the resolution needed to fully visualize biological structures as it is limited by the diffraction of light. Non-optical methods like atomic force microscopy can reach very high resolution but the intermittent contact of the probe with the sample and the use of force while imaging can distort delicate biological structures. All other techniques are either invasive or require extensive sample preparation. SICM has emerged as a non-optical, high resolution imaging technique for imaging topography of biological cells [2–4]. It has been extensively used to image topography of a large variety of biological cells. In this chapter we will limit our discussion to the applications of SICM in cardiac biology. The Gorelik group has extensively used SICM to study cardiomyocyte structures over the last decade. This group has combined SICM with several other techniques to get important insights in the role of biological structures in cell physiology.

2 The Technique

Originally invented as a form of scanning probe microscopy by Hansma and colleagues [1], SICM was improved for imaging biological cells by Korchev and colleagues [3]. In its original form the technique was limited in its ability to image complex and rough biological surfaces such as microvilli structures. In its refined

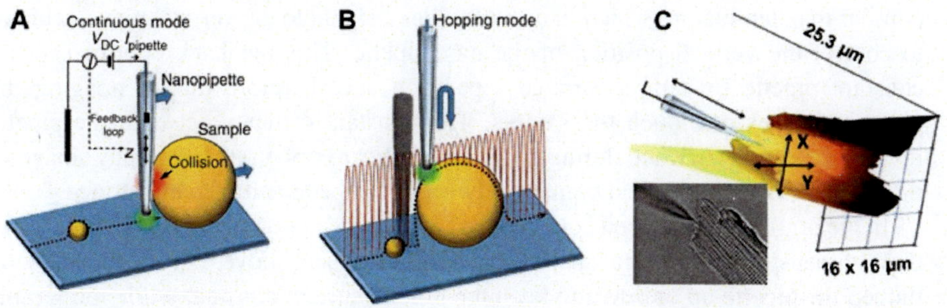

Fig. 1 Modes of SICM. (**a**) Originally described continuous scan mode of SICM depicting the possibility of the nanopipette colliding with a taller spherical object. (**b**) Hopping probe ion conductance microscopy (HPICM) depicting the nanopipette position with red lines showing the path of the nanopipette. (**c**) Angular mode of HPICM showing the nanopipette at an angle at the intercalated disk of an adult rat cardiomyocyte. The inset shows optical image of an adult rat cardiomyocyte and the nanopipette under an upright microscope. Figure modified with permissions from [5, 6]

form as hopping probe ion conductance microscopy (HPICM), the technique is now well suited to image taller biological structures with high resolution (Fig. 1a, b) [5].

The SICM setup consists of standard electrophysiological patch-clamp electronics, with an added SICM controller. As with the usual patch-clamp setup, SICM also uses an inverted microscope with a computer-controlled three axes piezo translation system mounted on the microscope stage. The piezo system typically has a range of 100×100 μm (xy) and 25 μm (z) for pipette movements. The glass nanopipette (similar to the glass patch pipette used in a patch-clamp setup) is mounted on the vertical piezo-actuator stage and is connected to the head stage of a patch-clamp amplifier. Electrical connections are also similar to the conventional electrophysiological circuit and are provided by Ag/AgCl electrodes in bath and patch pipettes. The electrode/nanopipette used in SICM is of prime importance as the resolution of imaging is dependent on the quality and size of the pipette tip. Nanopipettes with a defined tip geometry are pulled using standard glass pullers from borosilicate glass capillaries. The ion current which flows through the pipette is induced by an applied DC voltage of $\approx$50–200 mV and is used in the feedback loop that in turn controls the nanopipette's vertical position so that at no point of time the pipette touches the sample. A striking difference between the nanopipette used in SICM and the patch pipette used in patch-clamp is the size of the tip of the pipette. In SICM, the pipette tip is usually $\approx$100 nm in diameter (to obtain a resolution of 50 nm) as opposed to the large pipette tip of $\approx$1 μm used in patch-clamp. The SICM nanopipette tip size can be further decreased to improve the resolution.

SICM is based on the principle that the ion current flowing through a nanopipette tip in solution decreases when the nanopipette is lowered towards the surface of any non-conducting substrate due to the restriction of the space through which the current can flow. Thus, the ion current through the nanopipette strongly depends on the distance between the pipette tip and the non-conducting surface (in this case the cell surface) [3]. This ion current is used by the feedback control system of SICM (similar to patch-clamp) which in turn provides a signal to the vertical piezo-actuator to control the vertical position of the nanopipette. This feedback control keeps a constant pipette tip-sample surface separation and therefore during imaging the pipette tip does not touch the sample. The constant distance between the pipette tip and the sample surface during scanning allows to obtain a topography image of the surface without physical contact between the pipette and the cell surface.

In the originally described conventional SICM, the nanopipette was mounted on a three-dimensional piezo translation stage as described above. The feedback loop moved the pipette up or down to keep the pipette current constant while the sample was scanned in the x and y directions. A continuous feedback mechanism was used as a result of which the pipette was always in a close proximity to the cell surface [1, 4]. This technique was successful for relatively flat cells; however, it failed in the case of cells with taller structures as the pipette tip often collided with the taller structures. This prompted researchers to refine this technique and therefore HPICM was developed [5]. In HPICM, instead of the continuous feedback mode, at each imaging point, the pipette approaches the sample from a starting position that is far up from the cell surface and thus well above any of the surface features of the biological cells. At each imaging point, the nanopipette then approaches the cell surface and the current reduces by a predefined value, which can be $\approx$0.25–1.0% of its initial value. The position of the pipette tip at which this amount of current reduction is achieved is recorded as the height of the sample at this imaging point. The pipette is then again withdrawn to the starting point, the sample is moved laterally to a new imaging point, and the pipette is lowered again to achieve the pre-set current reduction. This avoids collision of the nanopipette with the sample and a 3D topography image of the sample is achieved (Fig. 1a, b) [5]. A most recent advancement in SICM for biological cells is the angle scanning (Fig. 1c) [6]. In this modification, the scanning nanopipette approaches the sample at an angle which can be as shallow as the approach angle of a patch-clamp pipette. This angle is adjustable. Angular mode SICM allows imaging of biological cells on optically non-transparent substrates [6]. This also allows scanning living cells on upright microscopes which was not possible before and thus presents the most advanced SICM mode for biological cells.

3 Revealing Nanoscale Structures Using SICM

SICM has emerged as a versatile technique. It was developed to image topography of live biological cells without affecting their structure. With SICM it is possible to generate 3D topographical images of cells with a nanoscale resolution which can be as high as 20 nm [5]. There are several features of SICM which set this technique apart and in front of other imaging techniques. To image topography with SICM, biological cells do not need any preparation such as fixing or labelling. Neither there is any requirement of special media – the cells can be imaged in the culture medium itself. As described in the previous section, since it is a non-contact microscopy, cells

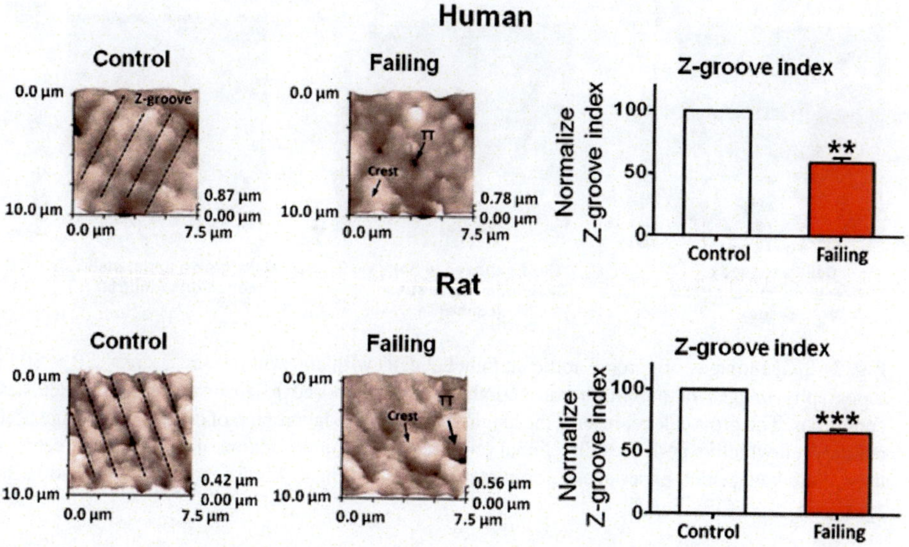

Fig. 2 Topography images obtained by SICM. SICM topography images of cardiomyocytes isolated from control (top and bottom left) and failing (top and bottom middle) human and rat hearts. Z-grooves, crest, and t-tubule (TT) openings are marked. The panels on the right show the Z-groove index compared between control and failing cardiomyocytes from human (top) and rat (bottom) hearts. Figure modified with permission from [7]

are not damaged and can be used for other purposes after imaging. For cardiac cells, SICM has emerged to be more beneficial since cardiomyocytes have specialized membrane structures, which were confirmed with electron microscopy and other techniques before the invention of SICM. However, all those conventional techniques required extensive sample preparation and the sample was destroyed by the end of the experiment. With the development of SICM, the Gorelik group has extensively characterized cardiomyocytes and other cells of the heart. A typical topography image of an adult cardiomyocyte generated by SICM is shown in Fig. 2 (left panels). Characteristic t-tubules of the cardiomyocytes are seen as holes in the image and the invaginations correspond to the Z-grooves. To be able to compare this structure across various conditions, Gorelik and colleagues have introduced a Z-groove index of structural integrity [8]. Z-groove index is a ratio between the length of Z-grooves on a topography image of a cardiomyocyte as obtained by SICM to the maximal length of Z-grooves possible on the given image. Later, this Z-groove index proved very useful in quantifying structural loss in cardio-pathological conditions. SICM imaging revealed that there is a considerable structural loss in cardiomyocytes under conditions of cardiovascular diseases like heart failure or myocardial infarction [9]. Prior to SICM, this structural loss was quantified using membrane dyes such as ANNEPS dyes; however, a clear picture of the loss of Z-grooves and t-tubules could be ascertained only by SICM. In Fig. 2, the topography of control and failing human and rat cardiomyocytes is compared, which revealed a significant loss of t-tubules and Z-grooves [7]. This structural loss is

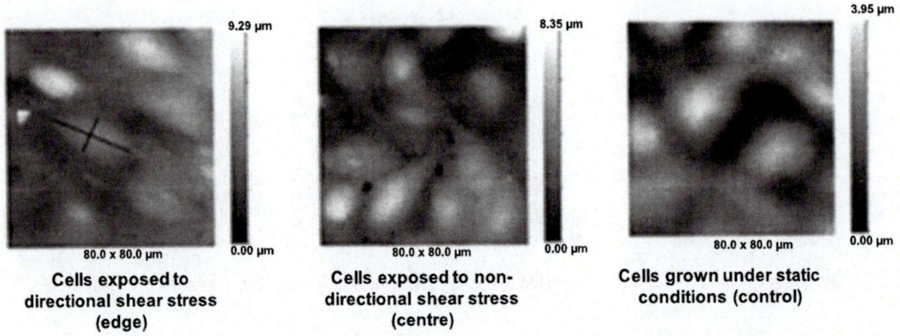

Fig. 3 SICM images of porcine aortic endothelial cells with and without shear stress. 80 × 80 µm topography images of porcine aortic endothelial cells exposed to shear stress and under static conditions. The greyscale represents the height of the cells. Edge: region of cells near the edge of the culture well which experiences directional pulsatile shear. Centre: region of cells near the centre of the culture well which experiences non-directional steady shear. Figure modified with permission from [11]

progressive and can be quantified temporally using SICM [10]. Since heart function is a collective effort of a variety of cells, researchers were prompted to obtain SICM images of endothelial cells and other structures of the heart. SICM images of endothelial cells from different aortic regions showed a very interesting pattern [11]. It is now well known that the heart is a very mechanical organ with the ability to adapt to several high and low pressure conditions. Similarly, aortic cells are known for their ability to adapt to shear stress. When isolated and subjected to shear stress in culture, aortic cells showed a unique organization with or without shear stress. In culture without shear stress, endothelial cells oriented randomly, however, they became aligned when shear stress was applied (Fig. 3) [11]. With SICM, not only cells, but cardiac structures such as intact valve preparations could be studied [12]. SICM was particularly useful in imaging these structures as the opaque tissue preparations are not useful for optical microscopy as they have few layers of cells. Analysing and studying valve structures could potentially help us to understand valve physiology to develop potential valve replacement therapies. With SICM, topographical differences could be observed in endothelial cells from the ventricular and atrial sides of the aortic valve (Fig. 4) [12]. SICM has also been used to image the topography of neonatal cardiomyocytes and human embryonic stem cell-derived cardiomyocytes (ESCMs) [13, 14]. In each case SICM has proved beneficial due to the advantages listed above (no sample preparation, no-contact), as these cells are fragile and sample preparation may destroy the precious samples such as stem cell-derived cardiomyocytes. In the case of ESCMs, since the culture was heterogeneous, SICM could identify ESCMs within the layer of mixed cell types and thus validated the suitability of ESCM culture for the discovery of antiarrhythmogenic drugs [14].

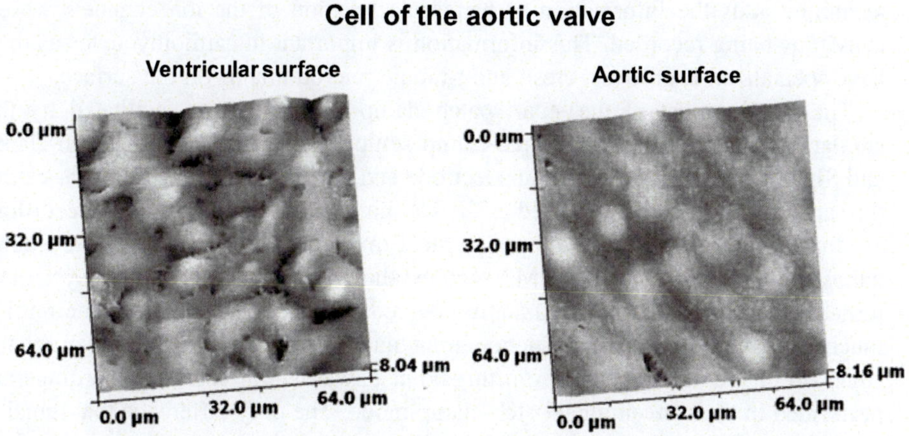

Fig. 4 SICM images of aortic valve endothelial cells. Topography of aortic valve endothelial cells on the ventricular and aortic surface of aortic valve cusps. Figure modified with permission from [12]

4 Correlating Structure to Function in Cardiac Cells

In order to have an in-depth understanding of biological systems one needs to understand function in relation to the structure. This becomes even more important in the case of very structural organs such as the heart. Emerging data suggest that a protein maintains its function in a specific environment and any disturbance in the environment, even without the changes in protein structure, may lead to the loss/ disruption of the function. Proteins such as receptors may change their signalling properties depending on their cellular locations [15, 16]. Also, key players in cell signalling such as protein kinases and phosphatases may be localized to certain microdomains of a cell to compartmentalize a signal [17]. The Gorelik group has developed and extensively used SICM for cardiac biology for a number of years now and in recent years they have combined SICM with an array of other techniques. The combination techniques described in the next few sections are helpful tools to study cell signalling in relation to cellular structural and functional microdomains in cardiomyocytes.

4.1 Patch-Clamp with SICM

Gorelik and colleagues combined SICM with patch-clamp electrophysiology to develop the "smart patch-clamp technique" [18]. Patch-clamp electrophysiology is conventionally used to record ion channel activity from biological cells. However, without SICM, the ion channel activity is recorded from the membrane without the knowledge of membrane microdomains or features. The smart patch-clamp technique adds the information of nanoscale location of the ion channels whose activity is being recorded. This information is important in cardiomyocytes as they have specialized Z-grooves, crest, and t-tubule microdomains on the surface.

The development of the smart patch-clamp technique was facilitated by the similarities in the SICM and patch-clamp setups. In particular, both patch-clamp and SICM use a glass pipette as an electrode and an amplifier. In smart patch-clamp the same glass nanopipette is used as the scanning probe and patch-clamp recording electrode. After obtaining the topographical image as described in the "revealing nanoscale structures using SICM" section, the feedback of the SICM system is turned off and the pipette is manually lowered towards the cell membrane until it touches the membrane. A light suction immediately results in a gigaseal owing to the small tip of the nanopipette. From this point onwards, ion channel recordings are performed in the conventional patch-clamp mode. The only difference that remains is the size of the tip of the patching pipette which is $\approx$1 μm in conventional electrophysiology vs $\approx$100 nm in smart patch-clamp. Using this method, researchers discovered how calcium and chloride channels are distributed in the cardiomyocyte membrane [2]. The size of the smart patch-clamp pipette, which is $\leq$100 nm (internal diameter), however, imposed a severe limitation on the throughput of the technique. This very fine-tipped scanning nanopipette, when used for patch-clamping, allowed a very small area to be covered in a single recording, thus severely lowering the chance of finding channels in the patch area. This prompted Gorelik and collaborators to further refine the technique of smart patch-clamp, in particular its throughput. The newer high throughput version of the technique is now called "super-resolution scanning patch-clamp" [13]. In super-resolution scanning patch-clamp, after obtaining the topography image with the fine-tipped nanopipette, the pipette tip is clipped using a precise movement controlled by the software. To achieve this, the pipette is taken off the cell first and the pipette tip is clipped. As a result of tip clipping, the size of the tip is increased, however, without affecting tip geometry. Since the topography image coordinates are already known, patch-clamp can now be performed with the clipped tip at the desired location. All the movements of the pipette are controlled by the piezo-actuators and therefore the repositioning of the pipette is quite precise with a lateral error of only $\pm$2 nm. For details on the tip clipping procedure, please refer Fig. 2 in reference [13]. This super-resolution scanning patch-clamp technique with increased throughput has so far been used to solve important biological questions in cardiovascular research. By now we know that the cardiomyocyte membrane possesses several microdomains where ion channels, receptors, and cytoskeletal proteins exist together and interact with each other. An ion channel distribution map created by super-resolution scanning patch-clamp revealed that the functional L-type calcium channels (LTCCs) and sodium channels have a specific microdomain preference on the surface of cardiomyocytes [13]. LTCCs are preferentially located in t-tubules in healthy ventricular cardiomyocytes and their activity can be rarely recorded from crest microdomains [13], however this is not the case in cardiovascular pathologies. Data from ventricular cardiomyocytes isolated from animal models and patients with distinct cardiovascular pathologies have now revealed that LTCCs may be located in different

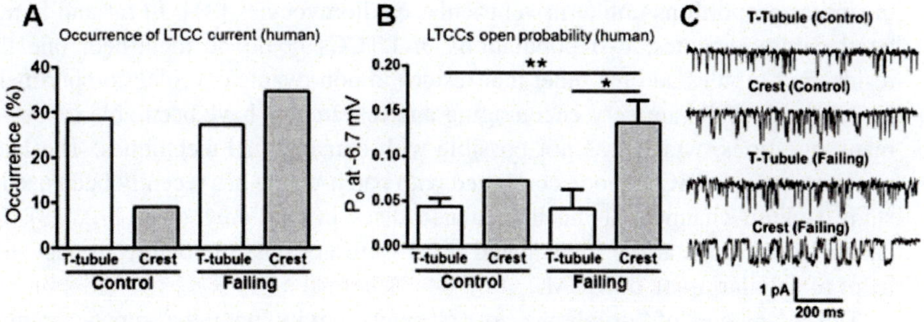

Fig. 5 LTCC localization and function in failing human cardiomyocytes. (**a**) Chance of obtaining LTCC current (% occurrence) in human control and failing cardiomyocytes (**$p < 0.01$ by Fisher exact test). (**b**) Open probability (P_o) at maximum activating voltage step (to −6.7 mV) in human control and failing cardiomyocytes (*$p < 0.05$, **$p < 0.01$). (**c**) Representative single-channel traces at −6.7 mV showing LTCC activity in human cardiomyocytes. Figure modified with permission from [7]

microdomains depending on the underlying pathology and that a shift from their "native" microdomain to any other microdomain can be detrimental. In failing human and rat ventricular myocytes, LTCC current could be recorded from crest microdomains (Fig. 5a, c) [7]. Interestingly, LTCCs in these failing cells showed aberrant activity (higher open probability) only in the crest microdomains, which was due to their hyper-phosphorylation by calcium-calmodulin kinase II in these non-native microdomains (Fig. 5b). A mathematical model predicted that these displaced LTCCs can be the potential source of re-entrant arrhythmias in heart failure [7]. The distribution and modulation of LTCCs in various cardiovascular pathologies also depended on the underlying pathology, for example in ischemic cardiomyopathy, LTCCs in the crest were hyper-phosphorylated by protein kinase A, whereas in dilated cardiomyopathy, LTCCs in the crest were hyper-phosphorylated by calcium-calmodulin kinase II [19]. Also, therapeutic interventions like LVAD implantation corrected the function of the relocated LTCCs (in this case corrected their activity back to normal) but could not change their location back to t-tubules. This is a very important result, which indicates that current therapies can only improve function but once the structure is lost, so far it has not been possible to revert the structural loss, which could be the reason for not being able to prevent heart failure. In an attempt to provide proof of concept for therapies aimed towards restoring the structure of cardiomyocytes in order to restore the function, it was demonstrated that over-expression of junctophilin-2 (a structural protein required for t-tubule tethering to the membrane, which is downregulated in heart failure) could restore the LTCC current, which is downregulated due to loss of t-tubular structures by keeping cardiomyocyte in culture [20]. Perhaps junctophilin-2 gene therapy could be a potential novel therapy in heart failure.

Interesting results are also obtained in the case of atrial cardiomyocytes using the super-resolution scanning patch-clamp technique. In atrial cardiomyocytes it was observed that functional LTCCs are present both in t-tubule and crest microdomains in similar proportions unlike in ventricular cardiomyocytes [21]. In rat and human atrial cardiomyocytes, two populations of LTCCs could be identified, one that resides in t-tubules and the other that resides in non-dyadic caveolar compartments [21]. These results are very encouraging and researchers have been able to answer many questions, which was not possible with conventional techniques. The latest angular scanning SICM mode combined with patch-clamp has recently been used to study sodium channels at the intercalated discs in cardiomyocytes [22, 23]. Ion channel recordings at the intercalated discs were not possible before the development of angular mode of SICM.

The advantage of visualizing ion channel activity with the super-resolution scanning patch-clamp technique over other traditional microscopic methods and over the conventional patch-clamp technique is the knowledge of functional channels in the specific microdomain. With only microscopy, we cannot ascertain whether the channel is active and is contributing to ion flux. With only patch-clamp, we cannot determine the spatial location of the ion channel being recorded on the membrane. The super-resolution scanning patch-clamp technique serves both the purposes, making it superior for ion channel biology in cardiomyocytes.

4.2 NanoFRET with SICM

The Förster Resonance Energy Transfer (FRET) technique has been extensively used in the field of biology to study protein–protein interactions and cell signalling. Therefore, to relate cell surface structures to cell signalling, it was combined with SICM. With the SICM-FRET combination technique, which was developed by the Gorelik group, it is possible to study signalling through individual microdomains on the surface. The first attempt using SICM-FRET was made to study β-adrenergic signalling in cardiomyocytes. Adrenergic signalling is one of the key signalling pathways in the heart and is also one of the prime therapeutic targets. β-adrenergic receptors (βARs), upon activation, generate cAMP and these β-adrenergic dependent cAMP signals in live cells can be detected by using highly sensitive fluorescent biosensors [24, 25]. These biosensors are based on FRET between two fluorophores (for example, CFP and YFP) fused to a single cyclic nucleotide-binding domain taken from proteins like Epac or PKA. Studies in cardiomyocytes isolated from transgenic mice expressing such biosensor revealed that β_1AR-mediated cAMP signals propagated through large areas in the cardiomyocytes, whereas β_2AR signals were locally confined [25]. However, all these previous studies could not reliably address the distribution and compartmentation of cAMP signals that are generated upon activation of βARs. SICM-FRET could capture for the first time cAMP signals from t-tubule and crest microdomains of the isolated cardiomyocytes that were transfected with EPAC based cAMP biosensor encoded by a viral vector [26]. Transfected cardiomyocytes were imaged for cell topography to reveal microdomains such as t-tubules, crests, and grooves. β-adrenergic agonist was then applied via the SICM pipette very locally. Local application of β-adrenergic agonist in this case was achieved by applying pressure through the nanopipette, which resulted in a jet of fluid coming out of the pipette for the duration of pressure applied. Also, cardiomyocytes were constantly superfused with the buffer so that any receptor stimulation that occurs stays local. A cAMP signal was observed upon stimulation of β_1ARs both in t-tubules and crests unlike β_2ARs where cAMP signal was observed only upon stimulation with the agonist in t-tubules [26]. These were fascinating results as they provided direct evidence of localization of β-adrenergic receptors in specific microdomains on the surface of cardiomyocytes, β1ARs being present in both t-tubules and crests and β_2ARs only in t-tubules.

Since heart failure is a structural disease, it was intriguing to determine the location of βARs in heart failure or other pathological conditions. It was observed that in failing cardiomyocytes, β_2AR dependent cAMP signal was no longer local meaning the signal could be observed in both t-tubules and crests and their pattern of expression was now similar to β1ARs (Fig. 6) [26, 27]. Therefore, it can be successfully concluded that the altered cAMP compartmentation with the signal going from local to global may contribute to alteration in heart function during failure. In order to understand the differences in cAMP signalling through β_1ARs and β_2ARs, perhaps it is important to look at the composition of the microdomains that they reside in. β_1ARs have been found in caveolar and non-caveolar membrane microdomains unlike β_2ARs, which have been found to be exclusively localized to caveolae [28]. Caveolae are plasma membrane structures that are enriched in cholesterol, glycophospholipids, and glycosylphosphatidylinositol-anchored proteins, etc. and can be visualized by high resolution techniques such as SICM and electron microscopy. Wright and colleagues showed that caveolin-3, which is the major component protein of caveolae microdomains, plays a crucial role in the localization of β_2ARs to the t-tubules and thus the compartmentation of β_2AR dependent cAMP signalling in healthy cardiomyocytes [29]. More interesting was the finding that the over-expression of caveolin-3 in failing cardiomyocytes could partially restore the local signalling through β_2ARs and thus may be therapeutically exploited [29]. On the other hand, cGMP signalling through β3ARs is known to be cardioprotective since long, however, due to lack of high resolution techniques, microdomain location and redistribution of β_3ARs during heart failure, if any, remained unknown until recently. By using SICM-FRET, it was demonstrated that β_3ARs dependent cGMP signal is largely obtained from t-tubules [30]. A very interesting finding of this study suggests that there is a phosphodiesterase 2 mediated cGMP to cAMP cross talk which appears to be disturbed in heart failure [30].

Regional differences have also been described in the heart leading to different properties of cardiomyocytes located in apex and base regions of the heart. It is also speculated that cardiomyopathies like Takotsubo could result from the differences in the signalling in these different regions of the heart. A recent study described that unlike described previously for cardiomyocytes in general, in apex cardiomyocytes, β_2AR dependent cAMP signals are produced from both t-tubules and crest microdomains whereas in base cardiomyocytes majority of the β_2AR dependent cAMP signals was observed only from t-tubules [31]. This difference in the β_2AR dependent cAMP signals contributed to the different contractile response by apex

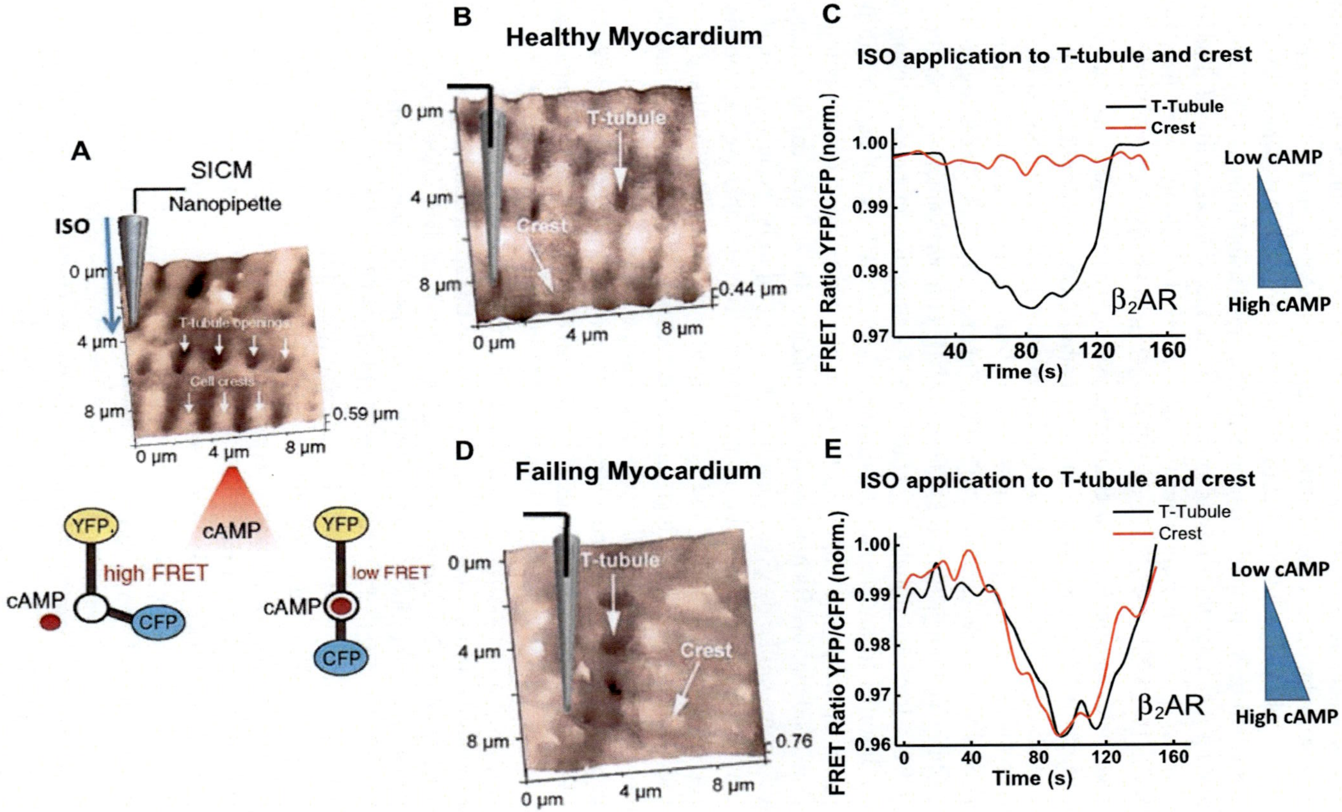

Fig. 6 βAR activated cAMP signalling in healthy and failing cardiomyocytes. (**a**) Schematics of NanoFRET (SICM-FRET). SICM topography image showing the application of isoprenaline (ISO) through the nanopipette above a t-tubule opening. βAR activated cAMP generation is measured by the FRET sensor Epac2-camps. (**b**) Topography of a healthy cardiomyocyte. (**c**) FRET ratio on application of ISO to either t-tubule (black line) or crest (red line). Low FRET indicates

high cAMP. (**d**) Topography of a failing cardiomyocyte showing structural loss. (**e**) ISO application produces cAMP signals both in t-tubules and crest. Figure modified with permissions from [26, 27] and base in the heart. The underlying mechanism seems to be the greater membrane organization including a greater number of t-tubules in base cardiomyocytes, which offered a tighter control over cAMP signalling as compared to apical cardiomyocytes [31].

Thus, SICM-FRET is a powerful technique to study compartmentation of cellular signals in relation to cell surface structures and provides new insights for novel therapeutic strategies.

4.3 Nanomechanics with SICM

Mechanosensation plays an important role in the function of heart. Mechanical properties of the individual cardiomyocyte's membranes and other cells of the heart contribute as a whole to the heart function. For years, measurement of mechanical properties of live cells was limited due to lack of techniques which can measure mechanics at high resolution without destroying the properties of the live cells. The development of SICM and its application in cardiac biology led to the easy determination of mechanical properties of single live cardiomyocytes and other heart cells. With SICM, it is now possible to simultaneously measure cell topography, membrane compliance, and modulus of elasticity. Simultaneous topography imaging also allows to measure properties of the membrane at selected microdomains in the membrane. Membrane properties are dynamic, can vary at nanoscale, and can be altered in diseased states or under the influence of signalling molecules. Therefore, SICM can prove very useful in comparing the membrane properties of cells in healthy and diseased states. Application of SICM to determine mechanics of the biological cells is possible due to the use of the nanopipette as a mechanical probe through which force can be applied. A positive pressure is applied through the glass pipette so that it ejects a stream of solution (force) strong enough to bend the membrane, however without damaging it. This force can be calibrated before so as to determine membrane compliance of live biological cells as was done by Sanchez and colleagues when they developed this technique [32]. They used an AFM cantilever which when bent by the pipette pressure allows to determine the corresponding change in the pipette position. The pressure applied can then be easily converted into force if we know the spring constant of the cantilever. From these parameters, compliance can be calculated as $C = d/P\ \sqrt{R}/\sqrt{R\mathrm{i}}$ where $C =$ compliance, $d =$ displacement of the membrane (μm), $P =$ pressure (kPa), $R =$ actual pipette resistance (Ω), $R\mathrm{i} =$ ideal pipette resistance (Ω). Using this technique, the membrane compliance of different cell types including cardiomyocytes, endothelial cells, and aortic valve cells has been measured [11, 33, 34]. Results suggest that the membrane properties can vary at nanoscale and are dependent on the cellular environment. In an interesting finding, it was revealed that the force through the pipette can indent specific membrane locations on cardiomyocytes and can initiate intracellular calcium release. More importantly, the propagation of this calcium whether localized or global depended on the health of the cells. In healthy cardiomyocytes the calcium propagation was local whereas in failing cardiomyocytes the initiation of calcium via mechanical interference propagated throughout the cell [33]. In order to investigate the mechanisms contributing to the mechanical properties of the cardiomyocytes, microtubule structures were investigated [35]. It is reported that post translational modifications such as acetylation or detyrosination in microtubules can affect membrane mechanical properties. Acetylation provides flexibility, whereas detyrosination makes the membrane rigid [35]. Experiments with peptide hormones like angiotensin II show that these hormones can also alter the membrane properties such as stiffness in a localized manner [34].

Therefore, SICM combined with mechanosensation seems to be a useful technique to relate properties of the membrane at the nanoscale level to the cell signalling.

4.4 NanoLCM with SICM

Laser confocal microscopy (LCM) is a useful technique to visualize labelled biological macromolecules and signalling molecules. LCM is also used to image calcium in the cells using fluorescent calcium indicators (e.g. Fluo-4). In cardiomyocytes, calcium plays an important role in cellular physiology and is the key to contraction. A massive release of calcium from the sarcoplasmic reticulum upon excitation brings about the contraction in cardiomyocytes. The Gorelik group has combined SICM with LCM to study contraction concurrently with the calcium release. This could be important under conditions of impaired contraction or calcium release. When an SICM pipette is positioned over a contracting cardiomyocyte, regular contractions of the cardiomyocyte result in the Z-displacement of the pipette. This is due to the feedback system which makes SICM a non-contact microscopy as described in the "The technique" section. This feedback system directs the pipette to go up and down following the membrane when the cardiomyocyte contracts to avoid touching of the membrane by the pipette. SICM-LCM combination technique can record the contraction of the cardiomyocytes via the pipette concurrently with recording changes in calcium concentration just beneath the membrane in a cardiomyocyte loaded with Fluo-4. To combine SICM with LCM, a laser is focussed on the tip of the pipette, and the sample is moved up and down by moving the stage to maintain a constant distance between pipette and cell surface [36]. This means that, during contraction, the fluorescence could be recorded at the same distance below the cell membrane. The distance moved by the stage and calcium fluorescence is simultaneously recorded. With this technique, it has been possible to probe intracellular calcium within 1 μm of the cell membrane (Fig. 7) [36].

SICM-LCM has been also successfully used to record contractions and calcium dynamics in human embryonic stem cell-derived cardiomyocytes (hESCs) [37]. In hESCs this technique could successfully characterize the beating clusters of stem cells and the calcium fluorescence in those beating cells. In principle, this SICM-LCM combination technique can be a useful method to study conditions or drugs

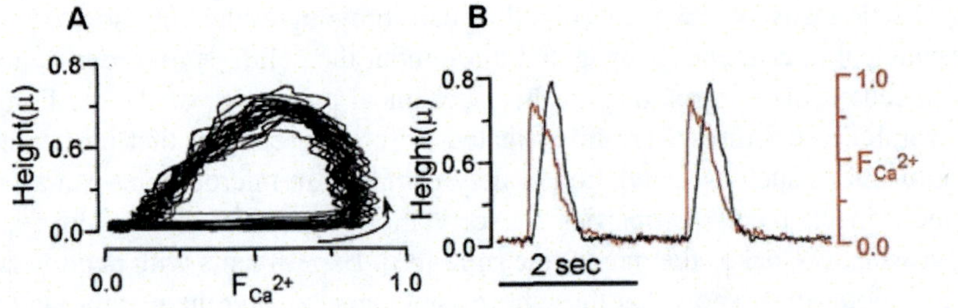

Fig. 7 Simultaneous measurement of calcium concentration and membrane movement using NanoLCM with SICM. (**a**) Cell height versus relative calcium concentration during contraction in 3-day-old cultured adult rat cardiomyocytes. (**b**) Two contractions are shown with overlay of cell height and calcium fluorescence. Figure modified with permission from [36]

which could alter contraction and/or calcium release in beating cardiomyocytes. With this technique, the amplitude of contraction and the duration of calcium transients can also be studied without the need of touching the cell.

5 Concluding Remarks

From our discussion in this chapter it is clear that SICM is the current state of the art super-resolution microscopy technique to study the topography of biological cells. It is even more advantageous when combined with other imaging and biophysical techniques and particularly useful in studying cells having distinct surface microdomains or surface structures. The combination techniques are exceptional tools to correlate structure with function at a resolution never achieved before. Its only limitation, that it can only image the surface of the cells and cannot provide insights into intracellular structures, is overshadowed by the advantages it offers. We hope that in the coming years researchers worldwide would be able to exploit this technique for better understanding of the biological processes.

References

1. Hansma PK, Drake B, Marti O, Gould SA, Prater CB (1989) The scanning ion-conductance microscope. Science 243(4891):641–643
2. Gu Y, Gorelik J, Spohr HA, Shevchuk A, Lab MJ, Harding SE, Vodyanoy I, Klenerman D, Korchev YE (2002) High-resolution scanning patch-clamp: new insights into cell function. FASEB J 16(7):748–750. https://doi.org/10.1096/fj.01-1024fje
3. Korchev YE, Bashford CL, Milovanovic M, Vodyanoy I, Lab MJ (1997) Scanning ion conductance microscopy of living cells. Biophys J 73(2):653–658. https://doi.org/10.1016/S0006-3495(97)78100-1

4. Korchev YE, Milovanovic M, Bashford CL, Bennett DC, Sviderskaya EV, Vodyanoy I, Lab MJ (1997) Specialized scanning ion-conductance microscope for imaging of living cells. J Microsc 188(Pt 1):17–23
5. Novak P, Li C, Shevchuk AI, Stepanyan R, Caldwell M, Hughes S, Smart TG, Gorelik J, Ostanin VP, Lab MJ, Moss GW, Frolenkov GI, Klenerman D, Korchev YE (2009) Nanoscale live-cell imaging using hopping probe ion conductance microscopy. Nat Methods 6(4): 279–281. https://doi.org/10.1038/nmeth.1306
6. Shevchuk A, Tokar S, Gopal S, Sanchez-Alonso JL, Tarasov AI, Velez-Ortega AC, Chiappini C, Rorsman P, Stevens MM, Gorelik J, Frolenkov GI, Klenerman D, Korchev YE (2016) Angular approach scanning ion conductance microscopy. Biophys J 110(10): 2252–2265. https://doi.org/10.1016/j.bpj.2016.04.017
7. Sanchez-Alonso JL, Bhargava A, O'Hara T, Glukhov AV, Schobesberger S, Bhogal N, Sikkel MB, Mansfield C, Korchev YE, Lyon AR, Punjabi PP, Nikolaev VO, Trayanova NA, Gorelik J (2016) Microdomain-specific modulation of L-type calcium channels leads to triggered ventricular arrhythmia in heart failure. Circ Res 119(8):944–955. https://doi.org/10.1161/CIRCRESAHA.116.308698
8. Gorelik J, Yang LQ, Zhang Y, Lab M, Korchev Y, Harding SE (2006) A novel Z-groove index characterizing myocardial surface structure. Cardiovasc Res 72(3):422–429. https://doi.org/10.1016/j.cardiores.2006.09.009
9. Lyon AR, MacLeod KT, Zhang Y, Garcia E, Kanda GK, Lab MJ, Korchev YE, Harding SE, Gorelik J (2009) Loss of T-tubules and other changes to surface topography in ventricular myocytes from failing human and rat heart. Proc Natl Acad Sci U S A 106(16):6854–6859. https://doi.org/10.1073/pnas.0809777106
10. Schobesberger S, Wright P, Tokar S, Bhargava A, Mansfield C, Glukhov AV, Poulet C, Buzuk A, Monszpart A, Sikkel M, Harding SE, Nikolaev VO, Lyon AR, Gorelik J (2017) T-tubule remodelling disturbs localized beta2-adrenergic signalling in rat ventricular myocytes during the progression of heart failure. Cardiovasc Res 113(7):770–782. https://doi.org/10.1093/cvr/cvx074
11. Potter CM, Schobesberger S, Lundberg MH, Weinberg PD, Mitchell JA, Gorelik J (2012) Shape and compliance of endothelial cells after shear stress in vitro or from different aortic regions: scanning ion conductance microscopy study. PLoS One 7(2):e31228. https://doi.org/10.1371/journal.pone.0031228
12. Miragoli M, Yacoub MH, El-Hamamsy I, Sanchez-Alonso JL, Moshkov A, Mongkoldhumrongkul N, Padala M, Paramagurunathan S, Sarathchandra P, Korchev YE, Gorelik J, Chester AH (2014) Side-specific mechanical properties of valve endothelial cells. Am J Phys Heart Circ Phys 307(1):H15–H24. https://doi.org/10.1152/ajpheart.00228.2013
13. Bhargava A, Lin X, Novak P, Mehta K, Khorchev Y, Delmar M, Gorelik J (2013) Super-resolution scanning patch-clamp reveals clustering of functional ion channels in the adult ventricular myocyte. Circ Res. https://doi.org/10.1161/CIRCRESAHA.111.300445
14. Gorelik J, Ali NN, Shevchuk AI, Lab M, Williamson C, Harding SE, Korchev YE (2006) Functional characterization of embryonic stem cell-derived cardiomyocytes using scanning ion conductance microscopy. Tissue Eng 12(4):657–664. https://doi.org/10.1089/ten.2006.12.657
15. Calebiro D, Nikolaev VO, Persani L, Lohse MJ (2010) Signaling by internalized G-protein-coupled receptors. Trends Pharmacol Sci 31(5):221–228. https://doi.org/10.1016/j.tips.2010.02.002
16. Wozniak M, Keefer JR, Saunders C, Limbird LE (1997) Differential targeting and retention of G protein-coupled receptors in polarized epithelial cells. J Recept Signal Transduct Res 17(1–3):373–383. https://doi.org/10.3109/10799899709036615
17. Scott JD (2003) A-kinase-anchoring proteins and cytoskeletal signalling events. Biochem Soc Trans 31(Pt 1):87–89. https://doi.org/10.1042/bst0310087
18. Gorelik J, Gu Y, Spohr HA, Shevchuk AI, Lab MJ, Harding SE, Edwards CR, Whitaker M, Moss GW, Benton DC, Sanchez D, Darszon A, Vodyanoy I, Klenerman D, Korchev YE (2002)

Ion channels in small cells and subcellular structures can be studied with a smart patch-clamp system. Biophys J 83(6):3296–3303. https://doi.org/10.1016/S0006-3495(02)75330-7
19. Sanchez-Alonso JL, Loucks A, Schobesberger S, van Cromvoirt AM, Poulet C, Chowdhury RA, Trayanova N, Gorelik J (2020) Nanoscale regulation of L-type calcium channels differentiates between ischemic and dilated cardiomyopathies. EBioMedicine 57:102845. https://doi.org/10.1016/j.ebiom.2020.102845
20. Poulet C, Sanchez-Alonso J, Swiatlowska P, Mouy F, Lucarelli C, Alvarez-Laviada A, Gross P, Terracciano C, Houser S, Gorelik J (2020) Junctophilin-2 tethers T-tubules and recruits functional L-type calcium channels to lipid rafts in adult cardiomyocytes. Cardiovasc Res. https://doi.org/10.1093/cvr/cvaa033
21. Glukhov AV, Balycheva M, Sanchez-Alonso JL, Ilkan Z, Alvarez-Laviada A, Bhogal N, Diakonov I, Schobesberger S, Sikkel MB, Bhargava A, Faggian G, Punjabi PP, Houser SR, Gorelik J (2015) Direct evidence for microdomain-specific localization and remodeling of functional L-type calcium channels in rat and human atrial myocytes. Circulation 132(25): 2372–2384. https://doi.org/10.1161/CIRCULATIONAHA.115.018131
22. Yang HQ, Perez-Hernandez M, Sanchez-Alonso J, Shevchuk A, Gorelik J, Rothenberg E, Delmar M, Coetzee WA (2020) Ankyrin-G mediates targeting of both Na(+) and KATP channels to the rat cardiac intercalated disc. eLife 9. https://doi.org/10.7554/eLife.52373
23. Leo-Macias A, Agullo-Pascual E, Sanchez-Alonso JL, Keegan S, Lin X, Arcos T, Feng Xia L, Korchev YE, Gorelik J, Fenyo D, Rothenberg E, Rothenberg E, Delmar M (2016) Nanoscale visualization of functional adhesion/excitability nodes at the intercalated disc. Nat Commun 7: 10342. https://doi.org/10.1038/ncomms10342
24. Nikolaev VO, Bunemann M, Hein L, Hannawacker A, Lohse MJ (2004) Novel single chain cAMP sensors for receptor-induced signal propagation. J Biol Chem 279(36):37215–37218. https://doi.org/10.1074/jbc.C400302200
25. Nikolaev VO, Bunemann M, Schmitteckert E, Lohse MJ, Engelhardt S (2006) Cyclic AMP imaging in adult cardiac myocytes reveals far-reaching beta1-adrenergic but locally confined beta2-adrenergic receptor-mediated signaling. Circ Res 99(10):1084–1091. https://doi.org/10.1161/01.RES.0000250046.69918.d5
26. Nikolaev VO, Moshkov A, Lyon AR, Miragoli M, Novak P, Paur H, Lohse MJ, Korchev YE, Harding SE, Gorelik J (2010) Beta2-adrenergic receptor redistribution in heart failure changes cAMP compartmentation. Science 327(5973):1653–1657. https://doi.org/10.1126/science.1185988
27. Lab MJ, Bhargava A, Wright PT, Gorelik J (2013) The scanning ion conductance microscope for cellular physiology. Am J Phys Heart Circ Phys 304(1):H1–H11. https://doi.org/10.1152/ajpheart.00499.2012
28. Rybin VO, Xu X, Lisanti MP, Steinberg SF (2000) Differential targeting of beta -adrenergic receptor subtypes and adenylyl cyclase to cardiomyocyte caveolae. A mechanism to functionally regulate the cAMP signaling pathway. J Biol Chem 275(52):41447–41457. https://doi.org/10.1074/jbc.M006951200M006951200
29. Wright PT, Nikolaev VO, O'Hara T, Diakonov I, Bhargava A, Tokar S, Schobesberger S, Shevchuk AI, Sikkel MB, Wilkinson R, Trayanova NA, Lyon AR, Harding SE, Gorelik J (2014) Caveolin-3 regulates compartmentation of cardiomyocyte beta2-adrenergic receptor-mediated cAMP signaling. J Mol Cell Cardiol 67:38–48. https://doi.org/10.1016/j.yjmcc.2013.12.003
30. Schobesberger S, Wright PT, Poulet C, Sanchez Alonso Mardones JL, Mansfield C, Friebe A, Harding SE, Balligand JL, Nikolaev VO, Gorelik J (2020) beta3-Adrenoceptor redistribution impairs NO/cGMP/PDE2 signalling in failing cardiomyocytes. eLife 9. https://doi.org/10.7554/eLife.52221
31. Wright PT, Bhogal NK, Diakonov I, Pannell LMK, Perera RK, Bork NI, Schobesberger S, Lucarelli C, Faggian G, Alvarez-Laviada A, Zaccolo M, Kamp TJ, Balijepalli RC, Lyon AR, Harding SE, Nikolaev VO, Gorelik J (2018) Cardiomyocyte membrane structure and cAMP

compartmentation produce anatomical variation in beta2AR-cAMP responsiveness in murine hearts. Cell Rep 23(2):459–469. https://doi.org/10.1016/j.celrep.2018.03.053
32. Sanchez D, Johnson N, Li C, Novak P, Rheinlaender J, Zhang Y, Anand U, Anand P, Gorelik J, Frolenkov GI, Benham C, Lab M, Ostanin VP, Schäffer TE, Klenerman D, Korchev YE (2008) Noncontact measurement of the local mechanical properties of living cells using pressure applied via a pipette. Biophys J 95(6):3017–3027. https://doi.org/10.1529/biophysj.108.129551
33. Miragoli M, Sanchez-Alonso JL, Bhargava A, Wright PT, Sikkel M, Schobesberger S, Diakonov I, Novak P, Castaldi A, Cattaneo P, Lyon AR, Lab MJ, Gorelik J (2016) Microtubule-dependent mitochondria alignment regulates calcium release in response to nanomechanical stimulus in heart myocytes. Cell Rep 14(1):140–151. https://doi.org/10.1016/j.celrep.2015.12.014
34. Swiatlowska P, Sanchez-Alonso JL, Mansfield C, Scaini D, Korchev Y, Novak P, Gorelik J (2020) Short-term angiotensin II treatment regulates cardiac nanomechanics via microtubule modifications. Nanoscale 12(30):16315–16329. https://doi.org/10.1039/d0nr02474k
35. Swiatlowska P, Sanchez-Alonso JL, Wright PT, Novak P, Gorelik J (2020) Microtubules regulate cardiomyocyte transversal Young's modulus. Proc Natl Acad Sci U S A 117(6): 2764–2766. https://doi.org/10.1073/pnas.1917171117
36. Shevchuk AI, Gorelik J, Harding SE, Lab MJ, Klenerman D, Korchev YE (2001) Simultaneous measurement of Ca2+ and cellular dynamics: combined scanning ion conductance and optical microscopy to study contracting cardiac myocytes. Biophys J 81(3):1759–1764. https://doi.org/10.1016/S0006-3495(01)75826-2
37. Harding SE, Ali NN, Brito-Martins M, Gorelik J (2007) The human embryonic stem cell-derived cardiomyocyte as a pharmacological model. Pharmacol Ther 113(2):341–353. https://doi.org/10.1016/j.pharmthera.2006.08.008

BIOREV (2022) 3: 159–186
https://doi.org/10.1007/11663_2021_12

Published online: 27 February 2022

Local Electrochemical Characterization Using Scanning Electrochemical Cell Microscopy

Marina V. Makarova and Yasufumi Takahashi

Contents

M. V. Makarova
Nano Life Science Institute, Kanazawa University, Kanazawa, Japan

Institute of Physics CAS, Prague, Czech Republic
e-mail: makarova@staff.kanazawa-u.ac.jp

Y. Takahashi (✉)
Nano Life Science Institute, Kanazawa University, Kanazawa, Japan
e-mail: yasufumi@se.kanazawa-u.ac.jp

Abstract *Scanning electrochemical cell* microscopy (SECCM) is a unique tool for nanoscale electrochemical imaging for characterizing sample reactivity. It uses a mobile meniscus state electrochemical cell between the nanopipette and the sample.

The chapter describes the method setup and principles as well as its reported applications in materials sciences. SECCM was successfully used to image various types of reactions such as hydrogen/oxygen evolution, oxygen/CO_2 reduction, corrosion, photocatalysis, and charge/discharge processes in battery materials. The studied objects included carbon phases: diamond and graphite with their derivatives; polymers; transitional metal dichalcogenides; metals; Li complex oxides. SECCM demonstrated site-dependent electrochemical activity, particularly, the role of defects and grain boundaries. Combination of SECCM with EBSD provided the information about the correlation between electrochemical reactivity and terminating facet in multiple kinds of 3D materials.

Moreover, SECCM can be used not only to image but also to build new patterned on-surface (sub)microstructures by loading the precursors into the pipette or by site-dependent electrooxidation/reduction of the surface. Thus, the advantages and goals of SECCM for various applications are described in the chapter.

Keywords Battery materials · BDD · Corrosion · Electrocatalysis · Graphite · Hydrogen evolution reaction · Nanopatterning · Photocatalysis · SECCM · TMDC

1 Introduction

The combination of electrochemical measurements and scanning probe microscopy is a promising tool for characterizing electrochemical reactions of inhomogeneous sample surfaces at the micro–nanoscale. Scanning tunnelling microscopy (STM) was the first probe microscopy-based electrochemical measurement tool that could visualize the atomic-scale topography and local density of states of the sample surface [1]. STM can be used for nanoscale electrochemical sensing, but it requires flat surfaces and is very sensitive to surface and tip contamination Scanning electrochemical microscopy (SECM) is a widely used technique for direct characterization of redox reactions using micro–nanoscale electrodes in a solution [2, 3]. Miniaturizing the electrode and controlling the electrode–sample distance are required to conduct nanoscale electrochemical imaging. This is because SECM measurements mostly employ a microscale electrode, and the probe scans the sample surface without controlling the probe–sample distance. This scanning method is called the constant-height mode. Atomic force microscopy [4, 5], shear force [6], impedance [7, 8], and ion current distance [9–11] control systems have been employed for SECM distance control, but it remains quite challenging.

Scanning electrochemical cell microscopy (SECCM) is a unique tool for nanoscale electrochemical imaging for characterizing sample reactivity. SECCM uses a mobile nanoscale glass pipette (nanopipette) to form a meniscus state electrochemical cell between the nanopipette and the sample for electrochemical imaging. It was developed by Professor Patrick Unwin and coworkers in 2010 [12]. The scanning micropipette contact method, which has a similar concept as SECCM, was reported in 2009 [13]. It was also categorized as a SECCM.

The nanoscale electrochemical cell for SECCM suppresses the capacitive current, and the ratio of faradaic and capacitive currents is significantly improved. Therefore, SECCM provides high-speed charge/discharge characteristics of battery materials [14], fast-scan cyclic voltammograms [15], and electrochemical imaging [16] of catalytic materials. Further, since the portion other than the measurement point is not immersed in the solution, the measurement can be performed without soiling the sample. The nanopipette for SECCM, with an aperture of less than 100 nm in diameter, is easy and reliable to fabricate. Some SECCM measurements are beyond the optical detection limit and provide super-resolution electrochemical images [15]. Initially, SECCM was used to characterize the reactivity difference in carbon nanotubes [17–19], edge and basal reactivity difference of HOPG [20–31], and facet-dependent reactivity difference on boron-doped diamond (BDD) [32–36].

Catalytic materials are also a suitable target for SECCM, for instance, facet-dependent reactivity difference in oxygen reduction reaction (ORR) [37], hydrogen evolution reaction (HER) [15, 38–47], CO_2 reduction [48–50], battery material [14, 51–56], corrosion [57–60], and photocatalysis [61, 62]. Recently, several review studies on SECCM and nanoscale electrochemical imaging have been reported [3, 63–70]. Here, we focus on important applications of SECCM.

1.1 SECCM Setup

In SECCM, a meniscus-shaped mobile nanoscale electrochemical cell is formed between the nanopipette and the sample for electrochemical imaging. Two types of nanopipettes are used in SECCM: double- and single-barrel nanopipettes. For the double-barrel nanopipette, each barrel (diameter <500 nm) of the tapered theta shape nanopipette is filled with an electrolyte solution, and an Ag/AgCl electrode is inserted. A potential is applied between the Ag/AgCl electrodes to generate an ionic current across the thin meniscus of the solution at the end of the theta nanopipette. The ionic current is used as a feedback signal to control the sample–pipette distance. Electrochemical measurements are performed at the meniscus-shaped nanoscale electrochemical cell on the sample surface. The sample is connected to a current amplifier, and one of the Ag/AgCl electrodes in the nanopipette is used as a quasi-reference electrode. Double-barrel nanopipette-based SECCM can analyze not only conductive samples but also semiconductive and insulating samples.

For the single-barrel nanopipette, the liquid meniscus at the nanopipette end closely links the nanopipette with the surface of the sample electrode. When the meniscus just contacts the surface, a small capacitive current is detected and the nanopipette is halted. The tip height at this point is recorded and electrochemical

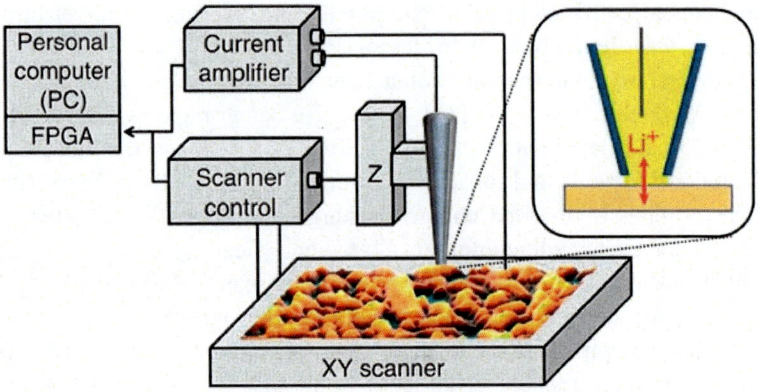

Fig. 1 Schematic of the SECCM setup for lithium-ion battery materials. Adapted with permission from Ref. [14], Copyright 2014 Springer Nature Publishing Group

measurements are collected. Single-barrel SECCM provides high-resolution images because the single-barrel glass capillary provides a small nanopipette aperture.

A SECCM system consists of five major components: a nanopipette probe with an inserted reference electrode and filled with an electrolyte solution, a current detector (current amplifier), a micro/nano-positioner, a data acquisition system, and an antivibration table (Fig. 1). In a typical SECCM measurement, a two-electrode setup is employed because the redox current is extremely small (pA–nA), and the *IR* drop is negligible. The nanopipette or sample is scanned by the micro/nano-positioner. A piezo stage is used for nanoscale imaging. Data are acquired by a field-programmable gate array board. The scanning area of the sample is selected by observing the sample position under a digital microscope.

To estimate reactivity from the redox current, it is important to keep the size of the electrochemical cell constant for SECCM measurement. As a method for estimating the size of an electrochemical cell, a footprint or electrochemically deposited Cu has been evaluated by SEM [48, 57]. Mineral oil has also been reported for eliminating the effect of hydrophilicity on the surface of the sample [58]. A method using a gel probe is also effective in eliminating the effect of hydrophilicity [71, 72]. However, even with these methods, it is difficult to keep the size of the electrochemical cell constant on very rough samples. To solve this problem, CV is measured at each measurement point, and the reactivity is evaluated from the half-wave potential [73].

2 Materials for Hydrogen Evolution Reaction (HER)

2.1 Transitional Metal Dichalcogenides

Recently, hydrogen has attracted increasing attention for energy storage and generation. It is currently generated predominantly from fossil fuels, but water splitting by electrocatalysis is proposed for the sustainable generation of H_2 since it would enable energy storage from various renewable sources. From the Nernst equation for HER, the overpotential value linearly decreases with a decrease in pH, so an acidic medium is most preferable. However, this also limits the choice of the catalysts to chemically stable conducting materials, such as platinum and its alloys, which are among the most effective but rare and expensive electrocatalysts. There is, therefore, a need to develop alternative electrocatalysts that can efficiently catalyze HER, sustaining high current densities at relatively low overpotentials with sufficient chemical stability. These include materials, such as noble metals, various carbon and phosphorus derivatives and compounds, and with special conditions, metal oxides [74].

Transitional metal dichalcogenides (TMD) are the most actively developing catalysts for HER [15, 38–40, 43–46]. Many of them have an anisotropic layered structure with a hexagonal lattice and (0001) basal plane. Their atomically flat surfaces, low hydrophilicity, and high chemical stability make them ideal for SECCM studies, and their semiconducting properties are desired in photocatalytic applications. However, the electrocatalytic properties of TMD are strongly affected by their microstructure, synthetic history, and local defect structure. The key features of the SECCM technique are that characteristic surface sites can be targeted and analyzed in detail, and the electrocatalyst area is known with good precision. A holistic view of the structure–activity relationship of (electro) materials results when local electrochemical maps and movies are combined with collocated microscopy/spectroscopy, such as EBSD [37, 42, 47, 48, 75–79], SEM and AFM [24, 38], micro-EDS, -XPS [43], and -Raman spectroscopy [24], local photoluminescence [45], optical spectroscopy [80], and experiments are further supported by theory (e.g., DFT calculations) [42].

Local electrocatalytic activity in TMD depends on the site position and crystal facet. From the bulk methods, it was obtained that MoS_2 HER activity is limited by the edge region, whereas the (0001) basal plane is inactive. However, it has been demonstrated by SECCM that the basal plane of bulk MoS_2 (2H crystal phase) possesses significant activity [38, 46], presumably due to the presence of point defects, such as S vacancies [15, 46], mechanical strain [46], and grain boundaries [15]. HER is greatly facilitated at the edge region (Fig. 2) (e.g., surface defects, such as steps, edges, or crevices) [38]. The latter is concerned with relatively freshly cleaved MoS_2, where cathodic polarization in acidic media has a cleaning effect on the surface, so the short-time aging has no apparent influence on electrocatalytic performance in contrast to SECM. However, after long-term aging, the edges and defect-rich areas oxidize more significantly [40, 45, 46], so the terraces demonstrate higher performance than the edges [15]. Similarly, in $Fe_{4.5}Ni_{4.5}S_8$, small variations in the surface Fe/Ni/S ratio, achieved by varying the synthesis conditions or material aging, can tremendously alter the catalytic HER activity [39].

Defect engineering is an important technique to increase electrocatalytic activity, including HER performance in TMDs. Studies on amorphous MoS_x have shown that the surface roughness and, especially, porosity are the main factors contributing to its spatially heterogeneous activity [43]. In crystalline phases, defects could also be

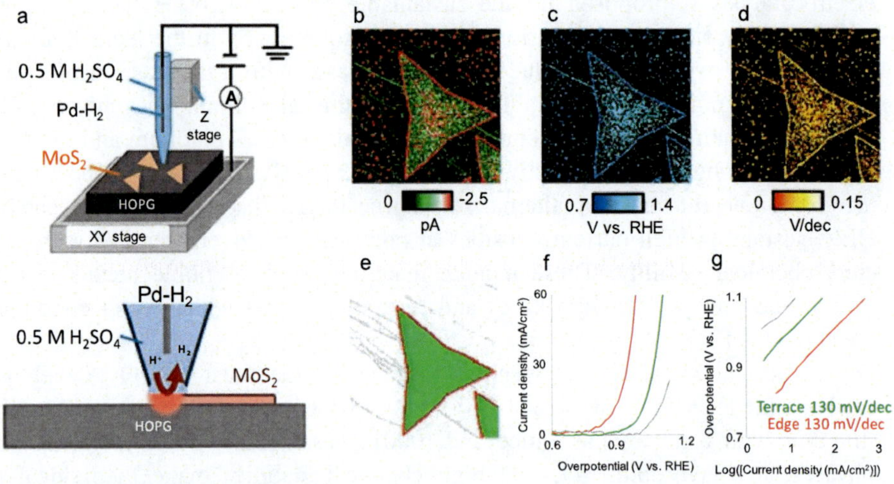

Fig. 2 (**a**) Illustration of SECCM measurements of MoS_2 nanosheets. Pd-H_2 used as a quasi-reference electrode. Nanopipette filled with 0.5 M H_2SO_4. (**b–d**) SECCM (**b**) current, (**c**) overpotential (30 mA cm^{-2}), and (**d**) Tafel slope images of 1H MoS_2 nanosheets on a HOPG substrate. Scan sizes 15 × 15 μm^2, scan rate 130 V s^{-1}, sweep voltage −1.3 V vs. RHE. (**e**) MoS_2 nanosheets edge (red), terrace (green), and HOPG edge (gray) tricolor images. (**f, g**) Graphs showing the (**f**) overpotential and (**g**) Tafel slope on the 1H MoS_2 edge (red), terrace (green), and HOPG edge (gray) regions. Adapted with permission from Ref. [15]. Copyright 2020 Wiley

generated by local anodization using SECCM [44, 45]. Artificial micron-sized holes are formed in the basal plane, increasing the density of monolayer high-step defects within the anodized regions, demonstrating improved catalytic behavior [44]. Furthermore, additional S vacancies are generated by applying a local negative potential to MoS_2 through SECCM, where HER activity enhancement is observed in electrochemically activated areas with potential not higher than $V = -1.4$ V vs. RHE. Crack-like defects are also generated by heating, which increases HER activity similar to the edge regions [15]. In the same study [15], aging processes, the effect of layer number, and heterostructure formation were investigated in detail.

Other layered materials, such as graphene, may also demonstrate increased HER activity of the edge atoms compared to the plane atoms, as determined by SECCM. Perforations were created in the graphene basal plane, which enriched the number of edges [29], as reported for TMD [44]. It was found that the synergetic effects of two important factors – edge structure of graphene and nitrogen/phosphorous cooping – enhance HER activity significantly. The SECCM results were confirmed by DFT, suggesting that the pyridinic nitrogen atom could be the catalytically active site, and phosphorus dopants promote the charge accumulation on the catalytically active nitrogen atoms [29].

Since classic TMDs are semiconductors, they have attracted attention as photoelectrocatalysts for light-energy conversion. However, defects, which enhance catalytic performance, may also serve as recombination centers for photogenerated charges, thereby reducing light-harvesting performance and HER and causing a trade-off between photovoltaic and electrocatalytic properties. SECCM data have shown that when large steps (>50 nm) are generally detrimental to HER, features much smaller than the optical penetration depth would not affect optical efficiency, maintaining a positive effect on HER performance [40].

It was reported that the number of layers does not significantly affect both HER overpotential and Tafel slope. As well, there are no synergistic effects when using MoS_2–WS_2 heterostructures; the underlying layer does not have any additional influence [15]. However, cathodic photocurrent in WSe_2 increases with an increase in the layer thickness in the range of 20–100 nm. This is attributed to the overlapping of space charge layers on each interface [40]. Therefore, photoelectrochemical applications require the tuning of other parameters in addition to pure electrochemical ones.

2.2 *Other Materials with HER Activity*

Besides TMD-like materials, the investigation of HER using SECCM could unveil important related effects in electrochemical systems. First, H_2 generation is a model for bubble nucleation in many devices, including fuel cells. Bubbles reduce electrocatalytic effectivity because they temporarily block the surface of the active catalyst. It was demonstrated that despite a heterogeneous distribution of the energy of nucleation for H_2 bubbles on a Pt surface, such heterogeneity is not correlated with crystal grains or grain boundaries but with nanosized surface features, such as crevices [41]. Also, HER served as the cathodic counterpart to anodic metal dissolution in acidic media and revealed corrosion processes in low carbon steel. SECCM, combined with EBSD, showed that HER activity only slightly depends on the grain facet index in the order: (100) > (111) > (101). However, a much more pronounced effect was observed for grain boundaries and MnS submicrometer inclusions, confirming that these sites are local HER "hot spots" and highly active corrosion centers (Fig. 3) [42]. For Zn, the low facet index significantly affects HER activity due to Zn corrosion, which correlates with both hydrogen adsorption energy and activity of the anodic process of Zn dissolution [78].

The HER process on Pd electrode was employed as a model system to investigate the effect of the surface structure on the ingress of hydrogen into metals, which is particularly important in corrosion areas (e.g., hydrogen embrittlement) and energy conversion/storage. Correlating spatially resolved cyclic voltammetric data from SECCM with colocated structural information from EBSD reveals a clear relationship between the crystal orientation and the rate of hydrogen adsorption–absorption. Grains closest to the low-index facets [i.e., (100), (101), and (111)] facilitate the lowest rates of hydrogen absorption, whereas those of high-index facets [e.g., (411)] promote higher rates. Enhanced kinetics are also observed at grain boundaries [47]. These experiments highlight the effectiveness of combined SECCM–EBSD in structure–activity measurements in electrochemical science.

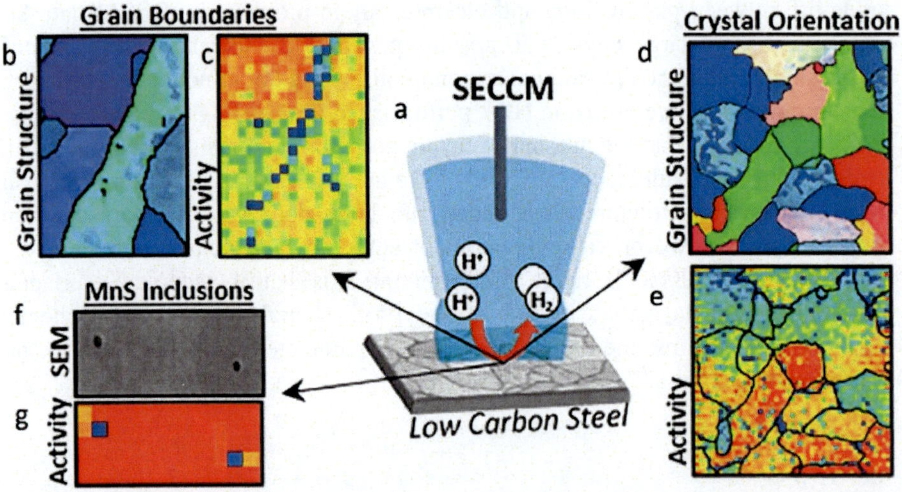

Fig. 3 (**a**) SECCM scheme for HER observation. (**b**) EBSD map of a "cathodically active" grain boundary, with the misorientation between the two neighboring grains. (**c**) Spatially resolved surface current (i_{surf}) map in the area of the surface shown in (**b**). Scale is between −90 and −55 pA. (**d**) EBSD map of the areas of the low carbon steel surface that were characterized with SECCM. The EBSD data correspond to (100) (red), (101) (green), and (111) (blue). (**e**) Spatially resolved i_{surf} map (2,500 pixels) over a 40 × 40 μm^2 area, corresponding to the areas shown above in (**d**). Scale is between −100 and −60 pA. (**f**) SEM image of the area of low carbon steel scanned with SECCM. The matrix of scanning points (pixels) can be seen as dark spots on the surface. The two darker spots are MnS inclusions. (**g**) Corresponding spatially resolved i_{surf} map. The i_{surf} maps were taken at $E_{app} = -1.337$ V vs. Ag/AgCl with a probe of 150 nm diameter containing 5 mM H_2SO_4. Adapted with permission from Ref. [42]. https://pubs.acs.org/doi/abs/10.1021/acs.jpcc.9b07216. Copyright 2019 American Chemical Society. Further permissions related to the material excerpted should be directed to the ACS

3 Metals

3.1 Effect of Grain Orientation on Electrochemical Activity

Metals are 3D-like materials with closest- and close-packing types; thus, they demonstrate many possible variants of grain orientation and facet indices. The unique advantage of SECCM over single-crystal measurements is the ability to investigate high-index facets, which are difficult to achieve but are common in practical applications [76]. In the previous part 2.2 the effect of grain orientation on Zn [78] and low carbon steel [42] corrosion has been described.

Noble metals, such as Pt and Pd, are used as model electrodes to study electrochemical processes, as described above for HER-related effects, hydrogen absorption [47], and gas bubble formation [41]. In addition to HER, other electrochemical processes are studied to assess the surface electrochemical effects. The main features of SECCM in these applications include (1) allowing pseudo-single-crystal experiments (in individual grains of a polycrystalline sample with high-index facets and grain boundaries) owing to the spatial confinement of the electrochemical cell;

(2) allowing access to fast surface kinetic effects owing to the higher mass transport rates generated; (3) enhancing the resolution to competing processes (such as surface oxidation) with the ability to fine-tune the time regime in which measurements are made [75].

$Fe^{2+/3+}$ redox couple is used as a model inner-sphere process to study electron transfer. SECCM showed that the electrochemical oxidation of Fe^{2+} is sensitive to the crystallographic orientation of the microcrystalline Pt grain orientation, regardless of the supporting electrolyte [75]. Similar activity dependence on grain orientation was observed for triiodide reduction in ionic liquid (Pt grains with a dominant (110) crystallographic character exhibited higher catalytic activity compared to those with a major (100) facet orientation [76]), similarly as it was reported for $Fe^{2+/3+}$ [75]. The obtained results are important for the efficiency enhancement in dye-sensitized solar cells (DSSCs).

The effect of electrode structure on ORR is of major interest in connection to electrocatalysis for energy-related applications [37]. The attributes of the approach herein stem from (1) the ease with which polycrystalline-substrate electrodes can be prepared; (2) the wide range of surface character open to study; (3) the possibility of mapping reactivity within a particular facet (or grain) in a pseudo-single-crystal approach and acquiring a high volume of data as a consequence; (4) the ability to measure the activity at grain boundaries; (5) experimental setups (SECCM) that mimic the three-phase boundary in low-temperature fuel cells. Pt grains having mixed (111) and (100) characters exhibit higher ORR activity than those with (100) and (110) features. However, ORR on ideal Pt (111) terraces is hindered by sulfate adsorption. Similar activity dependence on the surface structure was observed for hydrazine oxidation: grains having more (111) features showed higher activity than those having more (100) ones. In addition, in a deaerated medium, the signal was much higher than that in air, which is attributed to the competition between hydrazine oxidation and ORR [77].

3.2 Effect of Grain Boundaries on Electrochemical Activity

For some reactions, grain boundaries are more important than grains. Enhanced kinetics on hydrogen absorption were reported on Pd grain boundaries [47] as well as the corrosion-related HER on a low carbon steel surface (Fig. 3) [42]. $Fe^{2+/3+}$ activity variations between individual grains are minor compared with the enhanced activity at certain grain boundaries [75]. SECCM, with submicrometer resolution, showed that grain-boundary surface terminations in gold electrodes are more active than grain surfaces for electrochemical reduction of carbon dioxide (CO_2) to carbon monoxide (CO) but not for the competing HER. The catalytic footprint of grain boundaries is commensurable with its dislocation-induced strain field [48]. The potential of zero charge (PZC), its correlation with HER, and grain orientation have been mapped on polycrystalline Pt using both SECCM and EBSD. The experiment revealed a negative correlation between PZC and the Tafel slope of

HER, whereas PZC depended on grain orientation. It was observed that within a potential range on the same electrode, grains with higher PZC can be negatively charged, whereas those with lower PZC are positively charged. Under these scenarios, a net electric field is expected across the boundaries of grains of different PZCs, which is similar to the case of contact electrification [79]. Understanding the distribution of PZC at grains and grain boundaries of polycrystalline materials would enable rational tuning of electrocatalytic activity and selectivity based on electrostatics.

When metal catalysts are supported by other phases, the interfacial effect significantly enhances the performance. Reduced graphene oxide (rGO) shows high CO_2 adsorptivity and can be used for CO_2 transfer to catalyst surfaces. On-site SECCM experiments directly revealed this additional supply route and the enhancement of the catalytic activity at the interface between Sn catalyst and rGO (Fig. 4) [49]. This cooperative mechanism on the rGO-supported-catalyst provides a general design direction for accelerating the electrolytic synthesis of not only formic acid but also alcohols and olefins, which share the initial step of CO_2 adsorption on the catalyst surface.

4 Nanoparticle Activity Probing

Nanosized materials are practically important in catalysis owing to their large surface area, high number of near-surface atoms, and properties significantly different from the bulk. Imaging of nanoparticle electrochemical activity is required to investigate nanosized catalysts or determine the effect of nanosized co-catalysts on the local (photo) electrocatalytic performance of the bulk catalysts. However, most of the reported pipettes have a diameter ~100 nm and larger, which is comparable to the analyzed objects. In such cases, SECCM deals with probing rather than imaging, despite that some nanoparticle SECCM imaging have been reported [49]. The electrochemical response of ~100-nm aggregates of Pt nanoparticles supported by carbon nanotubes was successfully mapped in the cathodic and anodic conditions [81]. $LiFePO_4$ particles (~100 nm) were probed with a pipette of ~1 μm diameter, and heterogeneity was observed between particles attributed to a shift in the oxidation and reduction potentials [82]. The photoelectrochemical activity of Au nanorods in hydrazine oxidation was probed by combining SECCM with optical hyperspectral imaging, which was named optically targeted electrochemical cell microscopy. In situ spectroscopic investigations of individual nanorods showed that surface reactions at Au, such as gold oxide formation and reduction, play a significant role in the observed response [80]. ORR was probed on Pt nanoclusters in combination with XPS and TEM, and a decrease in catalytic activity due to Pt poisoning was observed [83]. A versatile, effective, high-throughput SECCM measurement to derive the electrochemical reactivity of oxygen evolution reaction (OER) of a single up to several individual Co-N/C composite nanoparticles in a single experiment under alkaline conditions has been reported [84]. SECCM CV mapping was employed to

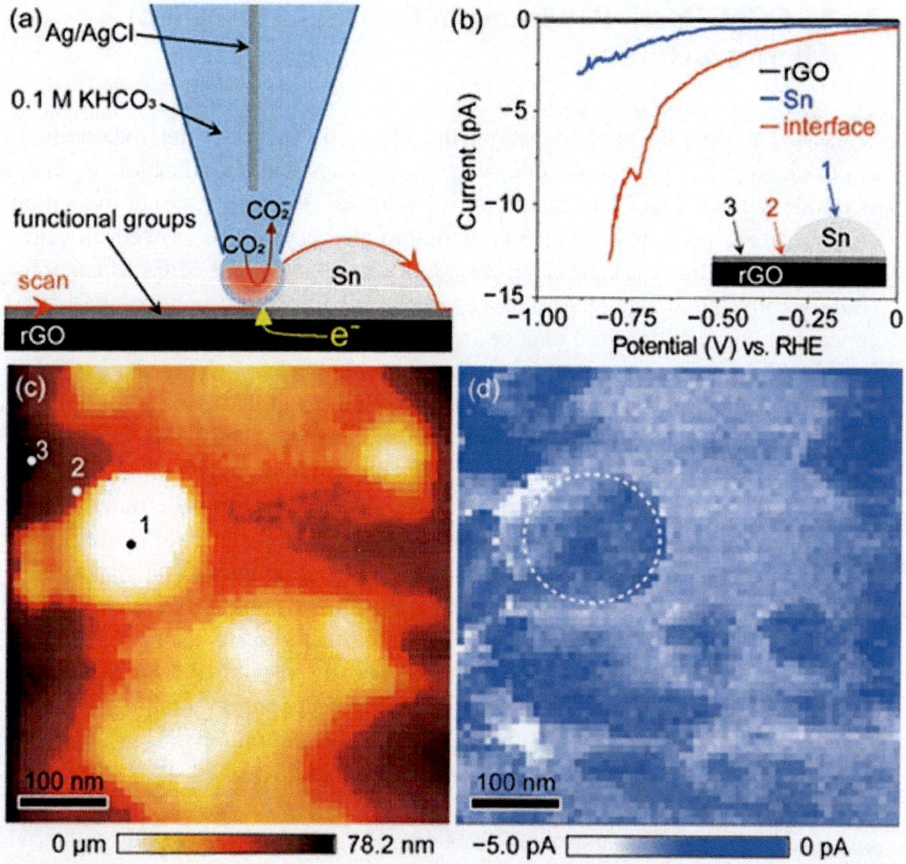

Fig. 4 SECCM electrochemical reduction current mapping. (**a**) Experimental setup and (**b**) typical linear sweep voltammograms for Sn/rGO in CO_2-saturated 0.1 M aqueous $KHCO_3$ electrolyte. The curves were obtained at the positions depicted in the inset and in (**c**). (**c**) Topography and (**d**) reduction current mapping of Sn/rGO in the CO_2-saturated electrolyte at −0.8 V (vs. RHE). The area of the electrochemical cell is 7,850 nm^2, which at a current of 1.00 pA gives a current density of 12.47 mA cm^{-2}. Adapted with permission from Ref. [49]. Copyright 2021 American Chemical Society

probe the facet-dependent electrocatalytic activity of individual nanoparticles, measuring the HER activity of a statistically meaningful set of Au nanocubes and octahedra with distinct types of crystal planes on their surfaces. The electrochemical activity of the nanocubes was increased compared to that of other particles [85].

5 SECCM-Based Patterning of the On-Surface (Sub) Microstructures

SECCM can also be used to pattern the required on-surface microstructures for electrochemical applications. One way is the chemical modification of existing electrode surfaces. SECCM formation of artificial defects by anodic oxidation to enhance HER performance [44] or photoluminescence [45], as well as cathodic generation of sulfur vacancies [15] was reported for TMD. The SECCM nanopipette could locally induce charge/discharge processes in $LiFePO_4$ and change local Li^+ concentration, resulting in a change in cell potential distribution with high control [14]. SECCM was used as a read–write probe for both electrochemical modifications of the local surface of boron-doped diamond (BDD) and its "reading" at the micron scale. Localized anodic oxidation converts the H-terminated BDD surface to –O, revealed by the change in current in $Ru(NH_3)_6^{3+}$ cathodic reduction [34].

Layered carbon-based materials are also promising for patterning. Patterned dopamine oxidation [22] and AQDS reduction [23] on the HOPG surface were employed to test the activity difference between the graphite (0001) plane and edge carbon atoms (Fig. 5). Local diazonium modification of pristine sp^2 carbon surfaces was performed with high control at the micrometer scale using SECCM. The properties of the modified surfaces were tuned by varying the applied potential and meniscus contact time. Such rehybridized defects not only tailor the chemical functionality of surfaces but may also generate an electronic bandgap in graphene, increasing the prospect of graphene-based electronic devices [33].

However, it is also possible to use a SECCM probe for on-surface electrochemical deposition of nano and microstructures. ZnO micro-spots with excellent electrochemical capacitance and electronic properties were patterned on an ITO surface by $Zn(NO_3)_2$ electrochemical reduction, where the Zn source was loaded into a pipette [86]. SECCM is also employed in the electropolymerization of patterned polymer structures, which has been performed for both nonconducting poly(HEAA) [87] and conducting PANI [88] and P3HT [61, 62] polymers. The former is important in biotechnology and surface modification, and the latter could be used to form electrical contacts between required objects, build micro- and nanoelectrodes and photoelectrochemical cells. Dual-barrel nanopipette was filled with a monomer-containing solution in water [87, 88] or acetonitrile [61, 62], which was delivered to an electrode surface through liquid meniscus at the pipette end. The surface could be additionally functionalized with initiating groups, such as a-bromoester for poly (HEAA) deposition [87]. By adjusting the potential of the surface, localized electropolymerization could be achieved. Thus, polymer objects of submicron and micron dimension, depending on the pipette diameter, were obtained. The thickness was controlled by varying the deposition conditions. SECCM meniscus growth of PANI allowed the building of fine 3D polymer structures with good electrical contacts even outside the electrode area [88].

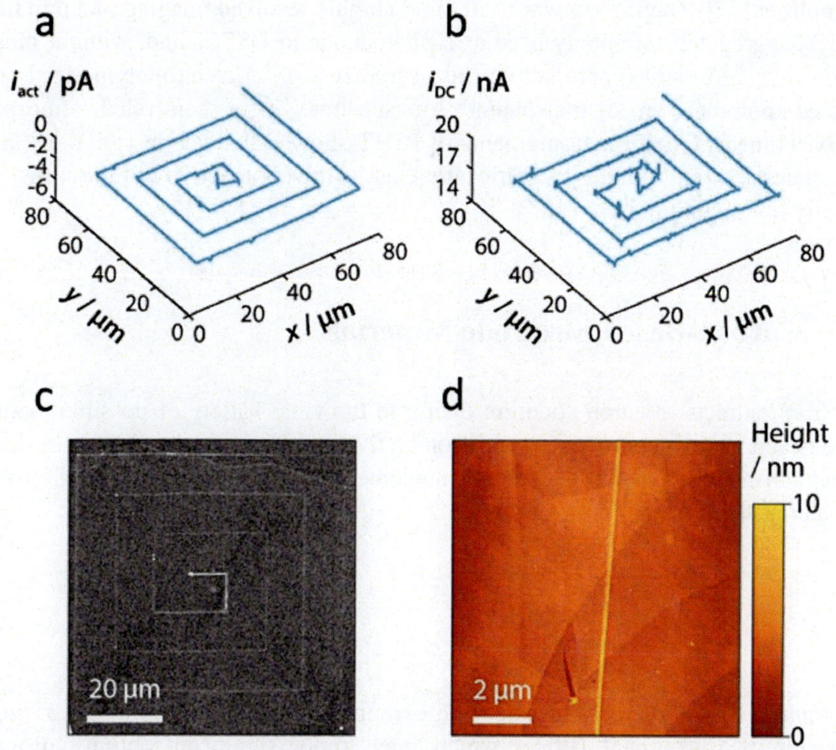

Fig. 5 SECCM maps of (**a**) surface activity i_{act} and (**b**) DC component of conductance current i_{DC} for the diffusion-limited reduction of 100 μM AQDS in 0.1 M HClO4 during reactive line patterning on AM HOPG, at a scan speed of 1 μm s^{-1}. (**c**) FE-SEM image showing the deposited AQDS line pattern. (**d**) AFM image of part of the pattern. Adapted with permission from Ref. [23]. Copyright 2014 American Chemical Society

6 Photoelectrochemical SECCM Imaging

The photoelectrochemical properties of TMD materials for HER electrocatalysis were described in the HER section [40, 44]. Through SECCM, it was obtained that charge photogeneration and electrochemical activity are not necessarily correlated parameters and may even contradict each other in some defect sites. Optical properties were tuned by dynamic electrochemical control of electrode potential. Spectroelectrochemistry showed that SECCM-induced electrochemical anodization of 2D MoS_2 results in (1) the 10-fold enhancement of photoluminescence (PL) and red-shift in the PL spectrum, attributed to the decrease in nonradiative decay rates of excitons, and (2) spatial heterogeneities in its photoelectrochemical activities, attributed to the spatially heterogeneous edge-type defect sites in the crystal [45].

SECCM was coupled with photo illumination to create a submicrometer spatial resolution cell for structure–(photo)activity mapping using LED intensity modulation with further detection of periodic photocurrent components using a lock-in amplifier [62]. The system was used for (1) highly resolved imaging of a thin film of TiO_2 aggregates commonly used as a photoanode in DSSCs and, without modulation, for (2) in situ (photo)activity measurements of an electropolymerized conjugated polymer on a transparent Au substrate in a controlled atmospheric environment. Ex situ measurements of P3HT showed that subtle variations in film thickness and morphology significantly change the photoactivity of the films, especially after aging [61].

7 Carbon-Based Electrode Materials

Carbon attracts research attention owing to the wide variety of possible modifications and available structures, with conductivity varying from dielectric to metallic and half-metallic. It is also an abundant element and forms environmentally friendly materials.

7.1 Graphite

Graphite has a layered structure with carbon in sp^2 hybridization with a strongly anisotropic hexagonal lattice, which may imply significant activity difference between the basal (0001) plane and the edge planes. Many non-SECCM studies revealed that the basal surfaces of sp^2 carbon materials are essentially inert, considering the electron transfer, and the electrochemical activity comes from the edges. However, SECCM data [21–23, 28] using different simple Red-Ox probes, such as $Ru(NH_3)_6{}^{3+/2+}$, $Fe(CN)_6{}^{4-/3-}$ [21], and $Fe^{2+/3+}$ [28], demonstrated that freshly cleaved, pristine basal surfaces support fast electron transfer which is not confined to step edges. The basal plane is the major electroactive site on cleaved HOPG. The edge state could be determined through the local electrochemical activity of a $Ru(NH_3)_6{}^{3+/2+}$ couple whose redox potential lies close to the intrinsic Fermi level of uncharged graphitic materials. On the other hand, for species whose redox potential is far from the intrinsic Fermi level (e.g., ferrocene derivatives), the edge state of graphene cannot be distinguished against the high activity of the basal surface [24].

The history of the HOPG surface strongly influences the electrochemical behavior because the local surface conductivity of HOPG decreases significantly with time [21]. An advantage of SECCM is that the measurements can be done shortly after sample preparation, minimizing history effects. Thus, different time-dependent electrochemical activity between the basal planes and step edges can be revealed, where higher step edge activity with $Ru(NH_3)_6{}^{3+/2+}$ is observed in "aged" graphite [24]. Such behavior after cleavage is not associated particularly with any surface contaminants but attributed to the spontaneous delamination of HOPG with time to create partially coupled graphene layers, which majorly affect the density of states of graphene and graphite edges [24]. This opposes the SECCM data for TMD, where the edge atoms lose their activity with aging due to oxidation [15]. It is also found that the time after the cleavage of HOPG surfaces affects the surface wettability (and surface contamination), and this slowly decreases the kinetics [28].

Besides simple outer-sphere redox couples, more complex electrochemical systems, such as dopamine, on the graphite surface have been studied using SECCM. Such systems undergo complex electron–proton-coupled electrooxidation, which is also affected by the adsorption of both the initial substance and side products. Therefore, there is a need to clarify the relative contributions of intrinsic electrode kinetics and reagent adsorption, which may be site-specific. However, for freshly cleaved graphite, dopamine oxidation [22] and AQDS reduction [23] demonstrated no significant difference for basal plane and edge electrochemical activities during electrochemical patterning (Fig. 5). Similarly, high basal-plane activity and independence on the step edge density were observed for NADH oxidation [25]. However, in all cases, the surface is easily contaminated by reaction products, which leads to its rapid deactivation [22, 23, 25].

For "aged" graphite, such as screen-printed carbon electrodes, the oxidation of dopamine was dependent on the location on the electrode surface. The most facile electrooxidation was observed in surface regions having graphite particles with a high degree of crystallinity, which also supports dopamine adsorption. The adsorbed layer may also play a role as an electrocatalyst for the oxidation of solution-phase dopamine [30].

Similar to TMD, the electrochemical performance of graphite can be further increased by an additional defect formation. Such defects include blisters, which are formed near grain boundaries and step edges during voltammetric cycling in $HClO_4$ due to ClO_4^- intercalation between the layers with further gas evolution. Linear sweep voltammograms (LSV)–SECCM data show that the blistered region exhibits a significant cathodic shift in the onset potential of hydrazine electrooxidation reactions. The improved electrochemical activity of the hollow structure could be attributed to the increased adsorption of protonated hydrazine at oxygenated defect sites, the ease of ion–solvent intercalation/deintercalation, and the reduced susceptibility to N_2 nanobubble attachment (as a product of the reaction) [26].

7.2 Graphene

Graphene, a few-layered form of graphite, has also been intensively studied as electrocatalytic material. However, sample history is important for electrochemical activity, and there are multiple preparation methods for graphene, including bottom–top (CVD, epitaxial growth, conversion of organic self-assembled monolayer, etc.) and top–bottom methods (scotch tape, chemical, and electrochemical exfoliations). The supporting substrate may also affect graphene properties. Hence, graphene electrochemical activity is affected by many parameters, complicating its studies.

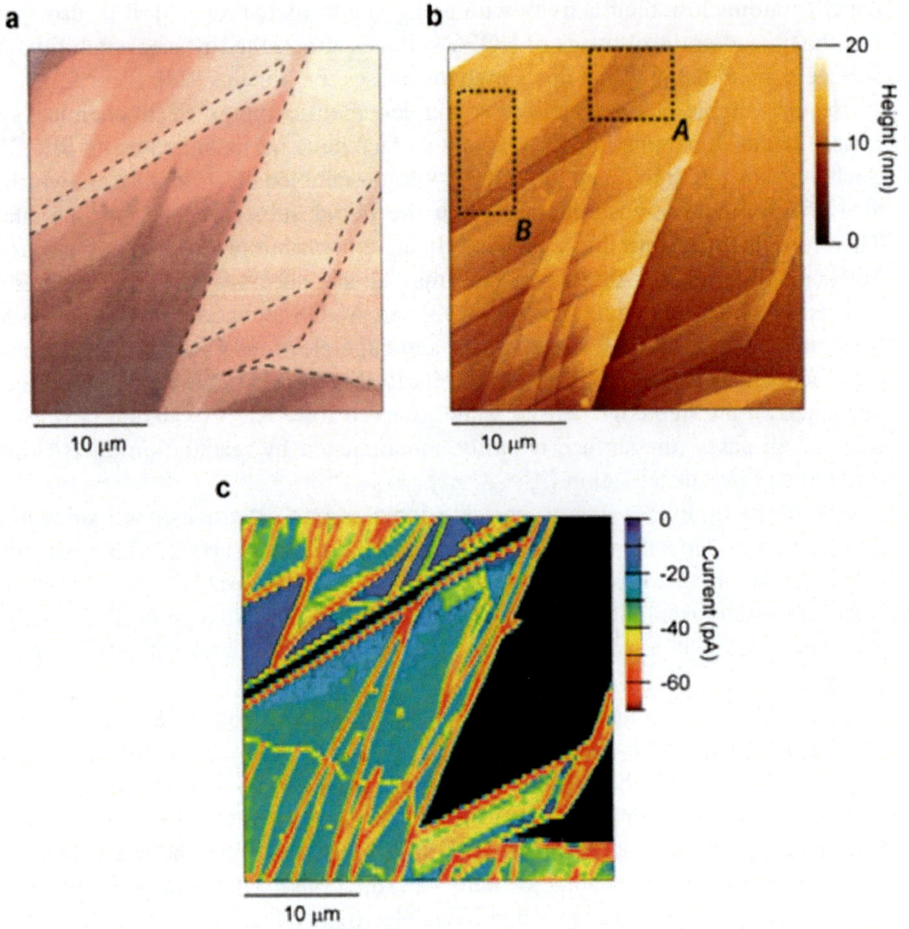

Fig. 6 Multimicroscopy of exfoliated graphene (from ZYA grade HOPG) on a Si/SiO_2 substrate. (**a**) Optical microscopy image (dashed lines define the boundaries between the graphene sample and Si/SiO_2 substrate), (**b**) AFM topographical image, and (**c**) SECCM electrochemical (EC) map for the reduction of $Ru(NH_3)_6^{3+}$ (5 mM in 25 mM KCl and phosphate buffer, pH = 7.2). The EC map was acquired at a potential $E_{sub} = -0.464$ V ($\eta = -0.198$ V) and 2 μm s^{-1} probe scan rate. For the EC current map, the color scheme was applied only when the EC current value is detectable above 3 times the background current (200 fA); otherwise, it was assigned the black color and coincides with the Si/SiO_2 substrate. In (**b**), the two areas labeled A and B exhibit the lowest EC currents. Adapted with permission from Ref. [24]. Copyright 2015 American Chemical Society

Various data on the edge state activities of graphene have been obtained by SECCM. As reported for freshly cleaved graphite, the electrochemical activity was uniform either along the edges of the flakes or at the (step edge) boundary between the flakes of the CVD-deposited graphene [20]. However, other studies using SECCM showed increased edge activity for "aged" exfoliated [24] and CVD-processed [29] graphene (Fig. 6). In the latter case the edge activity may be additionally enhanced by nitrogen or/and phosphorus doping. The difference between edge activities in different reports, however, could be attributed to the different used redox probes. Ferrocenemethanol [20] redox potential is far from the expected intrinsic Fermi level of graphene to detect edge state compared with $Ru(NH_3)_6{}^{3+/2+}$.

However, there is a strong dependence of electrochemical activity and electron transfer kinetics on the number of graphene layers, with monolayer graphene having the slowest rate. The rate increases for bilayer graphene, and ultrathin graphite has the fastest kinetics. This is common for both CVD-processed [20] and exfoliated [24] graphene with different redox couples and may be related to the underlying substrate effect, which is, to some extent, screened by the additional graphene layer beneath. This may increase the charge carrier mobility but does not enhance electron transfer kinetics compared to the intrinsic activity of monolayer graphene. The effect of the substrate has also been proven to be negligible as suspended graphene and graphene on Cu substrates demonstrated similar electrochemical activity with some differences in wettability due to meniscus compressing and pushing of suspended graphene [27].

SECCM provides a useful technique for evaluating the open-circuit voltage of 2D-coated semiconductor photoelectrodes, such as p-Si/graphene and p-Si/h-BN, using several one-electron outer-sphere redox couples. The measurements showed that the polycrystalline nature of 2D materials had little effect on the results and confirmed that the macroscale measurements reflected the junction behavior at the nanoscale [31].

7.3 Nanotubes

Carbon nanotubes are another form of sp^2 carbon, which could be considered as a rolled graphene layer. Here, we introduced SECCM-based studies on single-wall nanotubes (SWNTs). SWNT structure is determined by the chirality vector, with the two main variants of graphene rolling: "armchair" (metal conductivity) and "zigzag" (semiconductor; metallic only at $n \propto 3$). Intermediate types ("chiral") are also semiconducting. Similar to graphite derivatives, SWNTs are prospective materials for electrochemical sensing and biosensing.

Spatially resolved SECCM combined with micro-Raman spectroscopy showed that metallic and semiconducting SWNTs exhibit a high kinetic rate for ferrocenemethanol oxidation [17, 18]. This is consistent with the similar density of states of metallic and semiconducting nanotubes at this positive potential. For $Ru(NH_3)_6{}^{3+/2+}$ mediator and cathodic process, metallic SWNT has significantly higher electrochemical activity than the semiconducting one (Fig. 7) [18, 19]. This is because the formal potential of $Ru(NH_3)_6{}^{3+/2+}$ lies in the charge-depletion region of some of the semiconducting SWNTs [17]. Thus, SECCM offers an easy way to distinguish the SWNT type, and it could be possibly extended toward chirality determination.

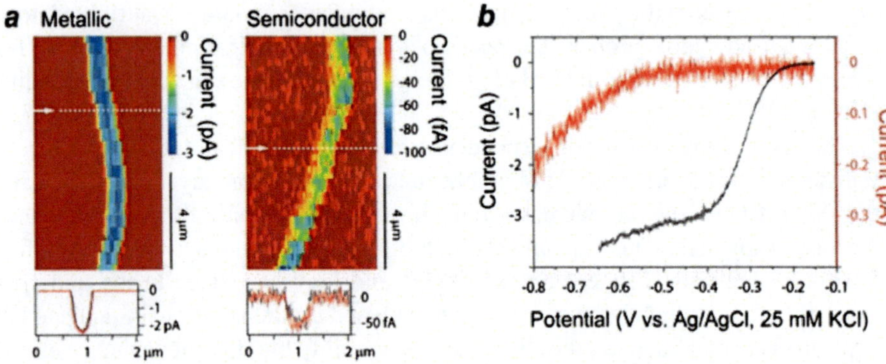

Fig. 7 (**a**) SECCM electrochemical (EC) current maps for the reduction of 0.5 mM $Ru(NH_3)_6^{3+}$ (25 mM KCl and phosphate buffer, pH = 7.2) at metallic and semiconducting SWNT devices. The EC map of the semiconductor SWNT was acquired at a much higher driving potential (V_{appl} = −0.65 V) than the metallic nanotube (V_{appl} = −0.35 V) to detect a response. Scan profiles, forward (red) and reverse (black) for each SWNT (marked with a white dashed line on the corresponding map) are shown. (**b**) Linear sweep voltammograms (starting potential V_{appl} = −0.15 V) obtained using SECCM setup in static mode with the meniscus centered over an individual SWNT (black line for metallic, left-hand y-axis; red line for semiconductor SWNT, right-hand y-axis) at 50 mV s^{-1}. All potentials are given with respect to Ag/AgCl, 25 mM KCl. Adapted with permission from Ref. [19]. Copyright 2014 American Chemical Society

Many authors have suggested that sidewalls are electrochemically inactive in SWNTs and only edge-plane-like sites and open oxygenated nanotube ends are active. However, highly resolved SECCM data demonstrated the opposite: most SWNTs are almost uniformly active on the length scale for two types of outer-sphere redox processes [17–19] as well as serotonin oxidation [18]. An important goal of the SWNT–SECCM platform for serotonin oxidation is that no electrode fouling with reaction products takes place while the concentrations are near-physiological. This is explained by the high mass transport toward and away from SWNT [18].

7.4 BDD

Diamond is another well-known form of carbon where carbon is in sp^3 hybridization and tetrahedral coordination in bulk. This results in strong covalent bonding in all three dimensions, in contrast to graphite consisting of stacked 2D layers. However, pure bulk diamond is an insulator with a large bandgap. Therefore, heterovalent doping is required to make it conducting. Among the possible dopants, boron is one of the most structurally similar to carbon, so it can effectively substitute it in the lattice.

Several factors affect the surface properties. First, the atomic structure of the facet, i.e., grain orientation, similar to that of metals (see Sect. 3.1). In contrast to strongly anisotropic graphite, boron-doped diamond (BDD) has multiple available facets, including (100), (110) and (111). Here, SECCM imaging is combined with EBSD to detect grain orientation. Second, the surface content of B, which also strongly depends on crystalline orientation [25, 32]. Third, the presence of surface defects [35]. Fourth, H- or O-termination [34, 36]. The surface of freshly grown BDD is hydrogen H-terminated and hydrophobic, but in the air (slowly) or under anodic conditions, it oxidizes to the oxygen O-terminated form. This makes the studies of spatial electrochemical activity of BDD extremely important for BDD electrode applications.

It was found that facets with higher boron content have higher heterogeneous electron transfer activity for outer-sphere, inner-sphere, and complex electrochemical reactions (serotonin oxidation). Also, the (111) facet may incorporate more B atoms than (100) [32, 35]. Half-wave potential, extracted from cyclic voltammetric and LSV-SECCM experiments, demonstrates that electron transfer is faster at the H-terminated (111) surface than at the H-terminated (100) one (Fig. 8) [73]. Voltammetric SECCM has revealed that the solvent window is directly affected by the local boron doping concentration, irrespective of the surface termination, with the solvent window for less-doped facets being wider than that for more-doped facets [36].

Generally, for a similar B-doping concentration, the solvent window is narrower for an O-terminated surface than for an H-terminated one. However, the onset potential for the anodic process is relatively insensitive to boron doping level for O-BDD (for the doping levels herein) but doping dependent on H-BDD. This is inferred to be partly due to a change in the mechanism to OER [36]. The conversion from H- to O-termination can be performed electrochemically using SECCM in anodic conditions as a “write” operation. The “read” process performed in cathodic conditions showed a decrease in the reduction current in the O-terminated area [34].

The activities of BDD (sp^3 hybridized C) and graphite (sp^2 hybridized C) toward NADH oxidation were compared. These two materials have a relatively similar density of states, but the electrooxidation of NADH is much faster on the basal plane of graphite than on BDD due to the better adsorption of NADH on sp^2 carbon materials. However, the reaction products also tend to exhibit strong adsorption, causing rapid deactivation of the electrode activity. BDD was found to be less susceptible to surface fouling, but it is not immune to the deterioration of the electrode response, and the reaction showed more sluggish kinetics on this electrode. Also, the electrochemical currents for NADH oxidation are strongly correlated with the local boron doping concentration in individual facets [25].

8 Applications of Battery Materials

Micro–nanoscale electrochemical measurements have advantages in characterizing the kinetic parameters of battery materials, such as suppressing the capacitive current. For example, an ultramicroelectrode (UME) has been used to distinguish the bottleneck process during charging/discharging and to estimate the diffusion

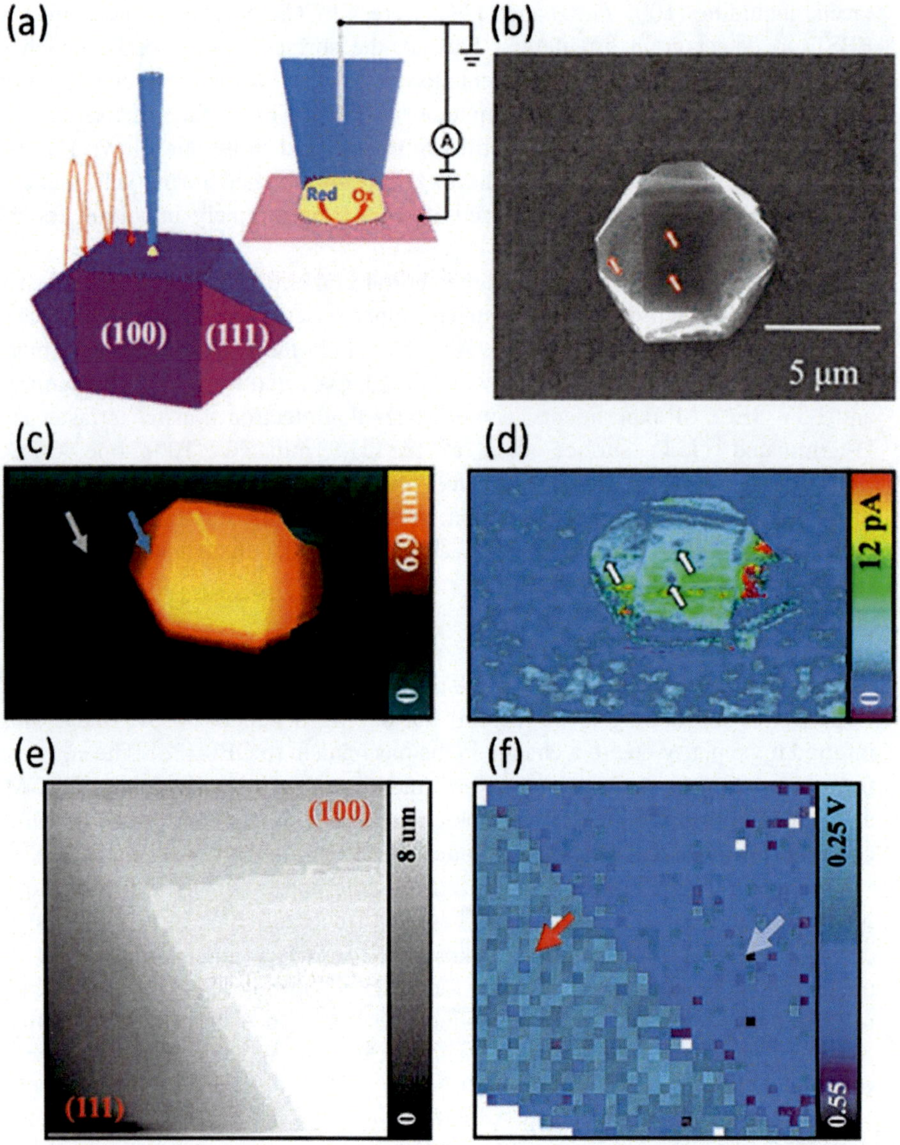

Fig. 8 (**a**) Illustration of the SECCM measurement of a H-BDD single particle. (**b**) SEM and SECCM (**c**) topography and (**d**) current images of a H-BDD particle with (100) face orientated upwards. Scan sizes (c, d) were 18 × 9 μm^2. Nanopipette of 50 nm nominal radius filled with 4 mM $K_4[Fe(CN)_6]$ in 0.1 M KCl was used. The applied voltage during the electrochemical measurement was +0.5 V vs. Ag/AgCl. The pixel number was 128 × 128. (**c, d**). (**e**) Topography and (**f**) half-wave potential images across (100) and (111) facets on a single-crystal H-BDD by SECCM. Scan sizes were 6 × 6 μm^2. The scan rate was 1 V/s. The pixel number was 32 × 32. Adapted with permission from Ref. [73]. Copyright 2021 American Chemical Society

coefficient of LIBs [89–92]. An UME in contact with single particles provides an ideal electron pass between a single particle and a UME and ion access owing to the formation of the sphere shape diffusion layer. SECM is a powerful tool for characterizing the reactivity and electrochemical properties of battery materials with spatial resolution. The SECM feedback mode has been used to monitor the solid–electrolyte interface formation [93–97]. For direct quantitative investigation of spatially resolved ionic processes, mercury-capped platinum ultramicroelectrodes have been developed and used for Li^+ imaging based on Li stripping [98].

For SECCM measurements, a nanopipette can be used as a Li^+ provider or collector to characterize battery materials with sub microscale spatial resolution. SECCM has been used to quantitatively visualize electrochemical phenomena on the composite cathode material $LiFePO_4$ (LFP) with a resolution of ~100 nm [14]. SECCM nanoscale electrochemical cells are effective for fast-scan CV to characterize the redox current. In this work, the LFP redox current was measured with 100 V/s by suppressing the capacitive current. Moreover, zero-current (essentially open circuit) cell potential mapping and local charge/discharge characteristics were also performed by SECCM.

Correlative analyses of SECCM and SEM are useful for characterizing the intrinsic properties of battery materials at the single-particle level and identifying variations in the individual single particles. Steen et al. characterized single LFP particles and found that most of them display single oxidation and reduction peaks at ca. 3.65 and 3.14 V vs. Li/Li^+, respectively [82]. However, heterogeneity between particles can be observed as a shift in the oxidation and reduction potentials. The authors also used ionic liquid for LFP characterization and improved the coulombic efficiency of the lithiation/delithiation reaction [52]. Unwin et al. also performed a correlative analysis of SECCM and SEM for single $LiMn_2O_4$ particles [55]. They found heterogeneous electrochemical behavior in individual $LiMn_2O_4$ particles attributable to the difference between particle size, composition, crystallinity, and orientation.

Cathode surface coating with metal-oxide thin layers has been intensively studied to improve the cycle durability of lithium-ion batteries. It was obtained that the cycle durability of lithium-ion batteries depends on side reactions occurring at the electrode/electrolyte interface. To improve the durability of LIBs, cathode surface coating with metal oxides, e.g., Al_2O_3, MgO, and ZrO_2, has been studied and validated [99]. To reveal the relationship between the morphology of a ZrO_2-coated layer and the charge/discharge property, SECCM was used to visualize Li^+ deintercalation current on $LiCoO_2$ (LCO) electrodes with different thicknesses of ZrO_2-coated layers [53]. SECCM current imaging, local CV, and galvanostatic charge/discharge experiment were performed to characterize the cycle durability and rate performance of ZrO_2-coated LCO thin-film electrodes and reveal the relationship between the ZrO_2 morphology and the rate performance (Fig. 9).

The crystal facet-dependent reactivity difference on thin-film electrodes for lithium-ion batteries has also been characterized by SECCM [56]. The SECCM analysis was performed in a glove box to obtain a well-defined argon gas atmosphere. The kinetic parameters of the anode material were studied. To estimate the

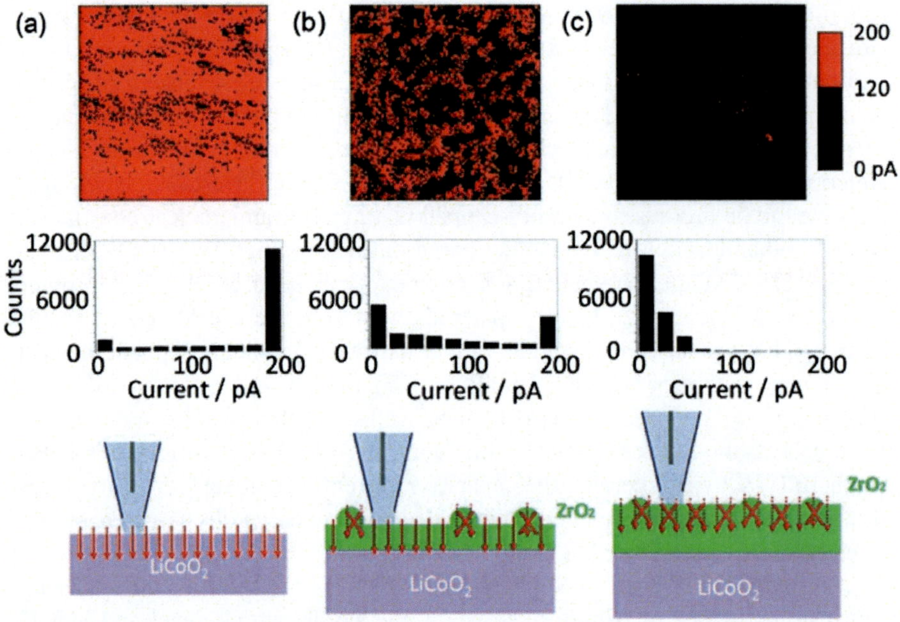

Fig. 9 SECCM current and binarization images and histograms for (**a**) $LiCoO_2$ thin-film electrodes, and $LiCoO_2$ thin-film electrodes coated by ZrO_2 for (**b**) 30 s and (**c**) 180 s. The threshold current was set to 120 pA. Scan size was 5×5 μm^2. Adapted with permission from Ref. [53], Copyright 2015, Royal Society of Chemistry

diffusion coefficient within the active materials using the Randles–Sevcik equation, the Li^+ diffusion state in solution should be faster than the diffusion state within the active materials. Micro/nanoscale electrochemical measurements can estimate high mass transfer rates because of the semi-sphere Li^+ diffusion state. This is an ideal situation for evaluating the Li^+ diffusion coefficient within the active material. The polycrystalline facet-dependent inhomogeneous diffusion coefficient of the $Li_4Ti_5O_{12}$ (LTO) thin-film electrode was visualized and evaluated as a 2D image using SECCM.

References

1. Binnig G, Rohrer H, Gerber C, Weibel E (1982) Surface studies by scanning tunneling microscopy. Phys Rev Lett 49(1):57–61
2. Polcari D, Dauphin-Ducharme P, Mauzeroll J (2016) Scanning electrochemical microscopy: a comprehensive review of experimental parameters from 1989 to 2015. Chem Rev 116(22): 13234–13278
3. Takahashi Y, Kumatani A, Shiku H, Matsue T (2017) Scanning probe microscopy for nanoscale electrochemical imaging. Anal Chem 89(1):342–357

4. Macpherson JV, Unwin PR, Hillier AC, Bard AJ (1996) In-situ imaging of ionic crystal dissolution using an integrated electrochemical/AFM probe. J Am Chem Soc 118(27): 6445–6452
5. Kueng A, Kranz C, Lugstein A, Bertagnolli E, Mizaikoff B (2003) Integrated AFM-SECM in tapping mode: simultaneous topographical and electrochemical imaging of enzyme activity. Angew Chem Int Ed 42(28):3238–3240
6. Turcu F, Schulte A, Hartwich G, Schuhmann W (2004) Label-free electrochemical recognition of DNA hybridization by means of modulation of the feedback current in SECM. Angew Chem Int Ed 43(26):3482–3485
7. Kurulugama RT, Wipf DO, Takacs SA, Pongmayteegul S, Garris PA, Baur JE (2005) Scanning electrochemical microscopy of model neurons: constant distance imaging. Anal Chem 77(4): 1111–1117
8. Zhao XC, Diakowski PM, Ding ZF (2010) Deconvoluting topography and spatial physiological activity of live macrophage cells by scanning electrochemical microscopy in constant-distance mode. Anal Chem 82(20):8371–8373
9. Takahashi Y, Shevchuk AI, Novak P, Zhang YJ, Ebejer N, Macpherson JV, Unwin PR, Pollard AJ, Roy D, Clifford CA, Shiku H, Matsue T, Klenerman D, Korchev YE (2011) Multifunctional nanoprobes for nanoscale chemical imaging and localized chemical delivery at surfaces and interfaces. Angew Chem Int Ed 50(41):9638–9642
10. Takahashi Y, Shevchuk AI, Novak P, Murakami Y, Shiku H, Korchev YE, Matsue T (2010) Simultaneous noncontact topography and electrochemical imaging by SECM/SICM featuring ion current feedback regulation. J Am Chem Soc 132(29):10118–10126
11. Comstock DJ, Elam JW, Pellin MJ, Hersam MC (2010) Integrated ultramicroelectrode-nanopipet probe for concurrent scanning electrochemical microscopy and scanning ion conductance microscopy. Anal Chem 82(4):1270–1276
12. Ebejer N, Schnippering M, Colburn AW, Edwards MA, Unwin PR (2010) Localized high resolution electrochemistry and multifunctional imaging: scanning electrochemical cell microscopy. Anal Chem 82(22):9141–9145
13. Williams CG, Edwards MA, Colley AL, Macpherson JV, Unwin PR (2009) Scanning micropipet contact method for high-resolution imaging of electrode surface redox activity. Anal Chem 81(7):2486–2495
14. Takahashi Y, Kumatani A, Munakata H, Inomata H, Ito K, Ino K, Shiku H, Unwin PR, Korchev YE, Kanamura K, Matsue T (2014) Nanoscale visualization of redox activity at lithium-ion battery cathodes. Nat Commun 5:5450
15. Takahashi Y, Kobayashi Y, Wang ZQ, Ito Y, Ota M, Ida H, Kumatani A, Miyazawa K, Fujita T, Shiku H, Korchev YE, Miyata Y, Fukuma T, Chen MW, Matsue T (2020) High-resolution electrochemical mapping of the hydrogen evolution reaction on transition-metal Dichalcogenide Nanosheets. Angew Chem Int Ed 59(9):3601–3608
16. Momotenko D, Byers JC, McKelvey K, Kang M, Unwin PR (2015) High-speed electrochemical imaging. ACS Nano 9(9):8942–8952
17. Guell AG, Ebejer N, Snowden ME, McKelvey K, Macpherson JV, Unwin PR (2012) Quantitative nanoscale visualization of heterogeneous electron transfer rates in 2D carbon nanotube networks. Proc Natl Acad Sci U S A 109(29):11487–11492
18. Guell AG, Meadows KE, Dudin PV, Ebejer N, Byers JC, Macpherson JV, Unwin PR (2014) Selection, characterisation and mapping of complex electrochemical processes at individual single-walled carbon nanotubes: the case of serotonin oxidation. Faraday Discuss 172:439–455
19. Güell AG, Meadows KE, Dudin PV, Ebejer N, Macpherson JV, Unwin PR (2014) Mapping nanoscale electrochemistry of individual single-walled carbon nanotubes. Nano Lett 14(1): 220–224
20. Guell AG, Ebejer N, Snowden ME, Macpherson JV, Unwin PR (2012) Structural correlations in heterogeneous electron transfer at monolayer and multilayer graphene electrodes. J Am Chem Soc 134(17):7258–7261

21. Patel AN, Collignon MG, O'Connell MA, Hung WO, McKelvey K, Macpherson JV, Unwin PR (2012) A new view of electrochemistry at highly oriented pyrolytic graphite. J Am Chem Soc 134(49):20117–20130
22. Patel AN, McKelvey K, Unwin PR (2012) Nanoscale electrochemical patterning reveals the active sites for catechol oxidation at graphite surfaces. J Am Chem Soc 134(50):20246–20249
23. Zhang GH, Kirkman PM, Patel AN, Cuharuc AS, McKelvey K, Unwin PR (2014) Molecular functionalization of graphite surfaces: basal plane versus step edge electrochemical activity. J Am Chem Soc 136(32):11444–11451
24. Guell AG, Cuharuc AS, Kim YR, Zhang GH, Tan SY, Ebejer N, Unwin PR (2015) Redox-dependent spatially resolved electrochemistry at graphene and graphite step edges. ACS Nano 9(4):3558–3571
25. Maddar FM, Lazenby RA, Patel AN, Unwin PR (2016) Electrochemical oxidation of dihydronicotinamide adenine dinucleotide (NADH): comparison of highly oriented pyrolytic graphite (HOPG) and polycrystalline boron-doped diamond (pBDD) electrodes. Phys Chem Chem Phys 18(38):26404–26411
26. Sharel PE, Kim YR, Perry D, Bentley CL, Unwin PR (2016) Nanoscale electrocatalysis of hydrazine electro-oxidation at blistered graphite electrodes. ACS Appl Mater Interfaces 8(44): 30458–30466
27. Zhang G, Guell AG, Kirkman PM, Lazenby RA, Miller TS, Unwin PR (2016) Versatile polymer-free graphene transfer method and applications. ACS Appl Mater Interfaces 8(12): 8008–8016
28. Zhang G, Tan SY, Patel AN, Unwin PR (2016) Electrochemistry of Fe(3+/2+) at highly oriented pyrolytic graphite (HOPG) electrodes: kinetics, identification of major electroactive sites and time effects on the response. Phys Chem Chem Phys 18(47):32387–32395
29. Kumatani A, Miura C, Kuramochi H, Ohto T, Wakisaka M, Nagata Y, Ida H, Takahashi Y, Hu KL, Jeong S, Fujita J, Matsue T, Ito Y (2019) Chemical dopants on edge of holey graphene accelerate electrochemical hydrogen evolution reaction. Adv Sci 6(10)
30. Martin-Yerga D, Costa-Garcia A, Unwin PR (2019) Correlative voltammetric microscopy: structure-activity relationships in the microscopic electrochemical behavior of screen printed carbon electrodes. Acs Sensors 4(8):2173
31. Thompson AC, Simpson BH, Lewis NS (2020) Macroscale and nanoscale photoelectrochemical behavior of p-type Si(111) covered by a single layer of graphene or hexagonal boron nitride. ACS Appl Mater Interfaces 12(10):11551–11561
32. Patten HV, Lai SCS, Macpherson JV, Unwin PR (2012) Active sites for outer-sphere, inner-sphere, and complex multistage electrochemical reactions at polycrystalline boron-doped diamond electrodes (pBDD) revealed with scanning electrochemical cell microscopy (SECCM). Anal Chem 84(12):5427–5432
33. Kirkman PM, Guell AG, Cuharuc AS, Unwin PR (2014) Spatial and temporal control of the diazonium modification of sp(2) carbon surfaces. J Am Chem Soc 136(1):36–39
34. Patten HV, Hutton LA, Webb JR, Newton ME, Unwin PR, Macpherson JV (2015) Electrochemical "read–write" microscale patterning of boron doped diamond electrodes. Chem Commun 51(1):164–167
35. Tomlinson LI, Patten HV, Green BL, Iacobini J, Meadows KE, McKelvey K, Unwin PR, Newton ME, Macpherson JV (2016) Intermittent-contact scanning electrochemical microscopy (IC-SECM) as a quantitative probe of defects in single crystal boron doped diamond electrodes. Electroanalysis 28(10):2297–2302
36. Liu D-Q, Chen C-H, Perry D, West G, Cobb SJ, Macpherson JV, Unwin PR (2018) Facet-resolved electrochemistry of polycrystalline boron-doped diamond electrodes: microscopic factors determining the solvent window in aqueous potassium chloride solutions. ChemElectroChem 5(20):3028–3035
37. Chen CH, Meadows KE, Cuharuc A, Lai SCS, Unwin PR (2014) High resolution mapping of oxygen reduction reaction kinetics at polycrystalline platinum electrodes. Phys Chem Chem Phys 16(34):18545–18552

38. Bentley CL, Kang M, Maddar FM, Li FW, Walker M, Zhang J, Unwin PR (2017) Electrochemical maps and movies of the hydrogen evolution reaction on natural crystals of molybdenite (MoS2): basal vs. edge plane activity. Chem Sci 8(9):6583–6593
39. Bentley CL, Andronescu C, Smialkowski M, Kang M, Tarnev T, Marler B, Unwin PR, Apfel UP, Schuhmann W (2018) Local surface structure and composition control the hydrogen evolution reaction on iron nickel sulfides. Angew Chem Int Ed 57(15):4093–4097
40. Hill JW, Hill CM (2019) Directly mapping photoelectrochemical behavior within individual transition metal dichalcogenide nanosheets. Nano Lett 19(8):5710–5716
41. Wang YF, Gordon E, Ren H (2019) Mapping the nucleation of H-2 bubbles on polycrystalline Pt via scanning electrochemical cell microscopy. J Phys Chem Lett 10(14):3887–3892
42. Yule LC, Shkirskiy V, Aarons J, West G, Bentley CL, Shollock BA, Unwin PR (2019) Nanoscale active sites for the hydrogen evolution reaction on low carbon steel. J Phys Chem C 123(39):24146–24155
43. Bentley CL, Agoston R, Tao BL, Walker M, Xu XD, O'Mullane AP, Unwin PR (2020) Correlating the local electrocatalytic activity of amorphous molybdenum sulfide thin films with microscopic composition, structure, and porosity. ACS Appl Mater Interfaces 12(39): 44307–44316
44. Hill JW, Fu ZG, Tian JF, Hill CM (2020) Locally engineering and interrogating the photoelectrochemical behavior of defects in transition metal dichalcogenides. J Phys Chem C 124(31):17141–17149
45. Strange LE, Yadav J, Garg S, Shinde PS, Hill JW, Hill CM, Kung P, Pan SL (2020) Investigating the redox properties of two-dimensional MoS2 using photoluminescence spectroelectrochemistry and scanning electrochemical cell microscopy. J Phys Chem Lett 11(9):3488–3494
46. Tao BL, Unwin PR, Bentley CL (2020) Nanoscale variations in the electrocatalytic activity of layered transition-metal dichalcogenides. J Phys Chem C 124(1):789–798
47. Yule LC, Daviddi E, West G, Bentley CL, Unwin PR (2020) Surface microstructural controls on electrochemical hydrogen absorption at polycrystalline palladium. J Electroanal Chem 872
48. Mariano RG, McKelvey K, White HS, Kanan MW (2017) Selective increase in CO_2 electroreduction activity at grain-boundary surface terminations. Science 358(6367):1187–1192
49. Tsujiguchi T, Kawabe Y, Jeong S, Ohto T, Kukunuri S, Kuramochi H, Takahashi Y, Nishiuchi T, Masuda H, Wakisaka M, Hu K, Elumalai G, Fujita J-I, Ito Y (2021) Acceleration of electrochemical CO2 reduction to formate at the Sn/reduced graphene oxide Interface. ACS Catal 11(6):3310–3318
50. Mariano RG, Kang M, Wahab OJ, McPherson IJ, Rabinowitz JA, Unwin PR, Kanan MW (2021) Microstructural origin of locally enhanced CO_2 electroreduction activity on gold. Nat Mater 20:1000–1006
51. Sharel PE, Kang M, Wilson P, Meng L, Perry D, Basile A, Unwin PR (2018) High resolution visualization of the redox activity of Li2O2 in non-aqueous media: conformal layer vs. toroid structure. Chem Commun (Camb) 54(24):3053–3056
52. Dayeh M, Ghavidel MRZ, Mauzeroll J, Schougaard SB (2019) Micropipette contact method to investigate high-energy cathode materials by using an ionic liquid. ChemElectroChem 6(1): 195–201
53. Inomata H, Takahashi Y, Takamatsu D, Kumatani A, Ida H, Shiku H, Matsue T (2019) Visualization of inhomogeneous current distribution on ZrO2-coated LiCoO2 thin-film electrodes using scanning electrochemical cell microscopy. Chem Commun 55(4):545–548
54. Kumatani A, Takahashi Y, Miura C, Ida H, Inomata H, Shiku H, Munakata H, Kanamura K, Matsue T (2019) Scanning electrochemical cell microscopy for visualization and local electrochemical activities of lithium-ion (de) intercalation process in lithium-ion batteries electrodes. Surf Interface Anal 51(1):27–30
55. Tao BL, Yule LC, Daviddi E, Bentley CL, Unwin PR (2019) Correlative electrochemical microscopy of Li-ion (De)intercalation at a series of individual LiMn2O4 particles. Angew Chem Int Ed 58(14):4606–4611

56. Takahashi Y, Yamashita T, Takamatsu D, Kumatani A, Fukuma T (2020) Nanoscale kinetic imaging of lithium ion secondary battery materials using scanning electrochemical cell microscopy. Chem Commun 56(65):9324–9327
57. Yule LC, Bentley CL, West G, Shollock BA, Unwin PR (2019) Scanning electrochemical cell microscopy: a versatile method for highly localised corrosion related measurements on metal surfaces. Electrochim Acta 298:80–88
58. Li YJ, Morel A, Gallant D, Mauzeroll J (2020) Oil-immersed scanning micropipette contact method enabling long-term corrosion mapping. Anal Chem 92(18):12415–12422
59. Liu SY, Shi M, Zhou YH, Li RX, Xie ZD, Hu DB, Zhang MQ, Hu G (2020) Scanning electrochemical cell microscopy: a powerful method to study the intergranular corrosions of archaeological silver artifacts. J Cult Herit 46:176–183
60. Yule LC, Shkirskiy V, Aarons J, West G, Shollock BA, Bentley CL, Unwin PR (2020) Nanoscale electrochemical visualization of grain-dependent anodic iron dissolution from low carbon steel. Electrochim Acta 332
61. Aaronson BD, Garoz-Ruiz J, Byers JC, Colina A, Unwin PR (2015) Electrodeposition and screening of photoelectrochemical activity in conjugated polymers using scanning electrochemical cell microscopy. Langmuir 31(46):12814–12822
62. Aaronson BDB, Byers JC, Colburn AW, McKelvey K, Unwin PR (2015) Scanning electrochemical cell microscopy platform for ultrasensitive photoelectrochemical imaging. Anal Chem 87(8):4129–4133
63. Ebejer N, Guell AG, Lai SCS, McKelvey K, Snowden ME, Unwin PR (2013) Scanning electrochemical cell microscopy: a versatile technique for nanoscale electrochemistry and functional imaging. Annu Rev Anal Chem 6:329–351
64. Barton ZJ, Rodriguez-Lopez J (2016) Emerging scanning probe approaches to the measurement of ionic reactivity at energy storage materials. Anal Bioanal Chem 408(11):2707–2715
65. Kang MY, Momotenko D, Page A, Perry D, Unwin PR (2016) Frontiers in nanoscale electrochemical imaging: faster, multifunctional, and ultrasensitive. Langmuir 32(32):7993
66. Bentley CL, Kang MY, Unwin PR (2017) Scanning electrochemical cell microscopy: new perspectives on electrode processes in action. Curr Opin Electrochem 6(1):23–30
67. Danis L, Gateman SM, Kuss C, Schougaard SB, Mauzeroll J (2017) Nanoscale measurements of lithium-ion-battery materials using scanning probe techniques. ChemElectroChem 4(1):6–19
68. Bentley CL, Unwin PR (2018) Nanoscale electrochemical movies and synchronous topographical mapping of electrocatalytic materials. Faraday Discuss 210:365–379
69. Wahab OJ, Kang M, Unwin PR (2020) Scanning electrochemical cell microscopy: a natural technique for single entity electrochemistry. Curr Opin Electrochem 22:120–128
70. Ventosa E (2021) Why nanoelectrochemistry is necessary in battery research? Curr Opin Electrochem 25:100635
71. Dang N, Etienne M, Walcarius A, Liu L (2018) Scanning gel electrochemical microscopy (SGECM): the potentiometric measurements. Electrochem Commun 97:64–67
72. Dang N, Etienne M, Walcarius A, Liu L (2020) Scanning gel electrochemical microscopy (SGECM): lateral physical resolution by current and shear force feedback. Anal Chem 92(9): 6415–6422
73. Ando T, Asai K, Macpherson J, Einaga Y, Fukuma T, Takahashi Y (2021) Nanoscale reactivity mapping of a single-crystal boron-doped diamond particle. Anal Chem 93(14):5831–5838
74. Rhatigan S, Michel MC, Nolan M (2020) Hydrogen evolution on non-metal oxide catalysts. J Phys-Energy 2(4)
75. Aaronson BDB, Chen CH, Li HJ, Koper MTM, Lai SCS, Unwin PR (2013) Pseudo-single-crystal electrochemistry on polycrystalline electrodes: visualizing activity at grains and grain boundaries on platinum for the Fe2+/Fe3+ redox reaction. J Am Chem Soc 135(10):3873–3880
76. Aaronson BDB, Lai SCS, Unwin PR (2014) Spatially resolved electrochemistry in ionic liquids: surface structure effects on triiodide reduction at platinum electrodes. Langmuir 30(7): 1915–1919

77. Chen CH, Jacobse L, McKelvey K, Lai SCS, Koper MTM, Unwin PR (2015) Voltammetric scanning electrochemical cell microscopy: dynamic imaging of hydrazine electro-oxidation on platinum electrodes. Anal Chem 87(11):5782–5789
78. Shkirskiy V, Yule LC, Daviddi E, Bentley CL, Aarons J, West G, Unwin PR (2020) Nanoscale scanning electrochemical cell microscopy and correlative surface structural analysis to map anodic and cathodic reactions on polycrystalline Zn in acid media. J Electrochem Soc 167(4)
79. Wang YF, Gordon E, Ren H (2020) Mapping the potential of zero charge and electrocatalytic activity of metal-electrolyte Interface via a grain-by-grain approach. Anal Chem 92(3): 2859–2865
80. Saha P, Hill JW, Walmsley JD, Hill CM (2018) Probing electrocatalysis at individual Au nanorods via correlated optical and electrochemical measurements. Anal Chem 90(21): 12832–12839
81. Lai SCS, Dudin PV, Macpherson JV, Unwin PR (2011) Visualizing zeptomole (electro)-catalysis at single nanoparticles within an ensemble. J Am Chem Soc 133(28):10744–10747
82. Snowden ME, Dayeh M, Payne NA, Gervais S, Mauzeroll J, Schougaard SB (2016) Measurement on isolated lithium iron phosphate particles reveals heterogeneity in material properties distribution. J Power Sources 325:682–689
83. Ustarroz J, Ornelas IM, Zhang GH, Perry D, Kang M, Bentley CL, Walker M, Unwin PR (2018) Mobility and poisoning of mass-selected platinum nanoclusters during the oxygen reduction reaction. ACS Catal 8(8):6775–6790
84. Tarnev T, Aiyappa HB, Botz A, Erichsen T, Ernst A, Andronescu C, Schuhmann W (2019) Scanning electrochemical cell microscopy investigation of single ZIF-derived nanocomposite particles as electrocatalysts for oxygen evolution in alkaline media. Angew Chem Int Ed 58(40): 14265–14269
85. Choi M, Siepser NP, Jeong S, Wang Y, Jagdale G, Ye XC, Baker LA (2020) Probing single-particle electrocatalytic activity at facet-controlled gold nanocrystals. Nano Lett 20(2): 1233–1239
86. Zhan DP, Yang DZ, Zhu YL, Wu XR, Tian ZQ (2012) Fabrication and characterization of nanostructured ZnO thin film microdevices by scanning electrochemical cell microscopy. Chem Commun 48(93):11449–11451
87. Oseland EE, Ayres ZJ, Basile A, Haddleton DM, Wilson P, Unwin PR (2016) Surface patterning of polyacrylamide gel using scanning electrochemical cell microscopy (SECCM). Chem Commun 52(64):9929–9932
88. McKelvey K, O'Connell MA, Unwin PR (2013) Meniscus confined fabrication of multidimensional conducting polymer nanostructures with scanning electrochemical cell microscopy (SECCM). Chem Commun 49(29):2986–2988
89. Dokko K, Fujita Y, Mohamedi M, Umeda M, Uchida I, Selman JR (2001) Electrochemical impedance study of Li-ion insertion into mesocarbon microbead single particle electrode: part II. Disordered carbon. Electrochim Acta 47(6):933–938
90. Dokko K, Mohamedi M, Fujita Y, Itoh T, Nishizawa M, Umeda M, Uchida I (2001) Kinetic characterization of single particles of LiCoO2 by AC impedance and potential step methods. J Electrochem Soc 148(5):A422–A426
91. Dokko K, Mohamedi M, Umeda M, Uchida I (2003) Kinetic study of Li-ion extraction and insertion at LiMn2 O 4 single particle electrodes using potential step and impedance methods. J Electrochem Soc 150(4):A425–A429
92. Munakata H, Takemura B, Saito T, Kanamura K (2012) Evaluation of real performance of LiFePO4 by using single particle technique. J Power Sources 217:444–448
93. Hui JS, Burgess M, Zhang JR, Rodriguez-Lopez J (2016) Layer number dependence of Li+ intercalation on few-layer graphene and electrochemical imaging of its solid-electrolyte interphase evolution. ACS Nano 10(4):4248–4257
94. Zampardi G, Ventosa E, La Mantia F, Schuhmann W (2015) Scanning electrochemical microscopy applied to the investigation of lithium (De-)insertion in TiO2. Electroanalysis 27(4): 1017–1025

95. Zampardi G, La Mantia F, Schuhmann W (2015) In-operando evaluation of the effect of vinylene carbonate on the insulating character of the solid electrolyte interphase. Electrochem Commun 58:1–5
96. Ventosa E, Schuhmann W (2015) Scanning electrochemical microscopy of Li-ion batteries. Phys Chem Chem Phys 17(43):28441–28450
97. Bulter H, Peters F, Schwenzel J, Wittstock G (2014) Spatiotemporal changes of the solid electrolyte interphase in lithium-ion batteries detected by scanning electrochemical microscopy. Angew Chem Int Ed 53(39):10531–10535
98. Barton ZJ, Rodríguez-López J (2014) Lithium ion quantification using mercury amalgams as in situ electrochemical probes in nonaqueous media. Anal Chem 86(21):10660–10667
99. Chen Z, Qin Y, Amine K, Sun YK (2010) Role of surface coating on cathode materials for lithium-ion batteries. J Mater Chem 20(36):7606–7612

BIOREV (2022) 3: 187–204
https://doi.org/10.1007/11663_2021_11

Published online: 21 October 2021

Comparison of Scanning Ion-Conductance Microscopy with Scanning Electron Microscopy for Imaging the Surface Topography of Cells and Tissues

Tatsuo Ushiki, Futoshi Iwata, Masato Nakajima, and Yusuke Mizutani

Contents

T. Ushiki (✉)
Division of Microscopic Anatomy, Graduate School of Medical and Dental Sciences, Niigata University, Niigata, Japan
e-mail: t-ushiki@med.niigata-u.ac.jp

F. Iwata
Research Institute of Electronics, Shizuoka University, Hamamatsu, Japan

M. Nakajima
Division of Surgery, Graduate School of Medical and Dental Sciences, Niigata University, Niigata, Japan

Y. Mizutani
Office of Institutional Research, Hokkaido University, Sapporo, Japan

Abstract Images of cells and tissues obtained by scanning ion-conductance microscopy (SICM) were compared with those obtained by scanning electron microscopy (SEM). Samples chemically fixed with glutaraldehyde were first observed using SICM in a phosphate buffer solution. Conductive staining was applied, followed by critical point drying; samples were then coated with metal and observed by scanning electron microscopy (SEM). Using this technique, SICM and SEM images of the same portions in different cells (COS-7, Mm2T.P3, and HeLa cell lines) and tissues (rat trachea, organ of Corti, and renal glomerulus) were compared to assess the quality of SICM and SEM images. The imaging quality of SICM was excellent and sufficient to investigate the three-dimensional surface structure of cells and tissues, although optimized imaging conditions (including the threshold and adjustment of pH) are essential to ensure the quality of SICM images. The main advantage of SICM imaging is that it can obtain topographic images of cells and tissues in an aqueous medium, which is not possible with conventional SEM imaging. SICM also has the advantage of avoiding SEM preparation artifacts, such as shrinkage and deformation produced by drying and metal coating. Although the imaging time needs to be improved, SICM is expected to be useful for observing cells and tissues to reveal their three-dimensional structure at a high resolution, similar to SEM.

Keywords Cells · Scanning electron microscopy · Scanning ion-conductance microscopy · Tissues

1 Introduction

Scanning ion-conductance microscopy (SICM) was introduced by Hansma et al. in 1989 [1] as a member of the family of scanning probe microscopy (SPM). This technique uses a glass micropipette as the scanning probe; the micropipette, filled with an electrolyte solution, is connected to an electrical amplifier via an electrode and is brought very close to the cell surface in an electrolyte solution by monitoring the ion current between the pipette electrode and bath electrode. Thus, SICM can obtain images of the sample surface topography in an aqueous medium. Because SICM uses an ion current to control the probe, SICM can obtain contact-free imaging. This is unlike imaging by atomic force microscopy (AFM), which uses the force between the probe and the sample. Thus, SICM has the potential to avoid deformation of soft samples, which would occur from the force of the AFM cantilever.

Since the first application of SICM to live-cell imaging was reported by Korchev et al. in 1997 [2], several investigators have been interested in imaging cells in a liquid environment [3–8]. The introduction of hopping-mode SICM further expanded the possible biological applications, as this technique can be used for imaging the complex surfaces of soft samples at the nanoscale [9, 10].

We previously investigated the application of SICM to study the surface topography of cells and tissues [11, 12]. Specifically, to assess the quality of SICM imaging, we compared SICM images with SEM images of the same portion of the same samples. Here, we introduce our data on SICM and SEM images of the same portions in different cells (COS-7 Mm2T.P3, and HeLa cell lines) and tissues (rat trachea, the inner-ear organ of Corti, and renal glomerulus) and show that the image quality of SICM is comparable to that of SEM. After discussing the merits and deficiencies of these two techniques, we also propose that SICM may become a useful tool for observing the three-dimensional structure of cells and tissues in a liquid environment as well as in a living state.

2 Materials and Methods

2.1 Sample Preparation

Cultivated Cells The COS-7 cell line, Mm2T.P3 line, and HeLa cells were used for cell imaging. These were obtained from RIKEN Cell Bank (Riken, Tsukuba, Japan). Cells were grown on cover slips in culture dishes (50 mm in diameter) or grown directly on culture dishes (35 mm in diameter) for 24–48 h in a CO_2 incubator at 37–38°C. They were then fixed with 1% glutaraldehyde in 0.1 M phosphate buffer (PB) at pH 7.4 overnight at 4°C, and observed by SICM.

Tissues Tissue samples were obtained from adult male Wistar rats (8–10 weeks of age). The rats were housed in a well-ventilated room with access to food and water ad libitum. Under deep anesthesia by ether and pentobarbital sodium, the animals were transcardially perfused with physiological saline followed by the fixative, 2% glutaraldehyde in 0.1 M phosphate-buffered saline (pH 7.4). After fixative perfusion, the trachea, kidney, and temporal bone including the inner ear were excised from the body and kept in the same fixative until use. These animal experiments were reviewed by the Committee of the Ethics on Animal Experiment at Niigata University and carried out according to the Guidelines for Animal Experiments at Niigata University (No. 209-2).

2.2 Scanning Ion-Conductance Microscopy

Pretreatment of Cultivated Cells Cultivated cells described above were kept at room temperature until the temperature stabilized before SICM imaging.

Pretreatment of Tissue Samples As for the trachea, the inner wall was washed by flushing the buffer solution with a hypodermic needle to clear mucus adherent to the luminal surface; the trachea was then carefully cut into small pieces under a dissecting microscope. The kidney was cut into two or three pieces, which were sectioned through the cortex at a thickness of 100–150 μm using a microslicer (Dosaka EM, Japan) to expose renal glomeruli on the sectioned surface. For observing hair cells in the inner ear, the temporal bone was dissected with sharp forceps under a dissecting microscope; the tectorial membrane of the cochlea was especially removed to expose the surface of the organ of Corti. These specimens were attached to the coverslip according to the technique already described in our previous paper [12]; they were mounted on a coverslip pre-warmed on a hotplate to 50–60°C and precoated with 5% low-melting-point agarose solution (Sigma, St. Louis, MO, USA). After the specimens were placed onto the warmed coverslip, the coverslip was removed from the hotplate and left at room temperature for several minutes. Thus, the specimens were affixed to the coverslip enough for imaging by SICM.

SICM Imaging SICM imaging was performed with a commercial scanning probe microscope (SPM; XE-Bio, Park Systems Corp., Suwon, Korea). As described in our previous papers [11, 12], this instrument has an SPM unit on the stage of an inverted optical microscope (Eclipse Ti, Nikon Corp., Tokyo, Japan). The SPM unit has an X–Y-axis flat scanner (100 μm × 100 μm) and a separate SPM head with a 25 μm Z-axis scanner. The glass micropipette attached to the SPM head was fabricated from borosilicate capillaries (inner diameter 0.6 mm, outer diameter 1.0 mm, Narishige, Tokyo, Japan) using a CO_2-laser-based micropipette puller (P-2000, Sutter Instruments, Novato, CA, USA); the inner diameter of the pipette used in the present study was about 100 nm. The pipette was filled with phosphate-buffered saline, and an Ag/AgCl electrode was connected. After the samples on the coverslip were attached to the cell chamber filled with physiological saline, SICM imaging was performed in hopping mode [9, 10]; the ion current was recorded while the pipette was moved vertically, repeatedly approaching and retracting from the sample surface. The threshold for stopping the pipette approach was set at 0.6–2% resistance increase with respect to the basal resistance.

2.3 *Scanning Electron Microscopy (SEM)*

After SICM imaging, the specimens were prepared for SEM. The method for SEM preparation was already shown in our previous papers [11, 12]. After SICM observation, glutaraldehyde-fixed cultured cells on the coverslip were postfixed with 1% osmium tetroxide (OsO_4) for 10 min and then treated with 1% tannic acid for 10 min and 1% OsO_4 for 10 min; this treatment was based on the conductive staining method of Murakami [13] which increases the electron conductivity of the specimens for SEM observation. Tissue samples after SICM observation were also conductive-stained with 1% tannic acid solution for 30 min and 1% OsO_4 for 30 min. These specimens were dehydrated with an ascending series of ethanol, transferred to isoamyl acetate, and dried in a critical point dryer using CO_2 (HCP-2; Hitachi High-technologies, Tokyo, Japan). During this preparation process, cultivated cells were preserved on a coverslip, but the tissue samples sometimes detached from the coverslip. The dried samples (or samples on coverslips) were then mounted on aluminum stubs with double-sided carbon tape, coated with platinum-palladium in an ion coater, and observed by SEM (either SU-3500 or S-3700, Hitachi High-technologies, Japan).

3 Observation of Cultivated Cells

The surface structure of cultivated cells varies depending on the cell type. Thus, we describe the data separately for each type of cell.

3.1 COS-7 Cells

The COS-7 cell line is an African green monkey kidney fibroblast-like cell line [14]. The COS-7 cells cultivated in the dish are flattened cells with a few short microvilli on the apical surface.

SICM imaging clearly showed the contour of the COS-7 cells with short lamellipodia (Fig. 1a). The line profile of the COS-7 cell is useful for measuring the height of the cell; the height of the nuclear elevation was approximately 10 μm, as shown in Fig. 1a. Higher magnification SICM images of the COS-7 cell surface also revealed the presence of short projections (microfolds); these projections ranging from 100 to 600 nm in height and were sparsely observed on the cell surface (Fig. 1c).

SEM imaging (Fig. 1b, d) of the same specimens after drying and coating enabled an accurate comparison between the SICM and SEM images. The findings revealed that the overall shape of the cell was well imaged by SICM, while the cells observed by SEM were often partially cracked or torn due to shrinkage during dehydration and drying for SEM preparation. At high magnification, the features of the microfolds were roughly matched in these two images, indicating that these structures were well pictured in the SICM images. The SICM images contain quantitative information on the sample height. Thus, the height of the microfold ranged roughly from 100 to 600 nm, as measured by SICM.

3.2 Mm2T.P3 Cells

Mm2T.P3 is a muntjac thymus cell line [15]. These cultivated cells are flattened, with a central round nuclear elevation. They often have complicated sheet-like projections (microplicae) on the cell surface.

SICM imaging clearly showed the features of the microplicae on the cell surface (Fig. 2a, c). The height of microplicae is roughly 1–3.5 μm. SEM imaging of the

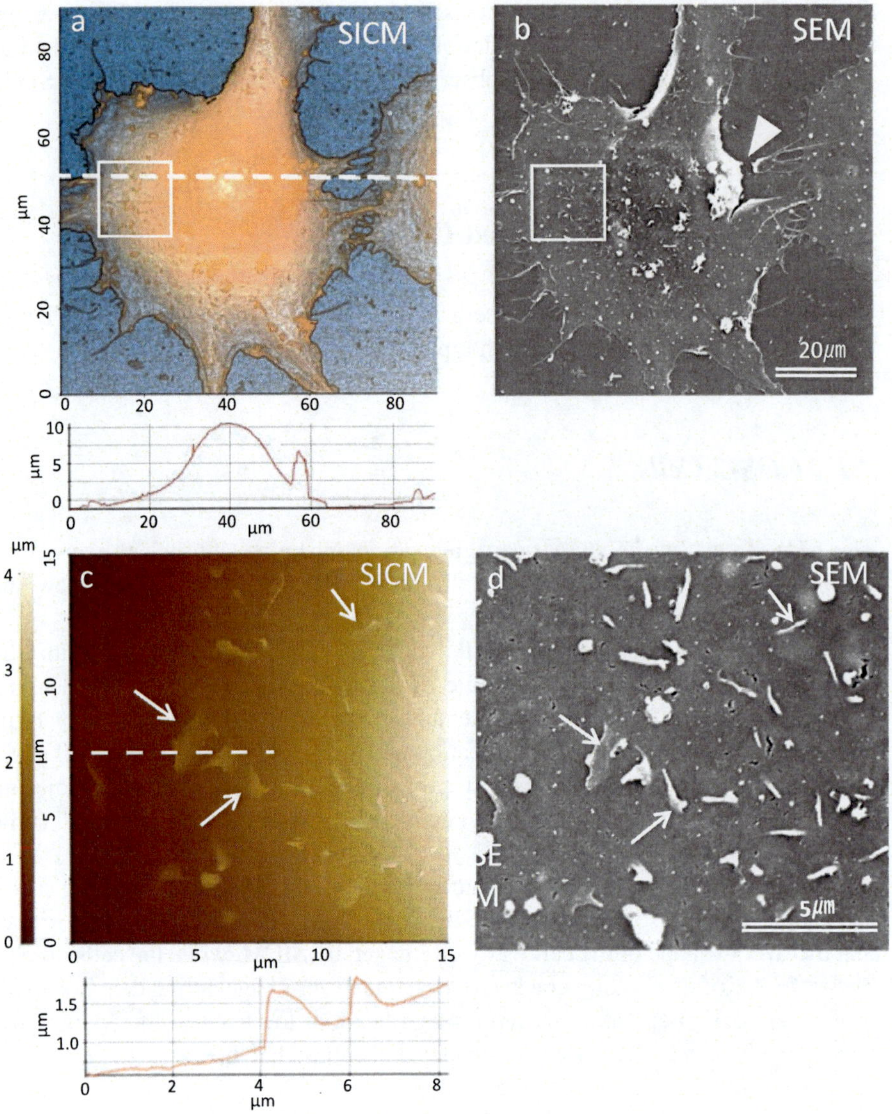

Fig. 1 SICM (**a** and **c**) and SEM images (**b** and **d**) of a COS-7 cell. After imaging by SICM, the sample was electron-conductive-stained, critical-point-dried, metal-coated, and observed by SEM. The shape of the COS-7 cell is well imaged by SICM (**a**), which is comparable to the corresponding SEM image (**b**). The boxed areas in a and b are magnified in (**c**) and (**d**), respectively. The arrowhead in the SEM image (**b**) indicates the artificially cracked portion of the cell process, probably due to shrinkage of the cell during SEM preparation. Arrows in (**c**) and (**d**) correspond to the cell processes observed in SICM and SEM images. The section profiles of the portions indicated by the broken lines in (**a**) and (**c**) are shown below the images

same portion (Fig. 2b, d) revealed that the shape of these microplicae was comparable between the two images, although they appeared somewhat distorted in the SEM images, probably due to the drying artifacts from the SEM preparation. These

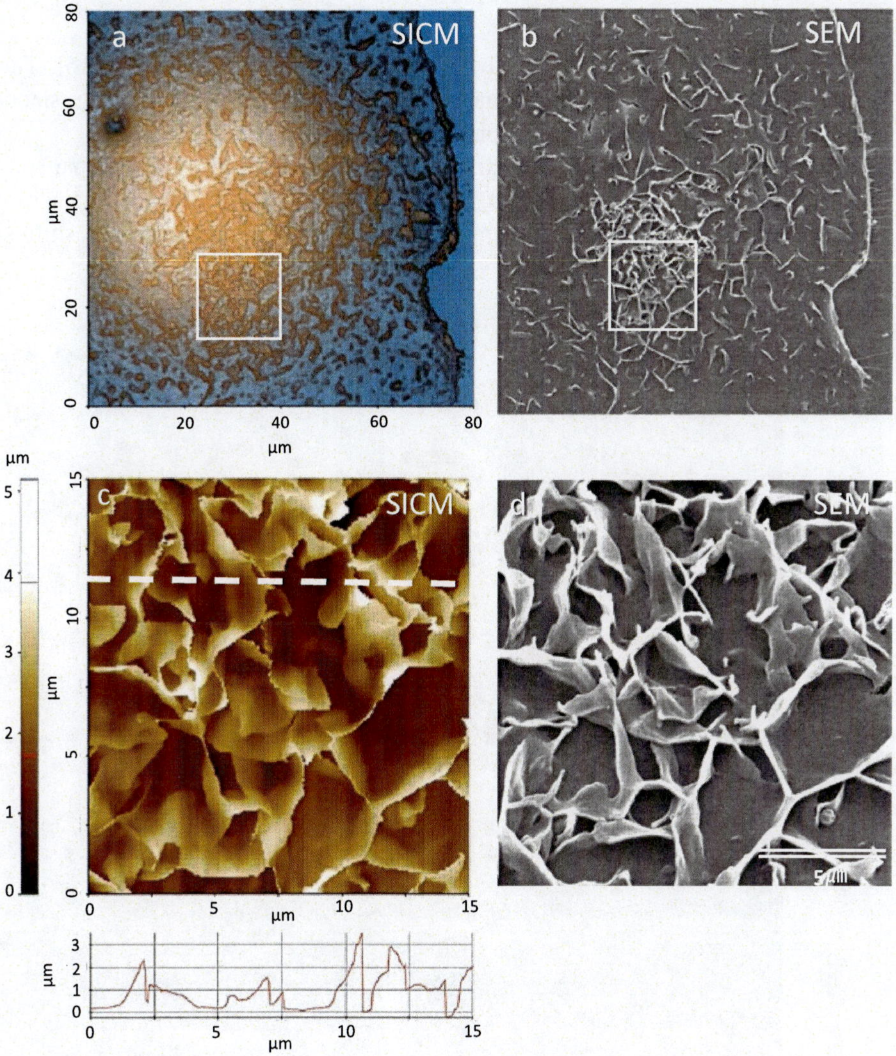

Fig. 2 SICM (**a** and **c**) and SEM images (**b** and **d**) of an Mm2T.P3 cell. This cell is flattened, with a round nuclear elevation at the center. The boxed areas in (**a**) and (**b**) are magnified in (**c**) and (**d**), respectively. The section profile shown below is obtained from the portion indicated by the broken line in (**c**)

findings indicate that the structure of the microplicae was better represented in the SICM images than in the SEM images.

3.3 HeLa Cells

The HeLa cell line is a human cell line derived from cervical cancer [16]. These cultivated cells are usually polygonal in a confluent monolayer and have slender, long filamentous microprojections on the cell surface.

Figure 3 shows the SICM and SEM images of HeLa cells. The comparison of SICM and SEM images at a low magnification showed that the overall shape of the HeLa cells was well represented in SICM images, in contrast to their shrunken

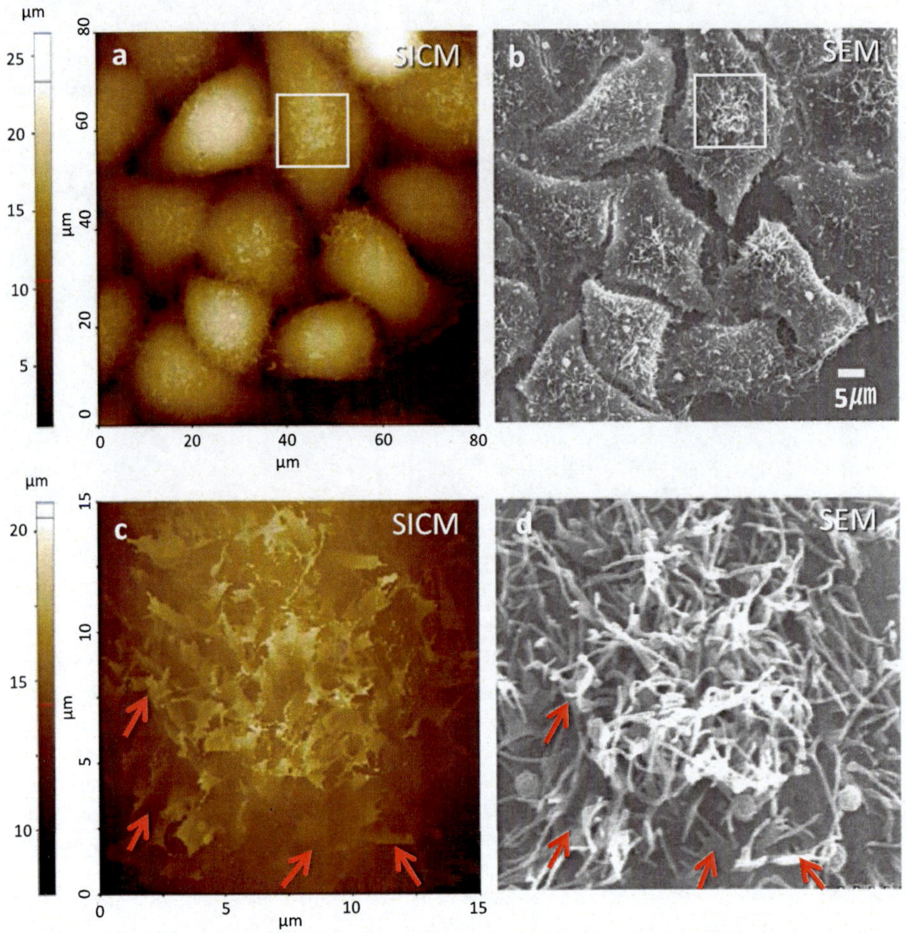

Fig. 3 SICM (**a** and **c**) and SEM images (**b** and **d**) of HeLa cells. HeLa cells are typically polygonal in a confluent monolayer. The comparison of SICM and SEM images at a low magnification shows that the overall shape of the HeLa cells is well represented in the SICM image, although the gap between the neighboring cells widened in the corresponding SEM image, probably due to the shrinkage of cells during SEM sample preparation. A comparison of the SICM and SEM images at higher magnification (**c** and **d**) shows that only the short processes (arrows in **c** and **d**) are represented in the SICM image, although long filamentous projections are observed on the cell surface in the corresponding SEM image

appearance in the SEM images, indicated by the appearance of gaps between cells in the confluent cell monolayer. The height of the HeLa cells was estimated using SICM images at approximately 15 μm.

Higher magnification images revealed that long filamentous projections were not always imaged in SICM images, as shown in Fig. 3c; slender long projections observed in the SEM image were not visualized in the SICM image, in which only the base of the projections was imaged.

Because the threshold level for stopping the pipette approach is important for determining the distance between the pipette tip and the sample in the SICM hopping mode, the level was changed from 0.6 to 2% for SICM imaging. As shown in Fig. 4a, b, some filamentous projections were well imaged in the 0.6% threshold images, while they sometimes disappeared or became discontinuous in the 2% threshold images. After obtaining these SICM images, we hardened this sample by treating with 1% OsO_4, 1% tannic acid, and again with 1% OsO_4. The portion observed before this treatment was again observed using SICM with a threshold level of 0.6%. Figure 4 c shows that filamentous projections were well imaged in this SICM image.

The comparison of the two SICM and SEM images also showed that the filamentous projections were not fully visualized in the SICM images, although the 0.6% threshold SICM image of the osmium-tannin-osmium treated sample most resembled the SEM image (Fig. 4c). In the SICM image, the height and thickness of these projections were approximately 1–2 μm and 200–250 nm, respectively.

4 Observation of Tissue Samples

4.1 *Trachea*

The luminal surface of the trachea is covered with ciliated epithelium, consisting of ciliated and non-ciliated cells. Because SICM can obtain images of samples with irregular surfaces, ciliated and non-ciliated cells were distinguishable by SICM images (Fig. 5a), which was comparable with the SEM images of the same portion (Fig. 5b).

As also shown in Fig. 6, it is important to consider the fact that the apparent cilium thickness changes depend on the threshold level in SICM images; the cilium is apparently thinner when the threshold level is $>0.6\%$.

4.2 *Hair Cells of Inner Ear*

The organ of Corti in the inner ear contains sensory cells called the inner and outer hair cells. These hair cells have auditory stereocilia on the apical cell surface, and the outer hair cells are characterized by the presence of W- or V-shaped bundles of stereocilia. The comparison of SICM and SEM images (Fig. 7) revealed that these

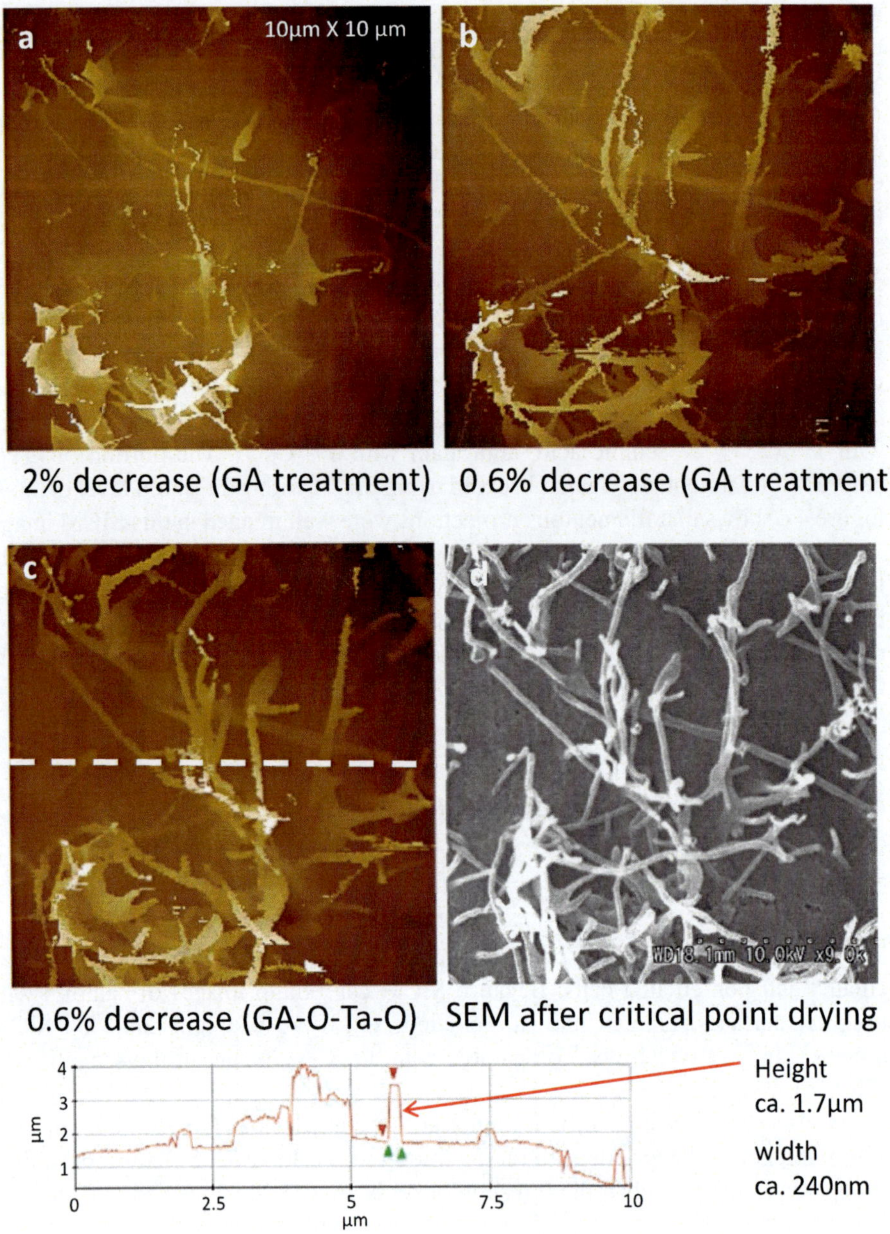

Fig. 4 SICM images (**a–c**) and SEM image (**d**) of HeLa cells. HeLa cells fixed with 2% glutaraldehyde (GA) were first observed by SICM at a threshold level of 2% (**a**) and 0.6% (**b**). After obtaining these images, the sample was treated with 1% OsO_4, 1% tannic acid, and 1% OsO_4 to harden the sample, then observed by SICM with a threshold level of 0.6% (**c**). They were then dried in a critical point dryer, metal-coated, and observed using SEM (**d**). The comparison of the SICM images in the different conditions shows that only the thicker processes as well as the filamentous processes lying on the cell surface are imaged in the SICM image in (**a**), but filamentous processes are more clearly visualized in (**c**). The section profile is obtained through the portion indicated by the broken line in (**c**)

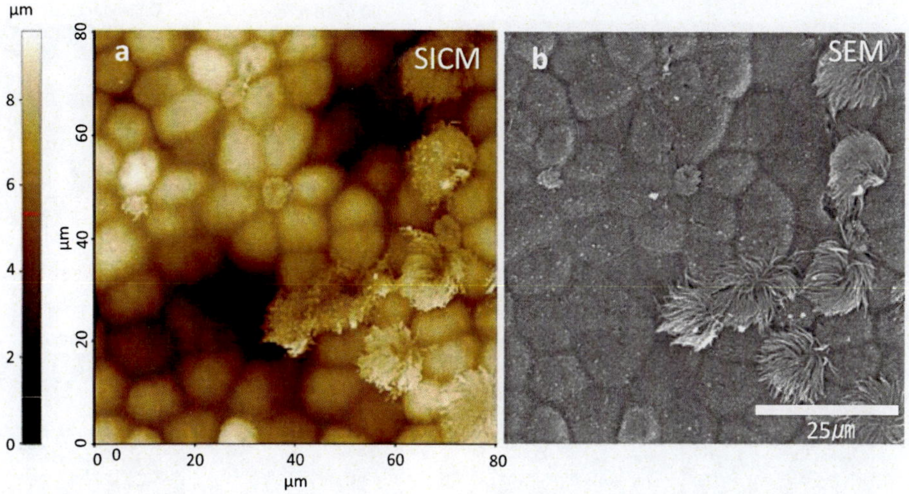

Fig. 5 SICM and SEM images of trachea. The luminal surface of the ciliated epithelium was first observed using SICM in phosphate-buffered saline (PBS) (**a**). The sample was electron-conductive-stained with the osmium-tannin-osmium method, dehydrated in ethanol, critical-point-dried, metal-coated, and observed by SEM (**b**). The SICM image is comparable to the SEM image (reproduced from Nakajima et al. [12])

bundles of stereocilia were well imaged in the SICM images. The width between the ends of the V-shaped bundles was 6.73 µm in the SICM image (Fig. 7a), but it was 5.07 µm in the SEM image. This indicates that the sample observed by SEM shrank during the dehydration of the SEM preparation. As shown in Fig. 7b, the SEM samples sometimes also cracked during preparation. Thus, the size of the sample is considered to be better preserved in the SICM images than in the SEM images, because SICM can obtain images in an aqueous environment.

4.3 *Renal Glomerulus*

The renal glomerulus is the ball of a capillary network surrounded by Bowman's capsule. These capillaries are covered by glomerular epithelial cells or podocytes, which are characterized by the presence of primary and secondary foot processes. Figure 8 shows that the SICM image of the renal glomerular capillary is comparable to the corresponding SEM image; primary and secondary foot processes were clearly observed in the SICM image, at a quality almost the same as in the SEM image. In addition, SICM images clearly showed that the thickness of the secondary interdigitated foot process was approximately 160–200 nm under liquid conditions.

Fig. 6 SICM images of ciliated cells in the trachea. The luminal surface of ciliated cells in the trachea was observed using SICM. In the left image (the threshold level at 0.6%), the individual cilia were clearly visualized, similar as in the SEM image in Fig. 5. The right four images show that the apparent thickness of the cilia in the SICM images changes depending on the threshold level. The section profiles of the portion indicated by red lines show that the thickness of the cilium is apparently thinner in SICM images than in SEM images when the threshold level was >0.6% (reproduced from Nakajima et al. [12])

5 Discussion

The three-dimensional (3D) ultrastructure of cells and tissues has been studied mainly by scanning electron microscopy (SEM) [17–20], as this technique can obtain the surface topography by collecting secondary electrons that are produced by scanning an electron beam over the sample surface. However, conventional SEMs are operated under high vacuum to preserve the electron beam coherency. Thus, for SEM imaging, a conductive sample must be observed under a vacuum environment. Given this constraint, cells and tissues are treated with conductive staining, and/or metal coating, and are dried with a special technique (critical point drying or freeze drying) to observe the surface structure of biological samples by SEM [21].

In this study, we showed that SICM can also obtain the surface topography of cells and tissues, with image qualities comparable to those obtained by SEM [11, 12]. The biggest advantage of SICM imaging for biological samples is that SICM can obtain sample images in a liquid environment, which is not possible using conventional SEM. This study indicates that SEM samples are sometimes artificially

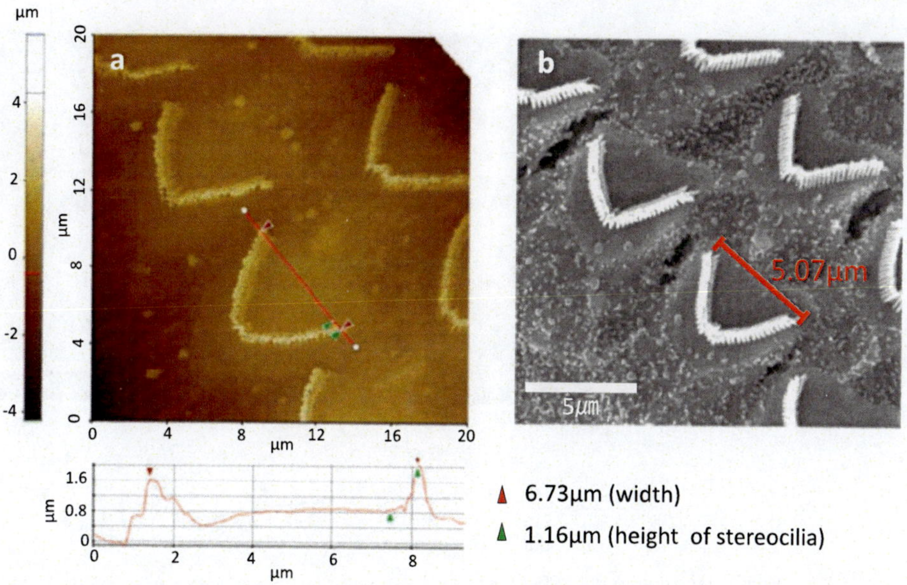

Fig. 7 SICM and SEM images of outer hair cells in the cochlea. The upper surface of the organ of Corti was first observed using SICM in PBS (**a**). After SICM observation, the sample was electron-conductive-stained, dehydrated, critical-point-dried, metal-coated, and observed by SEM (**b**). A section profile through the red line in the SICM image (**a**) is shown below. The end-to-end length of the V-shaped row of auditory stereocilia in the SICM image was 6.73 μm and the height of the cilia was 1.16 μm. Note that the end-to-end length was 5.07 μm in the SEM image (**b**)

cracked because of shrinkage during dehydration and drying for SEM preparation. In this regard, SICM imaging is superior to SEM imaging; SICM enables the direct observation of cells and tissues in a liquid medium without any additional treatment, which is mandatory for SEM imaging. Thus, SICM is expected to be used to study the surface topography of cells and tissues in a physiological environment.

Theoretically, SICM can be free from the interaction force between the probing tip and sample surface, in contrast to AFM, which uses the interaction force for imaging. However, as shown in this study, the threshold for stopping the pipette approach is important for determining the distance between the pipette tip and the sample in the SICM. This is because the pipette is controlled by the ion current decrease, and the decrease occurs as the pipette opening closely approaches the sample surface. This means that the pipette touches or pushes the sample when the threshold is too high. Our study showed that the threshold should be $<2\%$, and the most suitable threshold is considered to be approximately 0.6% for SICM imaging of soft samples. We also showed that it was rather difficult to image long filamentous projections even with a threshold of 0.6%; this is likely partly due to the projections being very soft and pushed away by the pipette, while there is also a possibility that the projections drift with changes in water flow caused by the pipette movement.

The time resolution and scanning area of SICM imaging should be considered in comparison with SEM imaging. For example, the height difference of the sample is

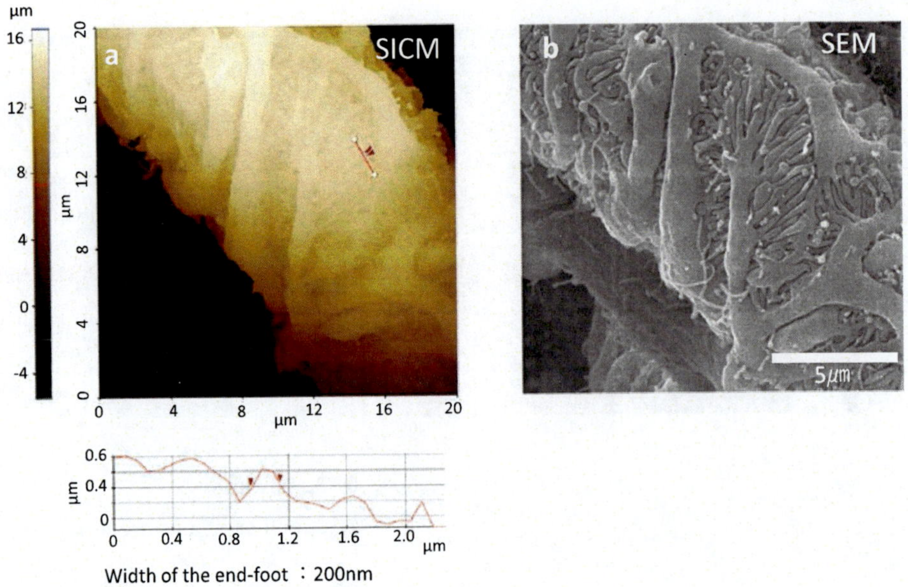

Fig. 8 SICM and SEM images of a capillary of rat renal glomerulus. A part of the renal glomerular capillary was first observed using SICM in PBS. After SICM observation, the sample was electron-conductive-stained, dehydrated, critical-point-dried, metal-coated, and observed by SEM (**b**). A section profile through the red line in the SICM image (**a**) is shown below. The thickness of the secondary process indicated by the arrowheads is approximately 200 nm

small when the size of the scanning area is small, but it increases when the size becomes larger. Thus, the scanning time for one image becomes longer when samples with large height differences are observed at a larger SICM scanning area. Although SICM has the potential for imaging live cells and tissues in liquid environments, the problem of long scanning duration should be improved to study the movement of larger-scale samples. Because several technologies have been developed for high-speed SICM [22–24], it can be expected to reduce the imaging time for larger-scale samples in the future.

Some biological samples are electrically charged in an aqueous environment. Several investigators have reported that the surface charge of the sample sometimes makes SICM imaging difficult [25–28]. Recently, we also clarified that SICM images are strongly influenced by the sample surface charge; it is difficult to obtain the topography if the charge is strong, as in chromosomes [29]. Figure 9a shows approach curves of ion current behavior on a metaphase chromosome obtained from Indian muntjac thymus cell lines. When the pipette potential is negative, the approach curve shows an increase in current as the pipette approaches the sample surface. This is because the surface of the chromosome is strongly negatively charged. As a result, under the negative potential of the pipette, the pipette tip cannot stop near the sample surface, and the surface topography of the chromosome cannot

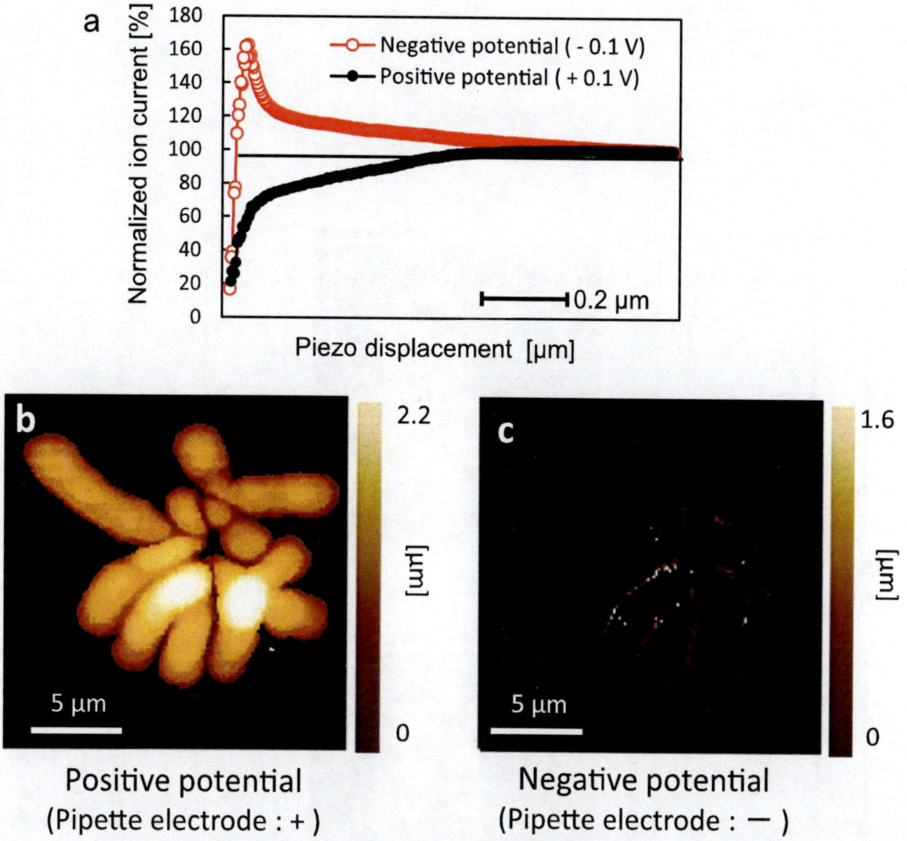

Fig. 9 Approach curves of ion current behavior (**a**) and SICM images (**b** and **c**) of metaphase chromosomes prepared from the muntjac thymus cell line. The vertical axis of the approach curves (**a**) shows the absolute value normalized by the ion current detected apart from the surface. The red curve shows the current behavior under the negative pipette potential, while the black curve shows the behavior under the positive pipette potential. The shape of chromosomes is clearly obtained in the SICM image under a positive potential of +0.1 V (**b**), but not obtained in the SICM Image under a negative potential of −0.1 V (**c**) (reproduced from Ushiki et al. [29])

be imaged as shown in Fig. 9c. In contrast, the chromosome can be imaged under the positive potential of the pipette.

To overcome this problem, techniques for observing charged samples have been proposed including an alternating current (AC) mode that vibrates the pipette vertically [25], a bias modulation mode [26] that applies a modulated bias potential to the pipette electrode, and a hopping mode with a low-potential direct current (DC) [27]. We also recently proposed a novel imaging method using a double-barrel nanopipette as an SICM probe [30]. Figure 10a shows the principle of this technique; the ion current flowing between the two apertures of the nanopipette would diminish the effect of the surface charge on imaging, thus facilitating the observation of charged samples by SICM. As shown in Fig. 10b, c, the approach curves decay near the sample surface under both the positive and negative pipette potentials. As a

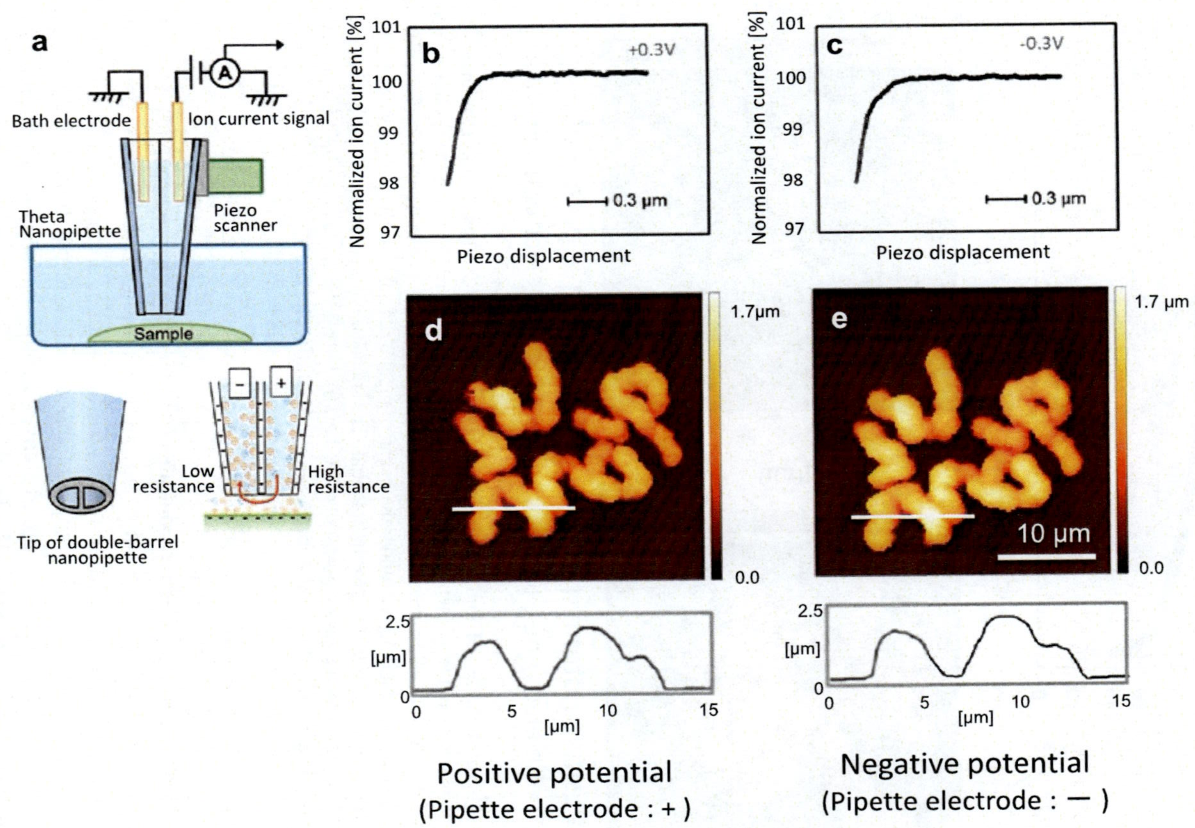

Fig. 10 A novel SICM imaging method using a double-barrel nanopipette (**a**). Approach curves were obtained using this method under the positive pipette potential of +0.3 V (**b**) and negative pipette potential of −0.3 V (**c**). The shape of the chromosomes was well preserved in both SICM images of chromosomes under the positive pipette potential of +0.3 V (**d**) and negative pipette potential of −0.3 V (**e**). The line profiles were obtained from the portions shown in the white line in (**d**) and (**e**) (reproduced from Iwata et al. [30])

result, the chromosome is successfully observed regardless of the pipette potential sign (Fig. 10d, e).

In conclusion, the present study compared SICM images of cells and tissues with SEM images of the same portions. The comparison has shown the applicability of SICM in studying the three-dimensional structure of cells and tissues in a liquid environment. By overcoming some drawbacks related to SICM, SICM is expected to become a useful tool for imaging biological samples beyond SEM.

Acknowledgements This study was supported in part by a Grant-in-Aid for Scientific Research (B) (no. 16H03868 for TU), Grant-in-Aid for Young Scientists (B) (no. 17K15541 for YM), and Grant-in-Aid for Scientific Research (C) (no. 20K07217 for YM), and Grant-in-Aid for Challenging Research (Exploratory) (19K22126 for FI) by the Japan Society for the Promotion of Science (JSPS).

References

1. Hansma PK, Drake B, Marti O, Gould SA, Prater CB (1989) The scanning ion-conductance microscope. Science 243:641–643
2. Korchev YE, Bashford CL, Milovanovic M, Vodyanoy I, Lab MJ (1997) Scanning ion conductance microscopy of living cells. Biophys J 73:653–658
3. Korchev YE, Gorelik J, Lab MJ, Sviderskaya EV, Johnston CL, Coombes CR, Vodyanoy I, Edwards RW (2000) Cell volume measurement using scanning ion conductance microscopy. Biophys J 78:451–457
4. Pastré D, Iwamoto H, Liu J, Szabo G, Shao Z (2001) Characterization of AC mode scanning ion-conductance microscopy. Ultramicroscopy 90:13–19
5. Happel P, Hoffmann G, Mann SA, Dietzel ID (2003) Monitoring cell movements and volume changes with pulse-mode scanning ion conductance microscopy. J Microsc 212:144–151
6. Gorelik J, Ali NN, Shevchuk AI, Lab M, Williamson C, Harding SE, Korchev YE (2006) Functional characterization of embryonic stem cell-derived cardiomyocytes using scanning ion conductance microscopy. Tissue Eng 12:657–664
7. Gorelik J, Zhang Y, Shevchuk AI, Frolenkov GI, Sanchez D, Lab MJ, Vodyanoy I, Edwards CRW, Klenerman D, Korchev YE (2004) The use of scanning ion conductance microscopy to image A6 cells. Mol Cell Endocrinol 217:101–108
8. Rheinlaender J, Geisse NA, Proksch R, Schäffer TE (2011) Comparison of scanning ion conductance microscopy with atomic force microscopy for cell imaging. Langmuir 27:697–704
9. Novak P, Li C, Shevchuk AI, Stepanyan R, Caldwell M, Hughes S, Smart TG, Gorelik J, Ostanin VP, Lab MJ, Moss GW, Frolenkov GI, Klenerman D, Korchev YE (2009) Nanoscale live cell imaging using hopping probe ion conductance microscopy. Nat Methods 6:279–281
10. Happel P, Dietzel ID (2009) Backstep scanning ion conductance microscopy as a tool for long term investigation of single living cells. J Nanobiotechnol 7:7
11. Ushiki T, Nakajima M, Choi M, Cho SJ, Iwata F (2012) Scanning ion conductance microscopy for imaging biological samples in liquid: a comparative study with atomic force microscopy and scanning electron microscopy. Micron 43:1390–1398
12. Nakajima M, Mizutani Y, Iwata F, Ushiki T (2018) Scanning ion conductance microscopy for visualizing the three-dimensional surface topography of cells and tissues. Semin Cell Dev Biol 73:125–131
13. Murakami T (1973) A metal impregnation method of biological specimens for scanning electron microscopy. Arch Histol Jpn 36:189–193

14. Gluzman Y (1981) SV40-transformed simian cells support the replication of early SV40 mutants. Cell 23:175–182
15. Takaoka T, Kasai M, Katsuta H (1982) A cell line derived from the Indian muntjac thymus. Dokkyo J Med Sci 9:133–138
16. Gey GC, Coffman WD, Kubicek MT (1952) Tissue culture studies of the proliferative capacity of cervical carcinoma and normal epithelium. Cancer Res 12:264–265
17. Ushiki T (1986) A scanning electron-microscopic study of the rat thymus with special reference to cell types and migration of lymphocytes into the general circulation. Cell Tissue Res 244:285–298
18. Souma T (1987) The distribution and surface ultrastructure of airway epithelial cells in the rat lung: a scanning electron microscopic study. Arch Histol Jpn 50:419–436
19. Iino N, Gejyo F, Arakawa M, Ushiki T (2001) Three-dimensional analysis of nephrogenesis in the neonatal rat kidney: light and scanning electron microscopic studies. Arch Histol Cytol 64:179–190
20. Kabata Y, Orime M, Abe R, Ushiki T (2019) The morphology, size and density of the touch dome in human hairy skin by scanning electron microscopy. Microscopy (Oxf) 68:207–215. https://doi.org/10.1093/jmicro/dfz001
21. Fujita T, Tanaka K, Tokunaga J (1981) SEM atlas of cells and tissues. Igaku-shoin, Tokyo
22. Shevchuk AI, Novak P, Taylor M, Diakonov IA, Ziyadeh-Isleem A, Bitoun M, Guicheney P, Lab MJ, Gorelik J, Merrifield CJ, Klenerman D, Korchev YE (2012) An alternative mechanism of clathrin-coated pit closure revealed by ion conductance microscopy. J Cell Biol 197:499–508
23. Simeonov S, Schäffer TE (2019) High-speed scanning ion conductance microscopy for sub-second topography imaging of live cells. Nanoscale 11:8579–8587
24. Watanabe S, Kitazawa S, Sun L, Kodera N, Ando T (2019) Development of high-speed ion conductance microscopy. Sci Instrum 90:123704
25. McKelvey K, Kinnear SL, Perry D, Momotenko D, Unwin PR (2014) Surface charge mapping with a nanopipette. J Am Chem Soc 136:13735–13744
26. Perry D, Al Botros R, Momotenko D, Kinnear SL, Unwin PR (2015) Simultaneous nanoscale surface charge and topographical mapping. ACS Nano 9:7266–7276
27. Page A, Perry D, Young P, Mitchell D, Frenguelli BG, Unwin PR (2016) Fast nanoscale surface charge mapping with pulsed-potential scanning ion conductance microscopy. Anal Chem 88:10854–10859
28. Klausen LH, Fuhs T, Dong M (2016) Mapping surface charge density of lipid bilayers by quantitative surface conductivity microscopy. Nat Commun 7:12447
29. Ushiki T, Ishizaki K, Mizutani Y, Nakajima M, Iwata F (2021) Scanning ion conductance microscopy of isolated metaphase chromosomes in a liquid environment. Chromosom Res 202. https://doi.org/10.1007/s10577-021-09659-0
30. Iwata F, Shirasawa T, Mizutani Y, Ushiki T (2021) Scanning ion-conductance microscopy with a double-barreled nanopipette for topographic imaging of charged chromosomes. Microscopy: dfab009. https://doi.org/10.1093/jmicro/dfab009

BIOREV (2022) 3: 205–230
https://doi.org/10.1007/11663_2022_13

Published online: 17 June 2022

Correlating Scanning Ion Conductance and Super-Resolved Fluorescence Microscopy

Patrick Happel, Annika Haak, and Astrid Gesper

Contents

Abstract The development of diffraction-unlimited fluorescence microscopy has paved the way for a detailed understanding of cellular dynamics since live-cell investigations with resolutions below 100 nm have become available. This allows studying labelled specimen such as proteins and their distribution within a cell at a new level of detail. However, examining the cell membrane as a site of interaction between the cell's internal processes and external signals with fluorescence microscopy (FM) leads to the difficulty that certain types of fluorescent dyes can alter the

Patrick Happel, (deceased), Happel wrote this chapter while at Nanoscopy Group, RUBION, Ruhr-Universität Bochum, Bochum, Germany

A. Haak (✉)
Nanoscopy Group, RUBION, Ruhr-Universität Bochum, Bochum, Germany

A. Gesper
Nanoscopy Group, RUBION, Ruhr-Universität Bochum, Bochum, Germany

Institute for Physiological Chemistry, Martin Luther University Halle-Wittenberg, Halle/Saale, Germany physicochemical properties of a cell. Investigating the cell membrane in a contact-free way and without any labelling or special pre-treatment is one of the greatest strengths of scanning ion conductance microscopy (SICM), which uses a fine pipette tip to determine the specimen's topography. Since FM and SICM can both be applied to living cells and provide a resolution below the diffraction limit, correlating data from these two techniques seems to be a promising approach to further the understanding of the cell membrane's organization and the mechanisms involved in transport processes across it. In the following, we briefly introduce the different approaches to super-resolved fluorescence microscopy (SRFM) and review applications of correlating SICM and FM (diffraction-limited) as well as SICM and SRFM (diffraction-unlimited). Finally, we discuss potential implementations of combined SRFM/SICM instruments and pitfalls that may arise during the development of such instruments.

Keywords Correlative microscopy · Diffraction-unlimited fluorescence microscopy · Photoactivated localization microscopy (PALM) · Scanning ion conductance microscopy (SICM) · Stimulated emission depletion (STED) microscopy · Super-resolution fluorescence microscopy · Super-resolved fluorescence microscopy

1 Introduction

Many cellular processes like exocytosis and endocytosis, cell growth, division, and migration, to name only a few, require the rearrangement, growth, or shrinkage of the cell membrane. Scanning ion conductance microscopy (SICM) [1] is a contact-free scanning probe technique providing information about dynamic changes, if applied to living cellular specimen, of the cell membrane's topography. However, the underlying dynamics of the molecular machinery, which play an important role in orchestrating cell membrane dynamics, cannot be investigated by using SICM. The latter can be studied by fluorescence microscopy (FM), which allows tagging a protein of interest with a fluorescent marker and observing its dynamics. Therefore, correlating FM and SICM data might help to unravel mechanisms underlying the previously mentioned cellular processes.

The resolution of SICM is determined by the inner opening radius r_i of the scanning pipette and can be approximated as 3 r_i (Fig. 1) [2, 3]. Thus, the resolution of SICM is only limited by the technical difficulties to produce pipettes with very small openings. Pipettes with radii of approximately 7 nm have been used for SICM imaging [4]. In contrast, the resolution in light microscopy is fundamentally limited by the diffraction of light. As Ernst Abbe found out in 1873, a beam of light cannot be focused to a spot smaller than the resolution limit d_{min} [5]:

$$d_{min} = \frac{\lambda}{2n\alpha} = \frac{\lambda}{2NA}$$

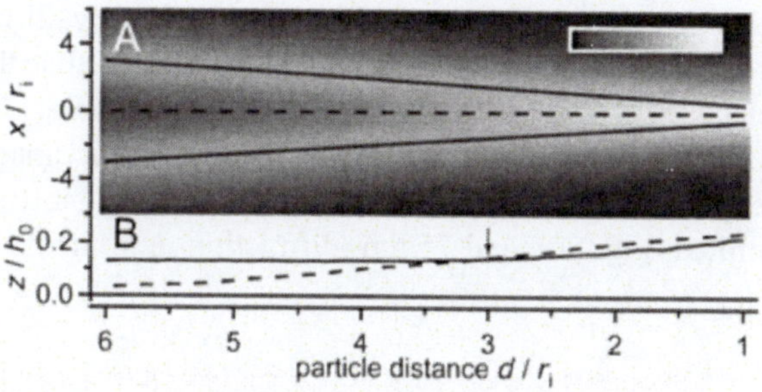

Fig. 1 Resolution of SICM. (**a**) Height profiles (plotted vertically) of modelled SICM recordings of two cylindrical particles (height and diameter: r_i) at varying distances. (**b**) Apparent height of the particles (solid line) and of the gap between them (dashed line) as determined by SICM. At distances $d \leq 3r_i$, no gap between the particles is detected and they cannot be resolved. Reprinted from Johannes Rheinlaender, Tilman E. Schäffer: *Image formation, resolution, and height measurement in scanning ion conductance microscopy*, Journal of Applied Physics 2009, 105(9), with the permission of AIP Publishing

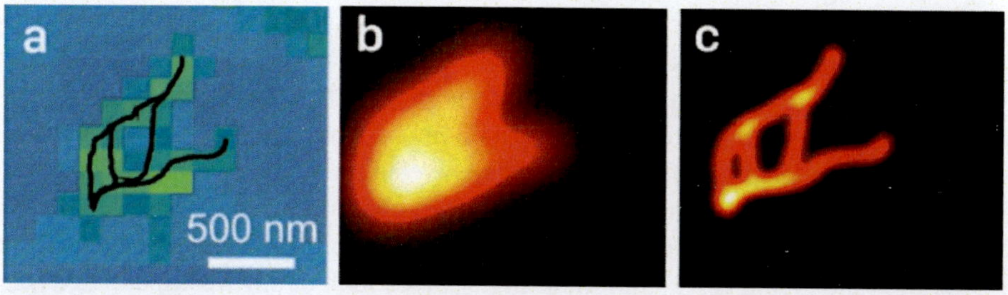

Fig. 2 Hypothetical fluorescence analysis of a protein forming membrane protrusions. (**a**) SICM recording of membrane protrusion and location of a putative protein within it. (**b**) Computed diffraction-limited fluorescence recording of the protein with a resolution of 250 nm. (**c**) Computed diffraction-unlimited fluorescence recording of the protein with a resolution of 75 nm. Reprinted with permission from Philipp Hagemann, Astrid Gesper, Patrick Happel: *Correlative stimulated emission depletion and scanning ion conductance microscopy*. ACS Nano 2018, 12, 5,807–5,815. Copyright 2018 American Chemical Society

Here, λ denotes the wavelength of the light used to generate the image, n the diffractive index of the medium between sample and objective lens, and α the half cone opening angle of the objective lens. The product $n \sin \alpha$ is called the numerical aperture NA of the objective. While the exact resolution limit is, as indicated by the previously introduced formula, dependent on λ and the NA of the objective, but as a rule of thumb, it is often stated that the resolution limit is in the range of 200 nm. In other words, the maximum achievable resolution limit is $\lambda/2$. This is approximately 14 times the size of the smallest structures observed in the SICM recordings, which were reported to be 14 nm [4].

To exemplify the impact of the limited resolution of light microscopy, assume that a protein (black line in Fig. 2a) forms a cellular protrusion. The SICM image showing the protrusion was part of a real scan series of a living HeLa cell's tip (see [6] Fig. 2, MIDDLE, second image, white arrow). A hypothetical, diffraction-limited recording of this protein would only show a blurred spot (Fig. 2b); while a recording with an improved resolution would clearly allow seeing that the location of the protein correlates with the position of the protrusion (Fig. 2c). Note that the resolutions used to compute the hypothetical fluorescent images correspond to resolutions experimentally obtained on corresponding instruments (250 nm (diffraction-limited; Fig. 2b), and 75 nm (diffraction-unlimited; Fig. 2c); [6]).

1.1 Fluorescence Microscopy

In order to understand how super-resolved fluorescence microscopy (SRFM) circumvents the diffraction limit, it is helpful to recapitulate the operating principle of fluorescence microscopy (FM). Fluorescence is the process of the absorption of a photon by a molecule, the fluorophore, followed by the spontaneous emission of a second photon. The transitions between the energetic states of the fluorophore are depicted in Fig. 3a. Due to the non-radiative loss of energy during the involved vibrational decays (black wave-like arrows in Fig. 3a), the energy $h\nu_{em}$ (h denotes the Planck constant) of the emitted photon is smaller than the energy of the absorbed photon $h\nu_{ex}$. Since the frequency ν of a photon is given by $c\,\lambda^{-1}$ (with c: speed of light, λ: wavelength of the photon), the wavelength λ_{em} of the emitted photon is larger than the wavelength λ_{ex} of the absorbed photon.

A typical absorbance/emission spectrum of an arbitrary fluorophore used as dye in FM is shown in Fig. 3b. In FM, a single wavelength or a wavelength range (green rectangle in Fig. 3b) suitable to excite the fluorophore used is chosen – by using a laser or LED or by selecting the wavelength range via a band-pass filter from a white

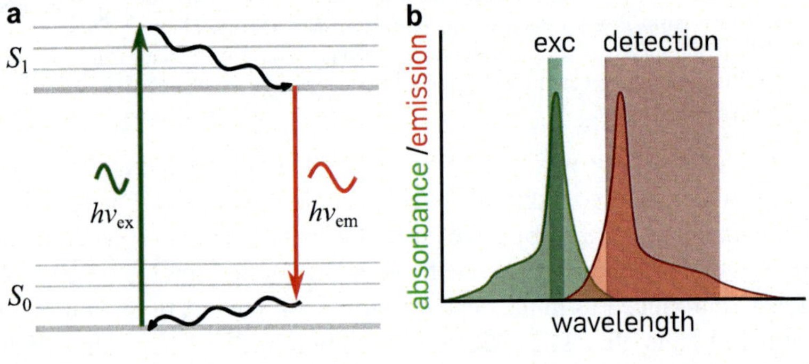

Fig. 3 Principle of fluorescence microscopy. (**a**) Jablonski diagram of fluorescence. A photon of energy $h\nu_{ex}$ gets absorbed and excites the molecule from its electronic ground state S_0 to a vibrational excited state of the electronic state S_1 (green arrow). After radiation-less relaxation (black wave-like arrow) into the vibrational ground state of S_1, the molecule returns to one of the vibrational states of S_0 (red arrow) by emitting a photon of energy $h\nu_{em}$, after which it relaxes to the lowest vibrational state of S_0. (**b**) Typical absorbance (green) and emission (red) spectra of a fluorescent dye molecule. Green and red rectangles indicate the wavelength range used for excitation (exc) and detection in FM

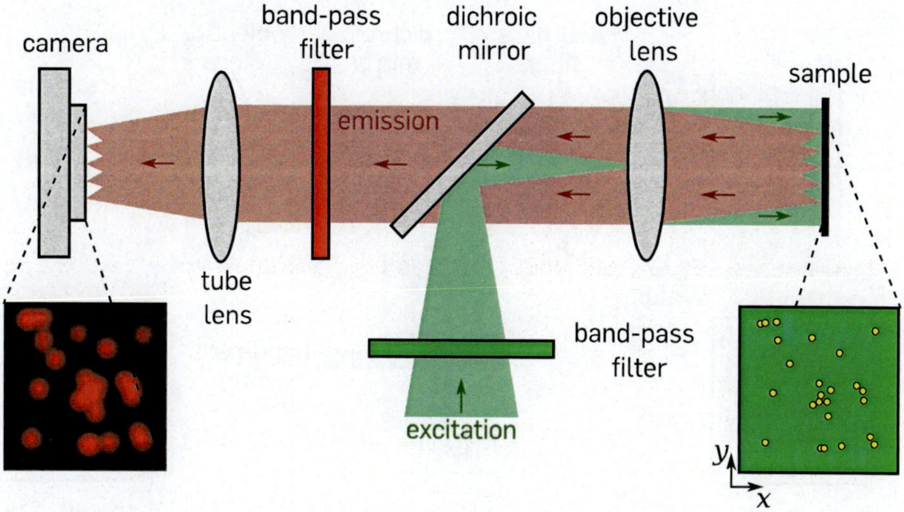

Fig. 4 Beam paths and major components of a wide-field fluorescence microscope. Excitation (green) and emission (red) beam paths

light source – and send through the objective lens of the microscope to the sample. In most FMs (the so-called *epifluorescence microscopes*), the emission is collected via the same objective lens. Since the emission is red-shifted with respect to the excitation wavelength, the emitted photons can be separated from the ones used to excite the fluorophores by a dichroic mirror. The wavelength range of a typical detection window is depicted by the red rectangle in Fig. 3b.

The two most common implementations of FM used in the life sciences are the wide-field and the confocal microscope. Both use a dichroic mirror to separate excitation and emission, but their implementations differ in the way in which the sample is excited and how the emission is recorded.

The beam paths and the major optical components of a wide-field FM are depicted in Fig. 4. In a wide-field FM, the entire sample is illuminated at the same time, and all fluorophores in the sample emit at the same time (yellow dots in the right insert in Fig. 4). The emission of the fluorophores (solid red shape in Fig. 4, only five beam paths are shown for clarity) is collected by the objective lens, passes the dichroic mirror, and is further cleared by a band-pass filter. The emission then is focused by the tube lens onto the camera, which records an image of the emission of all fluorophores at the same time. Each fluorophore in the sample generates a diffraction-limited spot on the camera. Spots which are not separated in space by at least the diffraction limit overlap and cannot be distinguished as single spots. The wavelength determining the resolution of a wide-field microscope is the emission wavelength λ_{em}.

The major disadvantage of a wide-field microscope is that the sample is not only illuminated in the focal plane of the objective lens, but also fluorophores above and below the focal plane are excited and their emission is collected (not shown for

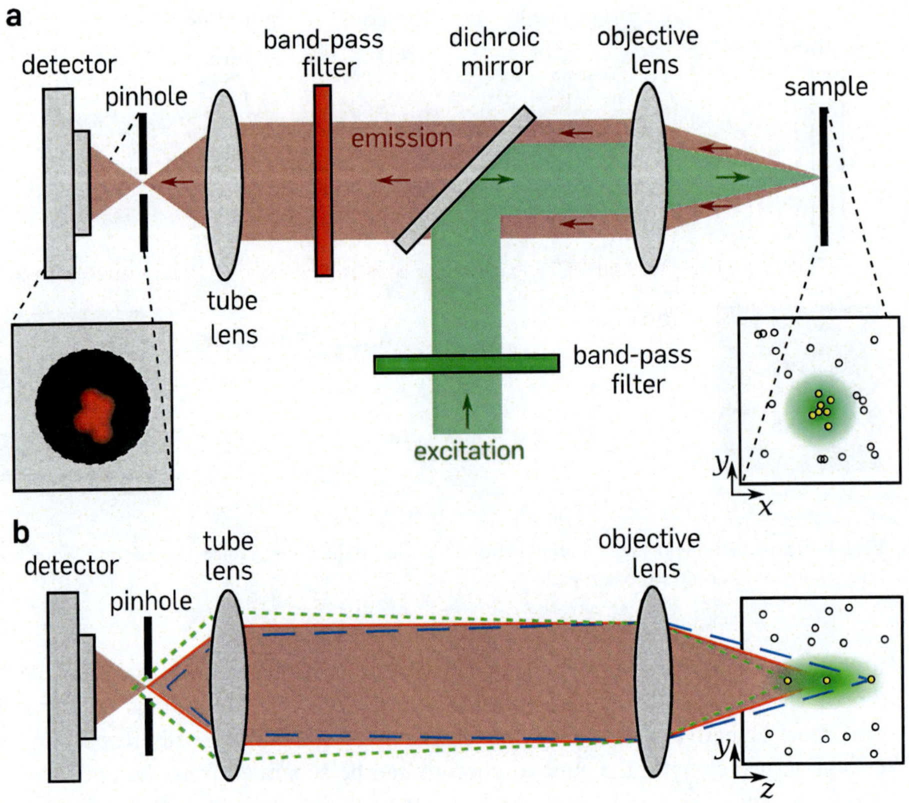

Fig. 5 Beam paths and major components of a confocal fluorescence microscope. (**a**) Excitation (green) and emission beam paths (red). (**b**) Emission beam path (filters omitted for clarity) of a confocal microscope showing how out-of-focus emission (green short-dashed line and blue long-dashed line) is suppressed by the confocal pinhole

clarity). Since they are not located in the focal plane of the objective lens, their emission does not result in a sharp spot on the camera, but instead appears as a blurred background.

This led to the invention of the confocal microscope [7, 8]. The beam path of a confocal microscope is shown in Fig. 5a. Here, the excitation light (green solid shape) is focused onto the sample by the objective lens and only the fluorophores within the resulting diffraction-limited spot are excited. Thus, the wavelength determining the resolution of a confocal microscope is the excitation wavelength λ_{ex}.

In contrast to the wide-field microscope, the emission of the fluorophores is focused onto a pinhole which is located in front of a single detector (left insert in Fig. 5a). Since this setup does not allow imaging the entire sample at once, the sample has to be scanned with respect to the excitation light beam and the image is collected pixel by pixel. Figure 5b illustrates the major effect of the pinhole: Emission from out-of-focus planes (green- and blue-dashed lines) is blocked by the pinhole since it is located in the same focal plane as the sample (thus the name *confocal*).

For a potential combination with SICM, the most important aspect of the two techniques is the way the fluorescence is recorded. Wide-field FM uses a camera to record the entire sample at once, whereas confocal FM is a scanning technique, which records the single pixels subsequently, either by moving the sample through the beam or vice versa.

1.2 Circumventing the Diffraction Limit

The general approach to circumvent the diffraction limit is to reduce the fluorophores within a diffraction-limited spot which contribute to the recorded signal that is to turn the fluorophores into an "OFF"-state. Only the fluorophores which contribute to the signal remain in the "ON"-state. However, a few more approaches exist, for example 4π-microscopy [9] or structured illumination microscopy (SIM) [10]. The first uses two objectives on opposite sides of the sample, which renders it impossible to be combined with SICM since the space required for the scanning pipette is already occupied by the second objective. In contrast, SIM is implemented on a wide-field microscope and could thus be combined with SICM. Therefore, combining SIM and SICM is similar to combining SICM with other wide-field based techniques (see Sect. 1.2.1). Since the gain in resolution is low, compared with the approaches explained in Sects. 1.2.1 and 1.2.2, we refrain from explaining SIM in detail here.

1.2.1 Stochastic Approaches

The stochastic approaches to SRFM use dye molecules, which can be switched between an inactivated ("OFF") and an activated ("ON") state by light of a specific wavelength. If only a sparse subset of molecules is activated, such that, on average, only one dye molecule per diffraction-limited spot emits, the position of the emitting molecule can be localized with much better precision than the diffraction limit. If this is repeated until all fluorophores have emitted, plotting the positions instead of the emission of all fluorophores yields an image with resolution beyond the diffraction limit (*localization microscopy*). The first techniques using this principle were photoactivated localization microscopy (PALM) [11] (using the Eos fluorescent protein [12] as marker), fluorescence photo-activation localization microscopy (FPALM) [13] (using photoactivatable green fluorescent protein (PA-GFP) as marker), and stochastic optical reconstruction microscopy (STORM) [14] (using the cyanine dyes Cy5 and Cy3).

As an example, the principle of PALM is illustrated in Fig. 6. Initially, all fluorophores within the sample are inactivated ("OFF"-state, gray dots in Fig. 6a). Next, light suitable to switch the fluorophores from the inactivated to the activated

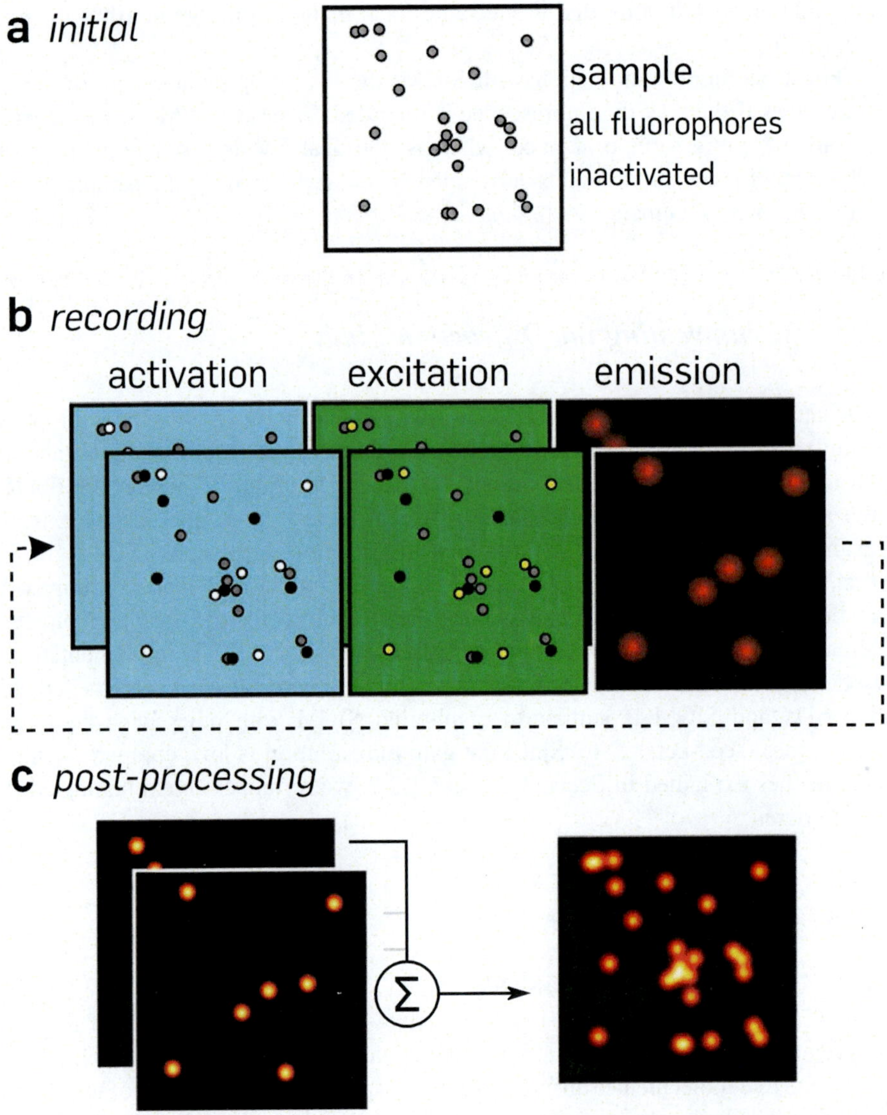

Fig. 6 Principle of PALM. (**a**) Initially, all fluorophores are in the inactivated state (gray dots). (**b**) (LEFT) A small subset of the fluorophores (dots) is activated (white dots; cyan background represents the activation light) and (MIDDLE) subsequently excited (yellow dots; green background represents the excitation light) until the fluorophores bleach (black dots). (RIGHT) The emission of the fluorophores is recorded. This series is repeated until all fluorophores have emitted. (**c**) The positions of the fluorophores in the recordings are determined, plotted as a Gaussian distribution with a standard deviation equal to the uncertainty of the localization, and the images are combined to a single image with improved resolution

("ON") state (white dots) is applied (the cyan background indicates the activation light, Fig. 6b (left)). The intensity of the activation light is selected such that only a small subset of the inactivated molecules is activated, at maximum one molecule per area of the diffraction-limited spot. The activated molecules are excited (green background indicating the excitation light, yellow dots, Fig. 6b (middle)) and their emission is recorded (Fig. 6b (right)). The excitation light is applied as long as all excited molecules have bleached (black dots, Fig. 6b (middle)). The series of activation, excitation and read-out is repeated until all fluorophores have been excited and bleached, which results in a set of images containing diffraction-limited emission spots of all fluorophores in the sample.

In a post-processing step (Fig. 6c), the positions of the fluorophores are determined by fitting (an approximation of) the instrument's point spread function to the data. The positions are plotted as Gaussian distributions with a standard deviation equal to the uncertainty of the determination of the position by the fit, finally yielding an image with an improved resolution (Fig. 6c, right).

The resolution provided by this approach scales with the number of photons N that are detected and theoretically is given by $\sigma_{PSF}/\sqrt{N}$ (with σ_{PSF} denoting the standard deviation of the point spread function of the microscope, which is in the range of 100 nm). However, in practice the resolution is in the range of 20 nm [11, 14] since the theoretical resolution cannot be reached due to background, noise, and other factors. A more detailed description of the resolution enhancement by localization is provided by [15].

This brief explanation is far from being exhaustive and does not consider the differences in the implementations of the various techniques. An in-depth overview is provided by several reviews of the techniques [16–19], of suitable dyes [20] as well as by detailed imaging protocols [21–23]. However, one aspect of the implementation which is important for the potential combination with SICM is the illumination scheme used by these methods. Since wide-field illumination is required to record the emission of the activated fluorophores, out-of-focus emission occurs, which increases the background in the recordings and affects the resolution. To avoid this, a total internal reflection fluorescence (TIRF) illumination scheme [24, 25] is used, which limits the excitation to the layer close to the interface of the microscopy slide and the sample.

1.2.2 Deterministic Approaches

In contrast to the stochastic approaches, which switch the fluorophores randomly in space, the deterministic approaches use a single-spot illumination and switch the fluorophores in the periphery of the excitation spot. The underlying principle is called reversible saturable optical fluorescence transitions (RESOLFT). Methods that use this principle are stimulated emission depletion (STED) microscopy [26, 27], ground state depletion (GSD) microscopy [28, 29], and RESOLFT with photo-switchable proteins [30]. The latter is often referred to as RESOLFT microscopy, although STED and GSD use the RESOLFT principle, too.

As an example for RESOLFT, the principle of STED microscopy is sketched in Fig. 7. To perform a scan (Fig. 7b), the excitation beam (2D-shape of the beam can

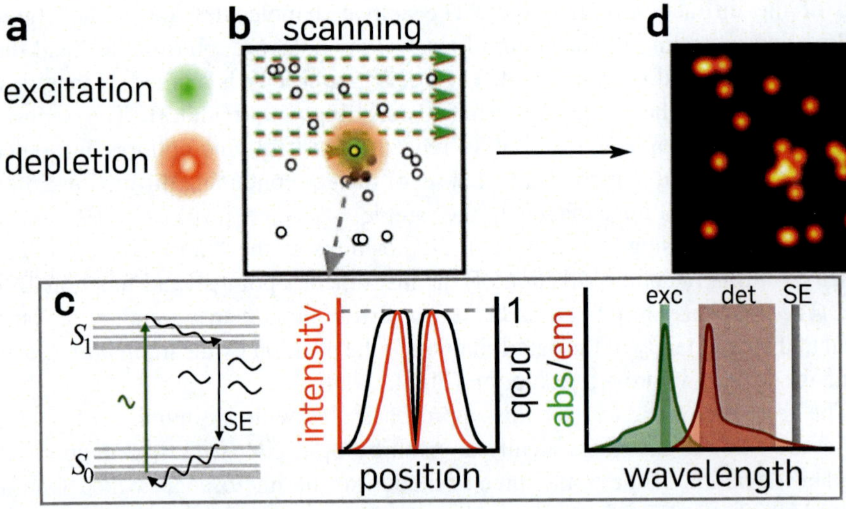

Fig. 7 Principle of STED microscopy. (**a**) Shape of excitation and ring-shaped depletion beam. (**b**) The two beams are superimposed and the sample is scanned. Fluorophores in the center are excited and emit spontaneously (yellow dot), while the depletion beam induces stimulated emission of the fluorophores in the periphery (black dots). (**c**) (LEFT) Jablonski diagram of stimulated emission (SE); (CENTER) Line profile of the intensity of the depletion beam (red) and probability (prob) to induce stimulated emission (black); (RIGHT) Wavelengths of excitation (ex), detection (det), and stimulated emission (SE) superimposed on the absorbance/emission spectrum of a fluorescent dye

be seen in Fig. 7a, top) is superimposed with a second, ring- or doughnut-shaped beam (2D-shape of the beam can be seen in Fig. 7a, bottom) and both are moved over the sample. The ring-shaped beam is also called the depletion beam. The fluorophores are excited by the excitation light, but only those located in the center of the minimum of the depletion beam are allowed to emit spontaneously (yellow dot in Fig. 7b). The fluorophores at the periphery are de-excited by the depletion beam and undergo stimulated emission (black dots in Fig. 7b).

The Jablonski diagram of stimulated emission is shown in Fig. 7c (left). An incoming photon of the depletion beam interacts with the excited fluorophore and causes it to de-excite to the ground state and to emit a second photon of the same energy as the incoming photon. Due to the non-linear relation between intensity and the probability to induce stimulated emission and the shape of the depletion beam (Fig. 7c, center), the size of the origin of spontaneous emission in the center of the depletion beam is decreased. The wavelength of the depletion beam and thus the wavelength of stimulated emission is selected such that it lies outside the detection window of the microscope and is therefore invisible to the detector (Fig. 7c, right). This switches the fluorophores which are undergoing stimulated emission de-facto to an "OFF"-state.

In summary, this directly yields a recording of the fluorophores with a resolution beyond the diffraction limit (Fig. 7d). The resolution obtained by STED microscopy depends on the intensity I of the depletion beam and is given by [31].

$$d_{\min,\mathrm{STED}} = \frac{\lambda}{2NA\sqrt{I/I_{\mathrm{sat}}}}$$

which is a modification of Abbe's original equation. Here, I_{sat} is the intensity required to induce stimulated emission in half of the fluorophores. The best resolution obtained by STED microscopy (approximately 2 nm) has been recorded on samples containing fluorescent color centers which were embedded in the crystal structure of a diamond [32].

Note that, by applying a depletion beam of a different shape, STED microscopy allows to improve the resolution along the optical axis [27, 33, 34]. A large number of review articles exist which summarize the use of STED (or in general, RESOLFT) microscopy in different fields in the life sciences, much too much to be listed here. We would like to refer the reader to exhaustive reviews, which, besides reviewing various applications, detail technical aspects of STED microscopy [35, 36] as well as to a review, which contains some anecdotal sections on how STED was developed [37].

2 Correlating SICM and Fluorescence Microscopy

Correlative SICM and FM, despite the mismatch in resolution, allows gathering information which cannot be obtained by using a single technique alone. However, the degree of correlation as well as the technical implementation differs, depending on the FM technique combined with SICM, as we will detail in the following sections.

Although recording SICM and FM data independently on *different samples* under the *same conditions* can support conclusions that could not be drawn from data from a single technique alone [38–40], we will focus on applications where SICM and FM data has been recorded from the *same sample*. Furthermore, we focus on far-field FM techniques and do not review the combination of SICM with near-field optical microscopy [41–44] or studies where the SICM pipette is used for targeted deposition [45–48] or for injection [49] of molecules, as a nano-sensor [50] or for manipulating the sample [51, 52].

2.1 Combining SICM and Wide-Field FM

Combining SICM and wide-field FM allows imaging the same region of the sample with both techniques (Fig. 8a). Because of the different fields of view, the different pixel sizes, and the different recording times required for acquisition of one image, the two data sets – due to their lack of synchronicity – cannot be correlated with one another pixel-by-pixel. Technically, the combination is simple, since the SICM only

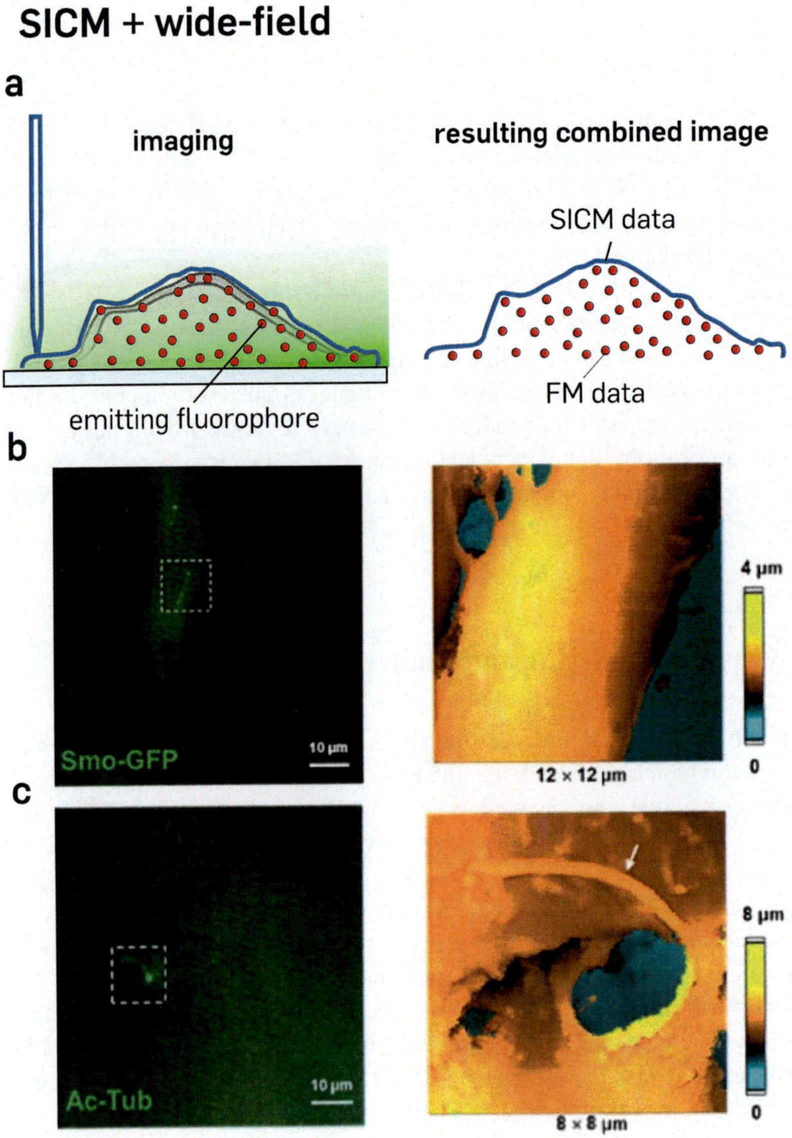

Fig. 8 Combining SICM and wide-field fluorescence microscopy. (**a**) Sketch of the data obtained by combining SICM and wide-field FM. (**b**) (LEFT) FM image of a cilium in an RPE-1 cell; (RIGHT) SICM recording of the area marked by the dashed rectangle. The cilium is not visible since it is located beneath the cell membrane (**c**) (LEFT) FM image of a cilium in an MDCK cell; (RIGHT) SICM recording of the region marked by the dashed rectangle. The cilium is clearly visible since it protrudes into the extracellular space. (**b**, **c**) Reprinted with permission from Yuanshu Zhou, Masaki Saito, Takafumi Miyamoto, et al.: *Nanoscale Imaging of Primary Cilia with Scanning Ion Conductance Microscopy*. Anal. Chem. 2018, 90, 4, 2,891–2,895. Copyright 2018 American Chemical Society

has to be mounted onto an inverted optical FM and their own software can control both instruments independently.

So far, the combination of SICM and wide-field FM has been used to find a region of interest such as an area containing active synapses in a culture of a neuronal network [53] or to identify the neuron which had been investigated by SICM-guided patch-clamp of synaptic boutons [54]. Furthermore, it has been used to first identify primary cilia in a cell culture and subsequently use SICM to reveal whether the selected cilium was extending into the extracellular space or was embedded in the cell (Fig. 8b, c), an information which could not be retrieved by using FM alone [55].

Cardiomyocytes exhibit a unique topography with transverse (T-)tubules and crests which can be displayed and investigated by using SICM [56, 57]. By combining SICM with Förster resonance energy transfer (FRET) measurements, the distribution of the β_1- or β_2-adrenergic receptor (AR) in healthy and in failing cardiomyocytes could be determined [56].

For this, the cell surface was first scanned via SICM. Afterwards the SICM pipette could be moved in a targeted manner directly toward the positions of the T-tubules or crests. This enables a local distribution analysis of the β_1- or β_2-AR. The recognition of a receptor was built on the application of an agonist against the receptor of interest and the simultaneous recording of the intensity ratio of cyan and yellow fluorescent protein (CFP and YFP) which were both coupled to a FRET-based cyclic adenosine monophosphate (cAMP) reporter. The measurement of the production of cAMP was used as evidence for receptor activity, or in other words, it was used to prove the presence of the receptor in this area. Therefore, it was shown that in healthy cardiomyocytes the distribution of the β_2-AR was exclusively connected toward T-tubules whereas the β_1-AR was found all over the surface. Interestingly, in failing cardiomyocytes the distribution of β_2-AR was altered [56].

The same approach was used to investigate the effect of partial mechanical unloading, showing that it reduces the cAMP response of β_2-ARs at the T-tubules [58]. To investigate the distribution of the β_3-AR, the approach was extended to use a FRET-based reporter of cyclic guanosine monophosphate (cGMP), showing that a redistribution of the β_3-AR occurs in failing cardiomyocytes, too [59].

2.2 *Combining SICM and Confocal FM*

Similar to a SICM and wide-field combination, a combination of SICM and confocal FM enables imaging of the SAME REGION AND SAMPLE with both techniques (Fig. 9a). Again, the technical implementation requires mounting a SICM onto an inverted but in this case confocal FM. Two different modes of operation are conceivable. In the first one, both microscopes record data independently from each other and thus can use their own soft- and hardware to control the single components. However, pixel-by-pixel correlation, if desired, might be impeded by different implementations to compensate for non-linear responses of the positioning piezoactors in the different software. This problem might become even more prominent if different positioning

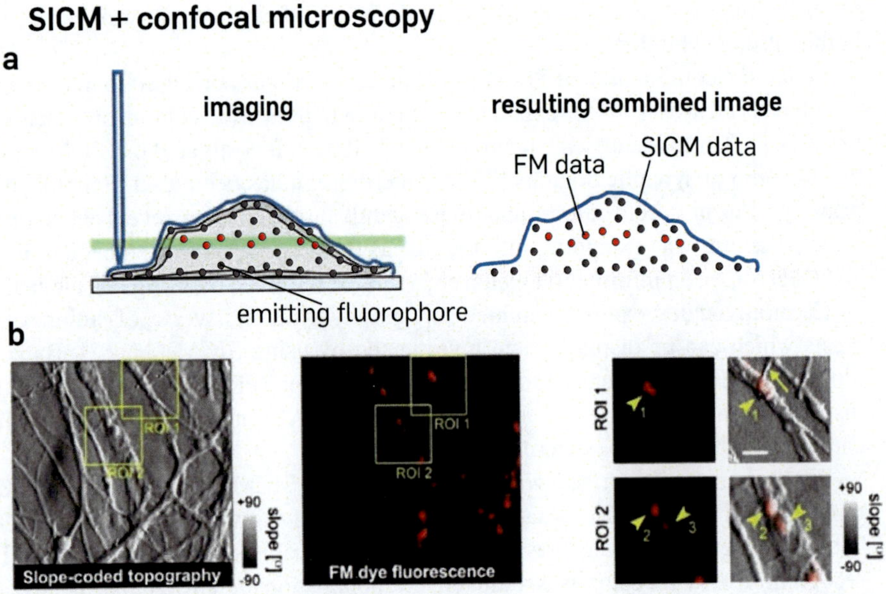

Fig. 9 Combining SICM and confocal fluorescence microscopy. (**a**) Schematic illustration of the data acquired by combining SICM and confocal FM. (**b**) (LEFT) Topography of a neuronal network; (CENTER) Fluorescence of the same neuronal network as depicted on the left; (RIGHT) Overlay of topography and fluorescence of two regions of interest (ROI 1, ROI 2) from the left and middle image. Reprinted with permission under the terms of the Creative Commons CC-BY license from P. Novak, J. Gorelik, U. Vivekananda, et al.: *Nanoscale-Targeted Patch-Clamp Recordings of Functional Presynaptic Ion Channels.* Neuron 2013, 79, 1067–1077

devices such as beam scanning for confocal FM and stage scanning for SICM are used.

The second mode of operation is to record the topography and the fluorescence at each *x*, *y*-position at the same time or with only a minimal temporal offset. Technically, this requires the synchronization of the SICM and the confocal microscope, which at least requires a distinct software.

SICM and confocal FM have been combined for the first time by Korchev et al. [41] in 2000. The group used this to verify volume calculations of renal tubular cells (A6 cell line) based on SICM measurements by comparing them with confocal images (rhodamine-phalloidin staining) obtained both in *x*,*y*- and in *z*-direction. The confocal images allow estimating the size of the cell and thereby provide a guide value for the expected order of magnitude. Just a year later, further development of this combination enabled simultaneous acquisition of Ca^{2+} imaging and topography data for investigating cardiac myocytes [42].

Seifert et al. investigated the morphological activity of platelets by time-lapse SICM, and subsequently obtained confocal recordings of cytoskeletal proteins of the investigated cells, allowing them to correlate the activity of the platelets and the structure of the cytoskeleton [60].

Another investigation which correlated SICM and subsequent confocal FM recordings studied cell–cell connections in co-cultures of cardiomyocytes and myofibroblasts and linked this to the expression of connexin-43 [61] which is one of the proteins able to constitute gap junctions [62, 63].

SICM has been shown to allow patch-clamp recordings from tiny neuronal structures like synaptic boutons [54]. The combination of SICM and confocal FM allowed to first record an overview image and to identify active synaptic boutons (Fig. 9b), which, in a second step, could be patch-clamped.

A second study on neurons investigated cargo transport in neuronal dendrites [64]. Here, the simultaneous combination of the two techniques (SICM and confocal FM) provided a hint that the observed transport processes were mitochondria mediated.

Furthermore, simultaneous recordings of both methods were used to study the formation of human immunodeficiency virus-like particles (VLPs) together with the dynamics of fluorescently labelled viral proteins like Gag or Viral Protein R [65]. In contrast to observing the virus formation with TIRF microscopy [66], the combination of SICM and confocal FM allowed investigations of the apical side of the cells. The release of VLPs at this side was observed to be a lot faster than the release at the basal cell side and the release time seems to be influenced by application of a fluorescent marker [65]. The total height differences of the topographies analyzed in this study were approximately in the range of the vertical resolution of the confocal data, and thus, the data sets were not only correlated in the horizontal, but also the vertical direction.

The applications reviewed above used only a single confocal layer. Yet, a combination of SICM and *z*-stacking confocal FM could yield more correlating data. After imaging the entire volume that is scanned by SICM, the data could be aligned and non-correlating data could be separated from correlating data. If stage scanning would be omitted, the two measurements could be performed simultaneously, but not synchronously.

2.3 Surface Scanning

Scanning surface confocal microscopy (SSCM) [67] is a combination of SICM and scanning confocal microscopy (SCM), which enables simultaneous SICM and FM imaging of the sample surface. In SSCM, the fluorescence data for each pixel is recorded at the position determined by SICM for that pixel. Up to now, two slightly different setups have been proposed (Fig. 10) [67, 68].

In both setups, a SICM is mounted onto an inverted confocal microscope and before simultaneous recording, the focus of the confocal FM objective is aligned with the tip of the SICM probe via micrometer screws and the confocal microscope's focusing mechanics. The difference between the two setups is the way in which the SICM pipette, the confocal focus, and the sample are positioned with respect to each other during imaging. The first setup (Fig. 10a) keeps the SICM probe at a fixed

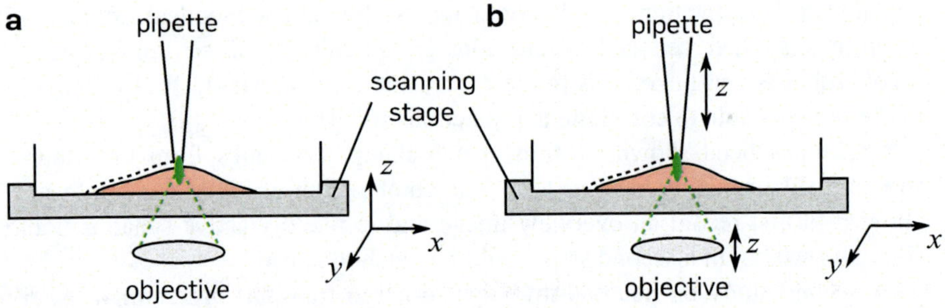

Fig. 10 Two setups for a scanning surface confocal microscope. (**a**) Diagram of the combined setup with an *x, y, z* sample scanner. (**b**) Diagram of the combined setup with an *x, y* sample scanner and additional *z*-scanners for probe and objective

position and scanning is performed by the sample stage, which positions the sample in all three dimensions (*fixed probe – variable stage (x,y,z)*).

In contrast, the second setup (Fig. 10b) uses the sample stage to position the sample only in *x*- and *y*-direction while additional piezoactors position objective and probe in *z*-direction (*variable probe (z) – variable stage (x,y)*). Thus, when the SICM pipette detects the surface, the position of the confocal spot must be adjusted by the objective positioning piezoactor subsequently.

Both setups do not allow using the controlling software of the confocal microscope since the fluorescence data has to be recorded after the SICM pipette has detected the surface. Thus, the soft- and hardware for controlling the SICM had to be expanded in order to at least enable reading from the detectors of the confocal microscope.

Compared to the *fixed probe – variable stage* setup the measuring time could be shortened by switching toward a *variable probe (z) – variable stage (x,y)* setup. This is because at the first setup the recording velocity is limited by the resonant frequency of the scanning stage. Additionally, usage of three piezoactors on the same platform led to cross-talk and consequently artifacts in the SICM data [68].

Additionally, both SSCM setups keep the probe-objective distance constant and enable the acquisition of a fluorescence image and the topography of the sample surface in the same scan. As the SICM and SCM data sets are recorded simultaneously, all data points in one set have a corresponding data point in the other set, as illustrated in Fig. 11a. This allows for determining the distribution of proteins in the proximity of the cell membrane (Fig. 11b).

First, SSCM has been used for investigating the dynamics of the interaction of Cy3-labelled polyoma VLPs with Cos-7 cells (simian (*Chlorocebus aethiops*) cell line [69, 70]) [67]. Furthermore, it has been utilized in studies investigating endocytotic pathways [71] (Fig. 11b), revealing an alternative closing mechanism for clathrin-coated pits by recording topography simultaneously with the fluorescence of tagged actin-binding protein and the tagged clathrin light-chain [72]. A further study on clathrin-mediated endocytosis revealed the effect of mutations in the Dynamin 2 gene on the maturation and internalization of clathrin-coated pits [73].

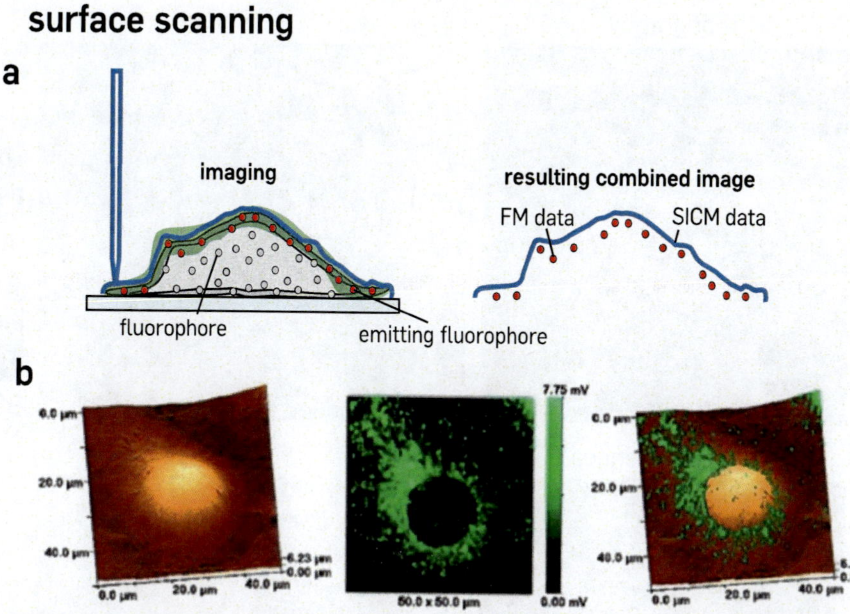

Fig. 11 Combining SICM and SCM. (**a**) Schematic illustration of the correlating data acquired by SSCM. (**b**) SSCM recordings of fixed clathrin-GFP transfected Cos-7 cells. (LEFT) Cell surface topography; (CENTER) Fluorescence recording of clathrin in the cell shown left; (RIGHT) 3D overlay of topographical and fluorescent recordings. Reprinted from Shevchuk, A.I., Hobson, P., Lab, M.J., et al.: *Endocytic pathways: combined scanning ion conductance and surface confocal microscopy study*. Pflugers Arch - Eur J Physiol 2008, 456, 227–235 with permission

SSCM was used to characterize the nanoparticle uptake by primary human alveolar epithelial type 2 (AT2) and by immortalized AT2 cells. Only the latter showed a strong tendency to internalize nanoparticles. Since, furthermore, the immortalized cells exhibited several characteristics of alveolar epithelial type 1 (AT1) cells, it was suggested that AT1 cells play a pivotal role in the translocation of nanometer-sized particles in the lung [74].

Figure 12 shows nanoparticles in the proximity of the surface of AT1-like cells at three different stages of the internalization: co-localized with membrane protrusions (yellow and white arrowheads, respectively), in the process of internalization (blue arrowheads), and fully internalized (green arrowheads). An improvement of the scanning velocity by the addition of a stiff, short-range piezoactor onto the pipette positioning piezoactor allowed following the internalization process with a temporal resolution of 15 s in detail [75].

Furthermore, SSCM has been applied to investigate exocytosis of insulin granules from pancreatic β-cells [76] and differentiation of stem cells [77]. To demonstrate the activation of the transient receptor potential cation channel subfamily V member 1 (TRPV1) after capsaicin stimulation, the topography of dorsal root ganglion (DRG) neurons was correlated with Ca^{2+} imaging [78]. Moreover,

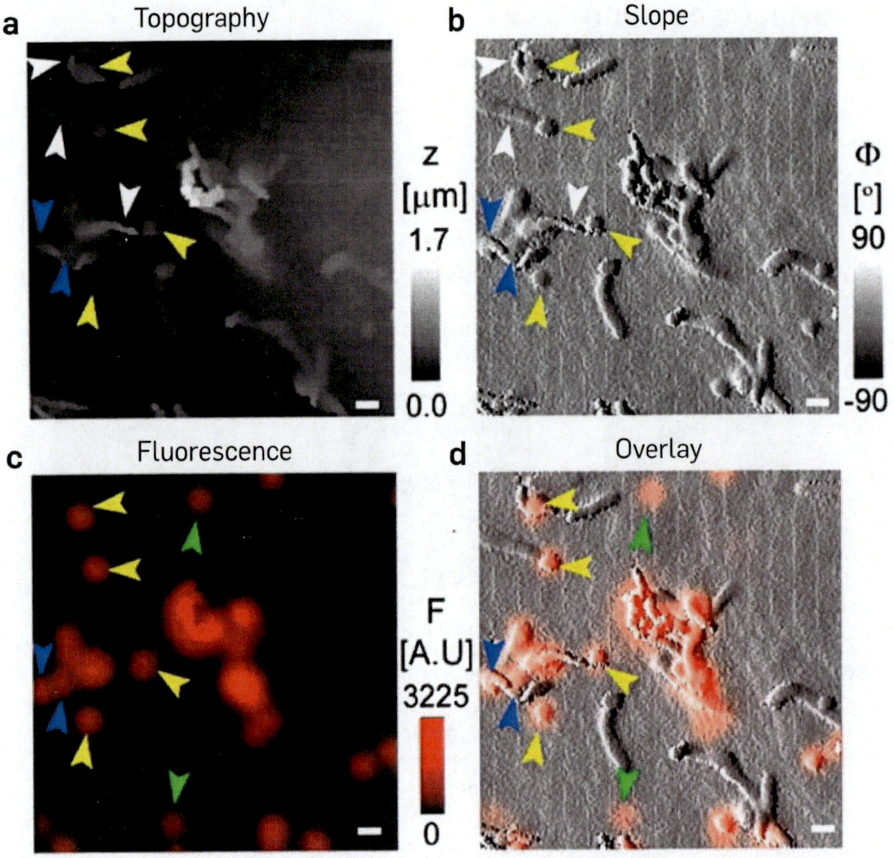

Fig. 12 Investigation on nanoparticle uptake by SSCM. (**a**) Height profile of a section of an AT1-like cell treated with 200 nm fluorescent nanoparticles. (**b**) Slope representation of the area shown in a. (**c**) Surface fluorescence of the same area. (**d**) Overlay of surface fluorescence and slope representation. Yellow arrowheads mark nanoparticles that co-localize with membrane protrusions (white arrowheads), blue arrowheads mark nanoparticles being in the process of internalization, and green arrowheads mark fully internalized nanoparticles. Reprinted with permission from P. Novak, A Shevchuk, P. Ruenraroengsak, et al.: *Imaging Single Nanoparticle Interactions with Human Lung Cells Using Fast Ion Conductance Microscopy.* Nano Lett. 2014, 14, 3, 1202–1207. Copyright 2014 American Chemical Society

SSCM was used for correlating topography and T-tubule distribution [79], or topography and mitochondria localization and appearance [80] in healthy and heart-failure cardiomyocytes.

In a current study, the uptake of the fluorescently labeled cell-penetrating peptide octaarginine (R8-Alexa Fluor 488) into HeLa cells was investigated by combining SICM and spinning disk confocal microscopy (SDCM) [81], a technique which enables confocal measurements with a temporal resolution of milliseconds [81, 82].

3 Correlating SICM and Super-Resolved Fluorescence Microscopy

The development of correlated SICM and SRFM is still in its very early stages. The first combination of SRFM and SICM has been the use of STORM to determine the number of fluorescent molecules deposited via a SICM tip [48]. However, recently a first proof-of-principle study combining STED and SICM has been published [6].

In the latter, the topography of a group of fixed HeLa cells has been investigated by SICM (Fig. 13a) and, subsequently, the distribution of the cytoskeletal protein actin was determined by confocal and STED microscopy (Fig. 13b) on a second instrument. The area of the sample imaged by both methods is marked in Fig. 13c (SICM: yellow rectangle, STED: gray rectangle).

Most interestingly, a protrusion (white arrowhead in Fig. 13a, b) extending from one cell was captured by both methods (magnified in Fig. 13d). The confocal recording showed a blurred spot within the protrusion (Fig. 13db), while the STED recording revealed that actin was arranged as a ring-shaped structure around two central spots (Fig. 13da). The authors speculate that this structure might be involved in the attachment or detachment of the cell from the cell culture dish. However, since the recordings were performed on fixed cells, time-lapse data that supports this speculation could not be obtained.

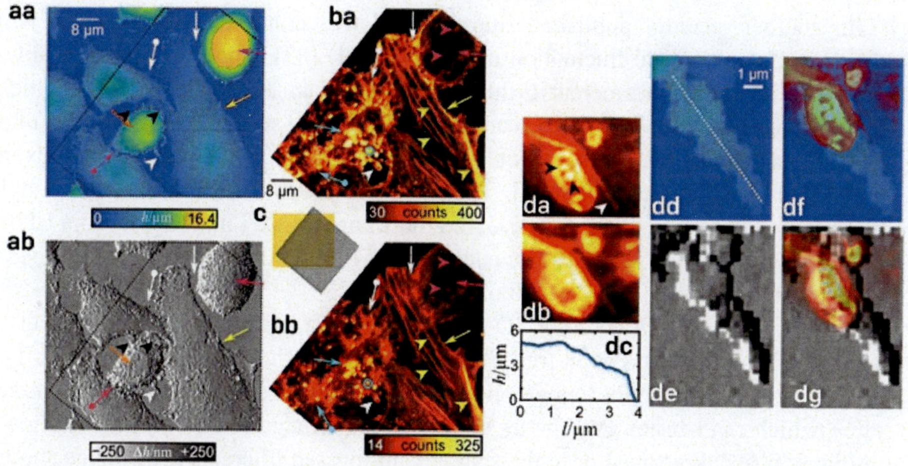

Fig. 13 Correlative SICM and STED recording of a group of HeLa cells. (**a**) Height (aa) and slope representation (ab) of a SICM recording of HeLa cells. (**b**) Confocal (ba) and STED (bb) recording of the region marked with a black dotted line in a. (**c**) Relative position of the SICM (yellow rectangle) and the confocal/STED (gray rectangle) images with respect to each other. (**d**) Magnification of the STED (da), confocal (db), SICM (dd), and slope (de) images. The magnified region is marked with a white arrowhead in **a** and **b**. df and dg show an overlay of the STED and the SICM data. dc shows the height profile along the white dotted line in dd. Reprinted with permission from Philipp Hagemann, Astrid Gesper, Patrick Happel: *Correlative stimulated emission depletion and scanning ion conductance microscopy.* ACS Nano 2018, 12, 5807–5815. Copyright 2018 American Chemical Society

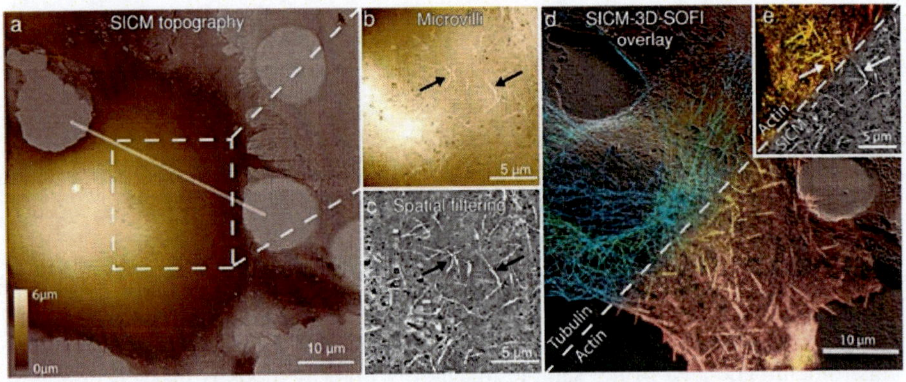

Fig. 14 Correlative SICM and 3D SOFI recording of a fixed Cos-7 cell. (**a**) Height representation of a SICM recording of a fixed Cos-7 cell. (**b**) Leveled zoom-in image of the area surrounded by a white dashed line in (**a**), revealing microvilli structures on the surface on the cell (marked with black arrows). (**c**) Spatial band-passed filtered image of the same area as shown in (**b**). (**d**) 3D-rendered overlay of a two-colored SOFI image and the SICM image shown in (**a**). The tubulin structures are displayed in colors from blue to green (color-coded height) and the actin structures are displayed in colors from red to yellow (color-coded height). (**e**) Correlative SICM/SOFI overlay of the same area as shown in (**b**). Reprinted with permission under the terms of the Creative Commons CC-BY license from V. Navikas, S.M. Leitao, K.S. Grussmayer, et al.: *Correlative 3D microscopy of single cells using super-resolution and scanning ion-conductance microscopy.* Nature Communications 2021, 12, 4565

In another recently published study, SICM was combined in one setup with super-resolution optical fluctuation imaging (SOFI) [83]. The principle of SOFI is based on independent stochastic fluctuations and can be obtained by wide-field microscopes outfitted with a CCD camera [84]. It is well suitable for live-cell studies [83, 85] and shows a four- to six-fold resolution improvement compared to conventional wide-field microscopes [84]. This was revealed by analyzing a microtubule staining with Abberior FLIP-565 dye: the fourth order SOFI image reached a lateral resolution of 72 nm ± 3 nm, 126/science.2464851 field microscope which reached a lateral resolution of 437 nm ± 106 nm [84].

A correlated SICM/two-color 3D SOFI image of a fixed Cos-7 cell with stained tubulin (blue fibers) and actin (red fibers) networks is shown in Fig. 14d [83]. The SICM image of the cell's topography (Fig. 14a) revealed several microvilli structures, which can be seen even better in the enlarged image of the cell surface shown in Fig. 14b (black arrows) or in the spatial band-passed filtered SICM image shown in Fig. 14c (black arrows). The correlated image indicated the presence of actin fibers inside the microvilli structures (Fig. 14e).

4 Outlook

In the previous sections, we have reviewed the combination of SICM and FM and have shown that this combination allows gathering new insights into the physiology of living cells. However, despite the mismatch in resolution, most of these studies combine SICM and diffraction-limited FM. Even other scanning probe techniques such as atomic force microscopy have been combined with SRFM only in a few studies [86–93].

In its most simple implementation, SICM and SRFM data could be recorded on two different instruments subsequently [6]. The major problem with this approach is finding the same region of the sample in both instruments. Here, strategies from correlating FM and electron microscopy (recently reviewed, for example, in [94]) might be adapted.

A drawback when using TIRF based methods such as PALM or STORM is the limited penetration depth of the excitation beam, which only allows recording fluorescence data close to the interface between sample and cell culture dish. However, approaches allowing 3D imaging by localization microscopy exist [95, 96] and might allow to circumvent this limit.

In the case of combining SICM with a scanning technique such as STED, there are two alignment options: The first option is aligning the pipette tip and the laser beam only in the *x*, *y*-plane. This is comparable, in terms of data recording, to the above-mentioned combinations of SICM and wide-field or confocal microscopy (Sects. 2.1 and 2.2). The second option, surface scanning (Sect. 2.3), requires additional alignment of the tip and, in case of STED, the two beams in the *z*-plane.

Especially surface scanning would require repeated focal plane adjustment of the optical setup. If this is implemented by moving the objective (Fig. 10b), this requires meticulous alignment of the excitation and depletion beam to retain the superimposition of the two beams during the adjustment of the focal plane. To minimize pixels requiring a focal plane adjustment, it is possible to only apply it, if a certain *z*-range, e.g. of more than 1 μm, has been exceeded [75].

However, the resolution of STED microscopy depends on the intensity of the depletion beam, which decreases along the optical axis. Thus, the range for unadjusted imaging might decrease and, if STED recordings with increased vertical resolution are targeted, will decrease even further, which in turn might lead to longer and less stable recordings.

Due to the increased horizontal resolution, even the *x*, *y*-alignment of the SICM pipette and the laser beams becomes more challenging since the tolerable offset between the two decreases. While it has been reported that fine adjustment screws are sufficient to align the pipette and the beam in the combination of SICM and confocal microscopes [67, 75, 97], additional piezoactors might be required for the alignment of STED and SICM. However, for the alignment of the SICM pipette and a confocal beam by piezoactors a remaining offset in the alignment has been reported [65].

If the depletion laser and the pipette tip are aligned in all three dimensions, the pipette tip might reflect the depletion beam. Since the intensity of the beam is large to achieve a high resolution, the reflected light might not be completely blocked by the filters in the emission beam path. In this case, the pipette would have to be retracted first before the recording of the fluorescence data could be performed.

This, in turn, requires a precise control of the data acquisition by the two imaging methods, which can be only obtained by a designated controlling soft- and hardware.

In summary, the development of combined SICM and SRFM instruments has just begun, and several technical hurdles have to be overcome during the further development. However, since there are no fundamental physical objections that would hinder the combination of SICM and SRFM, we expect this field of research to grow in the near future.

Acknowledgements The authors acknowledge support from the Volkswagen Foundation (grant 88 390) and the German Research Foundation (grant 411517989).

References

1. Hansma PK, Drake B, Marti O, Gould SA, Prater CB (1989) Science 243(4891):641. https://doi.org/10.1126/science.2464851
2. Rheinlaender J, Schäffer TE (2009) J Appl Phys 105(9):094905. https://doi.org/10.1063/1.3122007
3. Rheinlaender J, Schäffer TE (2015) Anal Chem 87(14):7117–7124. https://doi.org/10.1021/acs.analchem.5b00900
4. Shevchuk AI, Frolenkov GI, Sánchez D, James PS, Freedman N, Lab MJ, Jones R, Klenerman D, Korchev YE (2006) Angew Chem Int Ed Engl 45(14):2212. https://doi.org/10.1002/anie.200503915
5. Abbe E (1873) Arch Mikrosk Anat 9:413
6. Hagemann P, Gesper A, Happel P (2018) ACS Nano 12(6):5807. https://doi.org/10.1021/acsnano.8b01731
7. Heimstädt O (1911) Zeitschrift für wissenschaftliche Mikroskopie 28:330–337
8. Minsky M (1988) Scanning 10(4):128. https://doi.org/10.1002/sca.4950100403
9. Hell SW, Lindek S, Cremer C, Stelzer EHK (1994) Opt Lett 19(3):222. https://doi.org/10.1364/ol.19.000222
10. Guerra JM (1995) Appl Phys Lett 66(26):3555. https://doi.org/10.1063/1.113814
11. Betzig E, Patterson GH, Sougrat R, Lindwasser OW, Olenych S, Bonifacino JS, Davidson MW, Lippincott-Schwartz J, Hess HF (2006) Science 313(5793):1642. https://doi.org/10.1126/science.1127344
12. Wiedenmann J, Ivanchenko S, Oswald F, Schmitt F, Rocker C, Salih A, Spindler KD, Nienhaus GU (2004) Proc Natl Acad Sci 101(45):15905. https://doi.org/10.1073/pnas.0403668101
13. Hess ST, Girirajan TP, Mason MD (2006) Biophys J 91(11):4258. https://doi.org/10.1529/biophysj.106.091116
14. Rust MJ, Bates M, Zhuang X (2006) Nat Methods 3(10):793. https://doi.org/10.1038/nmeth929
15. Mortensen KI, Churchman LS, Spudich JA, Flyvbjerg H (2010) Nat Methods 7(5):377. https://doi.org/10.1038/nmeth.1447
16. Patterson G, Davidson M, Manley S, Lippincott-Schwartz J (2010) Annu Rev Phys Chem 61(1):345. https://doi.org/10.1146/annurev.physchem.012809.103444
17. Sengupta P, van Engelenburg SB, Lippincott-Schwartz J (2014) Chem Rev 114(6):3189. https://doi.org/10.1021/cr400614m
18. Liu Z, Lavis LD, Betzig E (2015) Mol Cell 58(4):644. https://doi.org/10.1016/j.molcel.2015.02.033
19. Sauer M, Heilemann M (2017) Chem Rev 117(11):7478. https://doi.org/10.1021/acs.chemrev.6b00667
20. Li H, Vaughan JC (2018) Chem Rev 118(18):9412. https://doi.org/10.1021/acs.chemrev.7b00767

21. Gould TJ, Verkhusha VV, Hess ST (2009) Nat Protoc 4(3):291. https://doi.org/10.1038/nprot.2008.246
22. Schermelleh L, Heintzmann R, Leonhardt H (2010) J Cell Biol 190(2):165. https://doi.org/10.1083/jcb.201002018
23. van de Linde S, Löschberger A, Klein T, Heidbreder M, Wolter S, Heilemann M, Sauer M (2011) Nat Protoc 6(7):991–1009. https://doi.org/10.1038/nprot.2011.336
24. Ambrose EJ (1956) Nature 178(4543):1194. https://doi.org/10.1038/1781194a0
25. Axelrod D (1981) J Cell Biol 89(1):141. https://doi.org/10.1083/jcb.89.1.141
26. Hell S, Wichmann J (1994) Opt Lett 19:780. https://doi.org/10.1364/ol.19.000780
27. Klar TA, Jakobs S, Dyba M, Egner A, Hell SW (2000) Proc Natl Acad Sci U S A 97(15):8206. https://doi.org/10.1073/pnas.97.15.8206
28. Hell SW, Kroug M (1995) Appl Phys B Lasers Opt 60:495. https://doi.org/10.1007/BF01081333
29. Bretschneider S, Eggeling C, Hell SW (2007) Phys Rev Lett 98(21):218103. https://doi.org/10.1103/PhysRevLett.98.218103
30. Hofmann M, Eggeling C, Jakobs S, Hell SW (2005) Proc Natl Acad Sci U S A 102(49):17565. https://doi.org/10.1073/pnas.0506010102
31. Harke B, Keller J, Ullal CK, Westphal V, Schönle A, Hell SW (2008) Opt Express 16(6):4154. https://doi.org/10.1364/oe.16.004154
32. Wildanger D, Patton BR, Schill H, Marseglia L, Hadden JP, Knauer S, Schönle A, Rarity JG, O'Brien JL, Hell SW, Smith JM (2012) Adv Mater 24(44):OP309. https://doi.org/10.1002/adma.201203033
33. Harke B, Ullal CK, Keller J, Hell SW (2008) Nano Lett 8(5):1309. https://doi.org/10.1021/nl073164n
34. Wildanger D, Medda R, Kastrup L, Hell SW (2009) J Microsc 236(1):35. https://doi.org/10.1111/j.1365-2818.2009.03188.x
35. Turkowyd B, Virant D, Endesfelder U (2016) Anal Bioanal Chem 408(25):6885. https://doi.org/10.1007/s00216-016-9781-8
36. Blom H, Widengren J (2017) Chem Rev 117(11):7377. https://doi.org/10.1021/acs.chemrev.6b00653
37. Sahl SJ, Hell SW (2019) High resolution imaging in microscopy and ophthalmology. Springer, pp 3–32. https://doi.org/10.1007/978-3-030-16638-0_1
38. Gesper A, Hagemann P, Happel P (2017) Nanoscale 9:14172. https://doi.org/10.1039/c7nr04306f
39. Lee Y, Jung GE, Cho SJ, Geckeler KE, Fuchs H (2013) Nanoscale 5(18):8577. https://doi.org/10.1039/c3nr02665e
40. Lyon AR, MacLeod KT, Zhang Y, Garcia E, Kanda GK, Lab MJ, Korchev YE, Harding SE, Gorelik J (2009) Proc Natl Acad Sci U S A 106(16):6854. https://doi.org/10.1073/pnas.0809777106
41. Korchev YE, Gorelik J, Lab MJ, Sviderskaya EV, Johnston CL, Coombes CR, Vodyanoy I, Edwards CR (2000) Biophys J 78(1):451–457. https://doi.org/10.1016/S0006-3495(00)76607-0
42. Shevchuk AI, Gorelik J, Harding SE, Lab MJ, Klenerman D, Korchev YE (2001) Biophys J 81(3):1759. https://doi.org/10.1016/S0006-3495(01)75826-2
43. Rothery AM, Gorelik J, Bruckbauer A, Yu W, Korchev YE, Klenerman D (2003) J Microsc 209 (Pt 2):94. https://doi.org/10.1046/j.1365-2818.2003.01122.x
44. Bruckbauer A, Ying L, Rothery AM, Zhou D, Shevchuk AI, Abell C, Korchev YE, Klenerman D (2002) J Am Chem Soc 124(30):8810. https://doi.org/10.1021/ja026816c
45. Ying L, Bruckbauer A, Rothery AM, Korchev YE, Klenerman D (2002) Anal Chem 74(6):1380. https://doi.org/10.1021/ac015674m
46. Bruckbauer A, James P, Zhou D, Yoon JW, Excell D, Korchev Y, Jones R, Klenerman D (2007) Biophys J 93(9):3120. https://doi.org/10.1529/biophysj.107.104737

47. Bruckbauer A, Dunne PD, James P, Howes E, Zhou D, Jones R, Klenerman D (2010) Biophys J 99(1):L1. https://doi.org/10.1016/j.bpj.2010.03.067
48. Hennig S, van de Linde S, Bergmann S, Huser T, Sauer M (2015) ACS Nano 9(8):8122. https://doi.org/10.1021/acsnano.5b02220
49. Hennig S, van de Linde S, Lummer M, Simonis M, Huser T, Sauer M (2015) Nano Lett 15(2): 1374. https://doi.org/10.1021/nl504660t
50. Piper JD, Clarke RW, Korchev YE, Ying L, Klenerman D (2006) J Am Chem Soc 128(51): 16462. https://doi.org/10.1021/ja0650899
51. Fuentes DE, Butler PJ (2012) Cell Mol Bioeng 5(2):143. https://doi.org/10.1007/s12195-012-0225-z
52. Wang K, Zhou L, Li J, Liu W, Wei Y, Guo Z, Fan C, Hu J, Li B, Wang L (2020) Nano Lett 20(9):6313. https://doi.org/10.1021/acs.nanolett.0c01735
53. Scheenen WJ, Celikel T (2015) Synapse 69(5):233–241. https://doi.org/10.1002/syn.21807
54. Novak P, Gorelik J, Vivekananda U, Shevchuk AI, Ermolyuk YS, Bailey RJ, Bushby AJ, Moss GWJ, Rusakov DA, Klenerman D, Kullmann DM, Volynski KE, Korchev YE (2013) Neuron 79(6):1067. https://doi.org/10.1016/j.neuron.2013.07.012
55. Zhou Y, Saito M, Miyamoto T, Novak P, Shevchuk AI, Korchev YE, Fukuma T, Takahashi Y (2018) Anal Chem 90(4):2891. https://doi.org/10.1021/acs.analchem.7b05112
56. Nikolaev VO, Moshkov A, Lyon AR, Miragoli M, Novak P, Paur H, Lohse MJ, Korchev YE, Harding SE, Gorelik J (2010) Science 327(5973):1653. https://doi.org/10.1126/science.1185988
57. Dutta AK, Korchev YE, Shevchuk AI, Hayashi S, Okada Y, Sabirov RZ (2008) Biophys J 94(5):1646–1655. https://doi.org/10.1529/biophysj.107.117820
58. Wright PT, Sanchez-Alonso JL, Lucarelli C, Alvarez-Laviada A, Poulet CE, Bello SO, Faggian G, Terracciano CM, Gorelik J (2018) Front Physiol 9. https://doi.org/10.3389/fphys.2018.01302
59. Schobesberger S, Wright PT, Poulet C, Mardones JLSA, Mansfield C, Friebe A, Harding SE, Balligand JL, Nikolaev VO, Gorelik J (2020) eLife 9. https://doi.org/10.7554/elife.52221
60. Seifert J, Rheinlaender J, Lang F, Gawaz M, Schäffer TE (2017) Sci Rep 7:4810. https://doi.org/10.1038/s41598-017-04999-6
61. Schultz F, Swiatlowska P, Alvarez-Laviada A, Sanchez-Alonso JL, Song Q, Vries AAF, Pijnappels DA, Ongstad E, Braga VMM, Entcheva E, Gourdie RG, Miragoli M, Gorelik J (2019) FASEB J 33(9):10453. https://doi.org/10.1096/fj.201802740rr
62. Fishman GI, Eddy RL, Shows TB, Rosenthal L, Leinwand LA (1991) Genomics 10(1): 250–256. https://doi.org/10.1016/0888-7543(91)90507-b
63. Beyer EC, Paul DL, Goodenough DA (1987) J Cell Biol 105(6):2621–2629. https://doi.org/10.1083/jcb.105.6.2621
64. Takahashi Y, Zhou Y, Miyamoto T, Higashi H, Nakamichi N, Takeda Y, Kato Y, Korchev Y, Fukuma T (2019) Anal Chem 92(2):2159. https://doi.org/10.1021/acs.analchem.9b04775
65. Bednarska J, Pelchen-Matthews A, Novak P, Burden JJ, Summers PA, Kuimova MK, Korchev Y, Marsh M, Shevchuk A (2020) Proc Natl Acad Sci 117(35):21637. https://doi.org/10.1073/pnas.2008156117
66. Jouvenet N, Bieniasz PD, Simon SM (2008) Nature 454(7201):236. https://doi.org/10.1038/nature06998
67. Gorelik J, Shevchuk A, Ramalho M, Elliott M, Lei C, Higgins CF, Lab MJ, Klenerman D, Krauzewicz N, Korchev Y (2002) Proc Natl Acad Sci U S A 99(25):16018. https://doi.org/10.1073/pnas.252458399
68. Shevchuk AI, Novak P, Velazquez MA, Fleming TP, Korchev YE (2013) J Opt 15(9):094005. https://doi.org/10.1088/2040-8978/15/9/094005
69. Jensen FC, Girardi AJ, Gilden RV, Koprowski H (1964) Proc Natl Acad Sci U S A 52(1): 53–59. https://doi.org/10.1073/pnas.52.1.53
70. Gluzman Y (1981) Cell 23(1):175–182. https://doi.org/10.1016/0092-8674(81)90282-8

71. Shevchuk AI, Hobson P, Lab MJ, Klenerman D, Krauzewicz N, Korchev YE (2008) Pflugers Arch 456(1):227. https://doi.org/10.1007/s00424-007-0410-4
72. Shevchuk AI, Novak P, Taylor M, Diakonov IA, Ziyadeh-Isleem A, Bitoun M, Guicheney P, Lab MJ, Gorelik J, Merrifield CJ, Klenerman D, Korchev YE (2012) J Cell Biol 197(4):499. https://doi.org/10.1083/jcb.201109130
73. Ali T, Bednarska J, Vassilopoulos S, Tran M, Diakonov IA, Ziyadeh-Isleem A, Guicheney P, Gorelik J, Korchev YE, Reilly MM, Bitoun M, Shevchuk A (2019) FASEB J 33(7):8504. https://doi.org/10.1096/fj.201802635r
74. Kemp SJ, Thorley AJ, Gorelik J, Seckl MJ, O'Hare MJ, Arcaro A, Korchev Y, Goldstraw P, Tetley TD (2008) Am J Respir Cell Mol Biol 39(5):591. https://doi.org/10.1165/rcmb.2007-0334OC
75. Novak P, Shevchuk A, Ruenraroengsak P, Miragoli M, Thorley AJ, Klenerman D, Lab MJ, Tetley TD, Gorelik J, Korchev YE (2014) Nano Lett 14(3):1202. https://doi.org/10.1021/nl404068p
76. Bednarska J, Novak P, Korchev Y, Rorsman P, Tarasov AI, Shevchuk A (2021) J Microsc 282(1):21–29. https://doi.org/10.1111/jmi.12972
77. Gorelik J, Ali NN, SheikhAbdul Kadir SH, Lab M, Stojkovic P, Armstrong L, Sviderskaya EV, Negulyaev YA, Klenerman D, Bennett DC, Lako M, Harding SE, Stojkovic M, Korchev YE (2008) Tissue Eng Part C Methods 14(4):311. https://doi.org/10.1089/ten.tec.2008.0058
78. Babakinejad B, Jönsson P, Córdoba AL, Actis P, Novak P, Takahashi Y, Shevshuk A, Anand U, Anand P, Drews A, Ferrer-Montiel A, Klenerman D, Korchev YE (2013) Anal Chem 85(19):9333–9342. https://doi.org/10.1021/ac4021769
79. Lyon AR, Nikolaev VO, Miragoli M, Sikkel MB, Paur H, Benard L, Hulot JS, Kohlbrenner E, Hajjar RJ, Peters NS, Korchev YE, Macleod KT, Harding SE, Gorelik J (2012) Circ Heart Fail 5(3):357. https://doi.org/10.1161/CIRCHEARTFAILURE.111.964692
80. Miragoli M, Sanchez-Alonso J, Bhargava A, Wright P, Sikkel M, Schobesberger S, Diakonov I, Novak P, Castaldi A, Cattaneo P et al (2016) Cell Rep 14(1):140–151. https://doi.org/10.1016/j.celrep.2015.12.014
81. Ida H, Takahashi Y, Kumatani A, Shiku H, Murayama T, Hirose H, Futaki S, Matsue T (2021) Anal Chem 93(13):5383–5393. https://doi.org/10.1021/acs.analchem.0c04097
82. Nakano A (2002) Cell Struct Funct 27(5):349–355. https://doi.org/10.1247/csf.27.349
83. Navikas V, Leitao SM, Grussmayer KS, Descloux A, Drake B, Yserentant K, Werther P, Herten DP, Wombacher R, Radenovic A, Fantner GE (2021) Nat Commun 12(1):4565. https://doi.org/10.1038/s41467-021-24901-3
84. Dertinger T, Colyer R, Iyer G, Weiss S, Enderlein J (2009) Proc Natl Acad Sci U S A 106(52): 22287–22292. https://doi.org/10.1073/pnas.0907866106
85. Geissbuehler S, Sharipov A, Godinat A, Bocchio NL, Sandoz PA, Huss A, Jensen NA, Jakobs S, Enderlein J, Gisou F, van der Goot EA, Dubikovskaya T, Lasser ML (2014) Nat Commun 5:5830. https://doi.org/10.1038/ncomms6830
86. Harke B, Chacko JV, Haschke H, Canale C, Diaspro A (2012) Opt Nanosc 1(1):3. https://doi.org/10.1186/2192-2853-1-3
87. Chacko JV, Zanacchi FC, Diaspro A (2013) Cytoskeleton 70(11):729. https://doi.org/10.1002/cm.21139
88. Chacko JV, Canale C, Harke B, Diaspro A (2013) PLoS One 8(6):1. https://doi.org/10.1371/journal.pone.0066608
89. Odermatt PD, Shivanandan A, Deschout H, Jankele R, Nievergelt AP, Feletti L, Davidson MW, Radenovic A, Fantner GE (2015) Nano Lett 15(8):4896. https://doi.org/10.1021/acs.nanolett.5b00572
90. Curry N, Ghézali G, Kaminski Schierle GS, Rouach N, Kaminski CF (2017) Front Cell Neurosci 11:104. https://doi.org/10.3389/fncel.2017.00104
91. Hirvonen LM, Cox S (2018) Meth Appl Fluoresc 6(4):045002. https://doi.org/10.1088/2050-6120/aad018

92. Hirvonen LM, Marsh RJ, Jones GE, Cox S (2020) Eur J Cell Biol 99(7):151106. https://doi.org/10.1016/j.ejcb.2020.151106
93. Cosentino M, Canale C, Bianchini P, Diaspro A (2019) Sci Adv 5(6):eaav8060. https://doi.org/10.1126/sciadv.aav8062
94. Begemann I, Galic M (2016) Front Synap Neurosci 8. https://doi.org/10.3389/fnsyn.2016.00028
95. Huang B, Jones SA, Brandenburg B, Zhuang X (2008) Nat Methods 5(12):1047. https://doi.org/10.1038/nmeth.1274
96. Huang B, Wang W, Bates M, Zhuang X (2008) Science 319(5864):810. https://doi.org/10.1126/science.1153529
97. Shevchuk AI, Hobson P, Lab MJ, Klenerman D, Krauzewicz N, Korchev YE (2008) Biophys J 94(10):4089. https://doi.org/10.1529/biophysj.107.112524